Multiple Literacy and Science Education:
ICTs in Formal and Informal Learning Environments

Susan Rodrigues
University of Dundee, Scotland

INFORMATION SCIENCE REFERENCE
Hershey · New York

Director of Editorial Content: Kristin Klinger
Senior Managing Editor: Jamie Snavely
Assistant Managing Editor: Michael Brehm
Publishing Assistant: Sean Woznicki
Typesetter: Michael Brehm, Kurt Smith
Cover Design: Lisa Tosheff
Printed at: Yurchak Printing Inc.

Published in the United States of America by
Information Science Reference (an imprint of IGI Global)
701 E. Chocolate Avenue
Hershey PA 17033
Tel: 717-533-8845
Fax: 717-533-8661
E-mail: cust@igi-global.com
Web site: http://www.igi-global.com/reference

Library of Congress Cataloging-in-Publication Data

Multiple literacy and science education : ICTs in formal and informal learning environments / Susan Rodrigues, editor.
p. cm.
Summary: "This book explores various learning mediums and their consequences within a classroom context to synchronize understanding within the schooling fields"--Provided by publisher.
Includes bibliographical references and index.
ISBN 978-1-61520-690-2 (hardcover) -- ISBN 978-1-61520-691-9 (ebook) 1. Science--Study and teaching--Technological innovations. 2. Computer literacy. 3. Information superhighway in education. 4. Information technology. I. Rodrigues, Susan.
Q197.M85 2010
507.1--dc22

2009043360

British Cataloguing in Publication Data
A Cataloguing in Publication record for this book is available from the British Library.

Editorial Advisory Board

Table of Contents

Detailed Table of Contents

learning science and learning to use digital technology). These are skills at both the operational and conceptual levels. The chapter will draw on these similarities to discuss how being digitally literate could better support the independent and personalized learning of science in the development of individuals who are scientifically literate. The use of multiple literacies pedagogy, the multimodal means of learning and communicating, as bridging the two literacies will be made. Specific examples will be used to illustrate the objective of this chapter. The chapter will conclude with a conceptual framework for the development of digital literacy in empowering students to become scientifically literate.

This chapter provides a teacher's view of the role and influence of multiple literacy in secondary school science. Multiple literacy from the author's perspective, (who is a secondary classroom science teacher) is the concept of doing things that teachers have always done in their lessons, but achieving them by incorporating new, engaging, ICT-rich strategies.

Large advances in technology in the last few years have made computers cheap and presentation technologies easily available in most secondary schools, at least in industrialised countries. Due to recent developments in software technology nearly anyone can create animations and visualisations. The Internet has helped to make the distribution of such graphic tools both wide and fast. Thus, using multimedia in science teaching is becoming more and more common. Today, integrating visualisations and animations from the Internet into the science classroom seems an obvious choice for enhancing science lessons. But are all of the animations offered on the Internet really helpful for promoting understanding? This chapter discusses what might occur while working with animations taken from the Internet and how these multimedia illustrations can potentially interact to reinforce rather than resolve students' misconceptions about chemical principles. Daniell's voltaic cell serves as a good example to illustrate the ways in which visual aids can be interpreted differently by experts and novices. The following discussion takes place in the form of an exaggerated example. It is meant to appear as a critical interjection making readers more aware of the myriad, often invisible, potential drawbacks which exist when first selecting promising-looking animated illustrations for classroom use.

In this chapter the authors examine three podcasts created by the same class. This particular class generated three different types of podcasts and this allows us to explore the contextual and staging dimensions of language found within these podcasts. The podcasts were created as part of a project investigating the use of podcasting in primary science. The teachers used class time over three-four weeks to allow the pupils to generate podcasts that were then uploaded onto a site for public dissemination. The authors examine the podcasts from various perspectives in order to explore the notion of multiple literacy.

In this chapter, the authors first discuss how Roger's theory of innovation diffusion can be incorporated into ICTs in formal and informal learning and teaching environments. The authors begin by presenting the use of ICT in education in general terms, then they introduce Rogers' diffusion of innovation (DoI) theory and the related literature. This is followed by a description of a project which explored the relationship between some characteristics of primary science teachers and their attitudes toward the use of ICT in education. A national project was funded by the Scientific and Technological Research Council of Turkey (TÜBİTAK), and Ege University, Science and Technology Application and Research Center. The last section involves a discussion of the diffusion of technological innovations into science education in the light of Rogers' DoI theory.

The purpose of this chapter is to show how information communication technologies (ICT) facilitated communication between primary pre-service teachers that enabled a 'community of learners' to develop children's scientific literacy. Cultural-historical theory was used to frame a study that sought to explicitly go beyond thinking as being individualistic, and to show how thinking can also be considered as a collective endeavour. In particular the study identifies how thinking forms part of a 'community of learners' both virtually and in reality within classrooms. The study was able to make visible child and pre-service teacher interactional sequences that brought together everyday concepts and scientific concepts to support concept formation in science. The study revealed the dialectical relations between everyday concepts and scientific concepts for moving from an interpsychological level to an intrapsychological level. The collective, rather than the individual orientation, made such a perspective possible. Importantly, the use of ICTs facilitated communication between members of the collective.

A project by James Cook University's School of Education created an online learning environment targeted at rural and regional schools in Far North Queensland. Pre-service teachers worked with practising teachers and children to develop learning activities which were shared through the BirdNet website. The site hosts a wide range of learning activities for bird identification, building school gardens, as well as professional learning tools such as lesson plans and integrated units of work. Project successes indicate that innovation, creativity and place-based learning can support high levels of both ICT and scientific literacy in all participants. The challenges faced included those resulting from technical issues, effects of distance, child-safety provisions for an on-line environment, and entry level of skills for participants. The value of informal learning by pre-service teachers, freed from the formal learning assessment regime, is endorsed as a valid sustainable, strategy which can be adopted by teacher educators.

This chapter reviews the Malaysian experience in implementing the Smart School Flagship initiatives, notably in the implementation of information communication technology (ICT) application in science and mathematics education. From a macro perspective, this chapter takes stock of the achievements of the Smart School Flagship in enabling ICT infrastructure and Internet connectivity in Malaysian schools. It attempts to appraise current trends and practice, clarifies emerging issues or challenges that schools face in trying to improve the ways in which ICT is applied to enhance teaching and learning, and identifies promising good practices so that general lessons may be drawn that are of interests to Malaysia and other countries. It does not claim to comprehensively cover every aspect of the initiatives but aims to contribute to current thinking about this topic by presenting a practical and pragmatic evaluation of some of its key features.

Universities and other tertiary institutions in developing nations around the world are facing major challenges in meeting the demand for the increasing access to higher education (HE): limitations imposed by inadequate funding, poor infrastructure and sometimes lack of political vision, added to the demographic explosion, make it almost impossible for some of these developing nations to ensure access to all to higher education solely through the conventional face-to-face mode. In this context, the information communication technologies (ICTs) are providing an alternative to face-to-face education. Moreover, they have the potential to significantly increase access to quality higher education, improve management of tertiary institutions, increase access to educational resources through digital libraries and open education resources, foster collaboration and networking between universities, foster collaboration between the private sector and tertiary institutions, enhance sub-regional and regional integration and facilitate the mobility of teachers and graduates. In Sub-Saharan Africa (SSA), the African Virtual University (AVU), a Pan African Inter-Governmental Organization initially launched in Washington in 1997 as a World Bank project, works with a number of countries toward reaching the goal of increasing access to quality higher education and training programmes through the use of ICTs. The AVU has been the first-of-its-kind in this regard to serve the Sub-Saharan African countries. In this chapter, the AVU's twelve years experience in delivering and improving access to quality higher distance education throughout Africa will be discussed. The AVU has trained more than 40,000 students since its inception; this is the proof that it is possible to achieve democratization of tertiary education in Africa despite many challenges.

The exponentially changing world of the Information Age is reflected in the emphasis on multiple literacies and the impact of information communication technology (ICT) in teaching and learning practices in global educational environments. Students' learning, teachers' curricula and teacher education programmes are being adapted to these changing circumstances. The concept of multiple literacies has had a powerful influence on classroom practice. Multimodal and multidimensional curricula have become the standard for students from a very young age to lifelong learners. While discipline-specific literacies such as scientific literacy are widely acknowledged as essential components of a multiple literacies concept, notions of 'information literacy' have taken centre stage in discussions of students' ability to access, retrieve and critically evaluate the information that floods the ICT driven delivery modes of the 21st century. However, it is important to remember that learning is a complex process, and that "Who is looking after our children?" is still an essential question to ask.

This chapter describes a small research project to place Internet-linked computers in a retirement complex in Melbourne, Australia (Murnane 2007, 2008). The aim was to research the existing computer skills of the residents, provide lessons in the use of email and general Internet and computer use, investigate the most appropriate type of lessons, document the problems encountered by very senior people encountering computing and the Internet for the first time, collect research on how they used the computers, the attitudes of financial management, nursing and occupational staff towards the activity, and the involvement of peers, family and friends.

This chapter proposes that the use of virtual worlds for science education is warranted and fits well with contemporary learning theory in the context of constructivist instructional approaches being desirable and that learners learn best when they are engaged in active mental processing. Over recent years, games have become increasingly social, supporting massively multi-player online game experiences and then evolving into virtual worlds, such as Second Life, which show significant promise for educational uses. This chapter introduces the field of virtual worlds, and then discusses relevant theory and research. The authors describe the potential of virtual worlds for education by emphasizing how they can be leveraged as an effective tool for constructivist teaching techniques. In addition, the authors present some of the literature that supports their use for science education. This chapter concludes with practical concerns and some possible solutions in the context of future research directions.

This chapter considers the enactment of competences in a particular science learning game Homicide, which is played in lower secondary schools. Homicide is a forensic investigation game in which pupils take on the role of police experts solving criminal cases in the space of one week. The game is designed to support work with genuine scientific inquiry and to meet the seventh- to tenth grade curriculum objectives for science and Danish education in Danish schools. This chapter presents the results of a long-term empirical study conducted with four school classes who played the game. The chapter considers how students construct visual representations of the cases they investigated and how they used these representations to establish hypotheses and evidence. The term 'representational inquiry competences' is developed; it refers to the students' ability to construct, productively use, transform and criticize visual representations as an integrated part of conducting an inquiry in the science game. The Homicide game-based learning environment was designed with the aim of supporting scientific inquiry through a simulation of elements of a professional forensic practice situation (Magnussen & Jessen, 2006). The goal of the current study has not primarily been to understand what science students learn in relation to formal science curriculum goals. It is an ethnographic study of what competences are enacted when a professional inquiry is played out in schools.

Preface

The idea for this book stemmed from various conversations I have had with my dad. My dad has a Papal knighthood, he was Under-secretary to both President Obote and President Amin, and he has worked as a Company Secretary for a large multinational company. So, he is used to working under pressure and in demanding circumstances, however, he finds using a computer exasperating. Sometimes when we are pursuing a computer related task, (whether it involves email, skyping, searching the Internet, using spreadsheets, word documents or video/photos editing), he says, 'but how did you know to do that?' Or, 'why did it do that?' Or, 'how can I get this to do this?' Most of his questions are a bit of a challenge for me, as I tend to prefer working with an Apple Mac, whereas my dad uses a PC. So taking what I know about how I use my Mac and finding a way to help my dad use his PC has made me more aware of the various facets that comprise this notion of multiple literacy when using information communication technologies. My dad and I are, (I think) subject competent (in our particular fields and with the English language), he is better than me at information processing, and our technology competence varies. I began to wonder about this balance between subject (discipline and language), information processing and technology competence, with respect to the use of ICT in science education. With this in mind, I asked my international colleagues to document their notion of multiple literacy in science education by describing their research or activity in the area. My international colleagues are using contemporary tools of communication within their specific, culturally dependent and context specific, environments and this book seeks to document rationales, theories and ideas that underpin their use of ICT in various formal and informal science education domains.

The book begins with a 'chapter' that discusses, more academically, the ideas signaled in the conversation I have had with my dad. The first chapter is a transcript of a conversation, between colleagues, about technology informed and warranted literacy. During the conversation a colleague said,

"I question whether there is one definition for digital literacy which is constant in time and across different societies. There are new technologies coming out all the time and hence digital literacy twenty/thirty years ago would have meant something quite different from what we might think of today. I also wonder whether there is a requirement for the phrase digital literacy, why do we have the word digital there? Are we not just talking about literacy using the contemporary tools of communication?"

So that is really the question. Are we just talking about literacy in science education when using contemporary tools of communication? To try to address this question, the book contains chapters that illustrate the nature and challenge of technology inclusion and infusion in formal science education (whether it be primary, secondary or tertiary education) and informal education (site visits, computer games).

The first few chapters illustrate the nature of multiple literacy found when information communication technologies are used in science education. For example, the chapter by Ng describes the promotion of scientific literacy through the use of digital literacy. Mackenzie's chapter follows Ng and provides concrete descriptions for the use of various technologies in science education, highlighting the need for multiple literacy in terms of attitude, science and technology. Mackenzie's chapter signals the promise of technologies and strategies for secondary school science from a Scottish school teacher's perspective. However, the need for caution is signalled in the next chapter, with a focus on simulations. Eilks, Witteck and Pietzner draw attention to the challenges when using multimedia aids to support the development of chemistry ideas and concepts. Rodrigues and Williamson add to the multiple literacy discussion by adding literacy and information literacy to the previously identified aspects of digital literacy and scientific literacy. They explore cognition and examine a social context and the resulting pupil engagement. The next batch of chapters focus on the development of multiple literacy within the context of science teacher education. The chapters document how teacher professional development to support digital pedagogies are attempting to move away from a focus on the addition of ICT tools to classroom practice, to a way of working in a digital world that requires competence with respect to a variety of literacies. The chapters describe projects involving teacher professional development within specific milieus. For example, the chapter by Cavas, Cavas, Karaoglan and Kisla describes a variety of 'adopter/adapter' characteristics and describes the challenge in developing ICT use in science education in Turkey. The chapter by Jane, Fleer and Gipps presents an Australian perspective, in which they explore a cultural-historical framework not only in terms of cognition, but also in terms of examining social contexts and engagements. Whitehouse and Hickey also present an Australian perspective of primary teacher science teacher education, and draw attention to the challenges faced when trying to promote various literacy competences (science, technical, social, etc.) in rural environments. The following four chapters show how the success and failure of ICT use in science education is significantly influenced by the multiple literacy skills of particular agencies (policy makers, teacher educators, cultural environments, institutions).

The chapters highlight the interrelationship between infrastructure, curriculum and assessment, and teacher professional development. For example, Yoong and Yuen Lew outline the Malaysian Smart School experience, a policy driven, heavily resourced, significant initiative. Yoong and Yuen Lew describe how an initiative may be ahead of the times, but still come unstuck if those implementing it have limited awareness of multiple literacy demands. In the case of Malaysia, technology issues and the fact that few teachers were proficient in ICT application, let alone proficient to teach science using the English medium for communication (most are trained in the Malay medium), as well as the fact that the learners first language was not English meant that the innovative initiative was beset with troubles. The Diallo, Traore and Fernadez chapter picks up ideas identified in the Yoong and Yuen Lew chapter, by documenting the challenges and successes in Africa in terms of infrastructure, curriculum and teacher development. Fehring addresses issues mentioned in both the Yoong and Yuen Lew chapter, and Diallo, Traore and Fernadez chapter. Fehring uses an Australian context to explore the requirements in teacher education and she shows the challenge teacher education faces in promoting the development of multiple literacy. Walker uses a Scottish study to illustrate the demands with regard to assessment practices. Walker's chapter illustrates the ad hoc development of competences to address online assessment, and the influence of this literacy on test anxiety. Marks describes the educational challenges from a sociologist's perspective. Thus far the chapters have focussed on formal learning environments based learning.

The next four chapters consider beyond or outside school learning environments: descriptions of field trips, the challenge for older lifelong learners and the use of recreational computer games. We start with a chapter by Lavonen, Laherto, Loukomies, Juuti, Kim, Lampiselkä and Meisalo who use a site visit to show how learning involving multiple literacy can occur in environments beyond the school locale. But their chapter also suggests that scientific literacy is more than knowledge *of* science, they suggest that it includes knowledge *about* science as well as understanding interconnections through reading, writing, speaking, and listening. Some of the argument involving the promotion of scientific literacy has at its centre a desire to promote public engagement in science with the wider community. Murnane shows how multiple literacy is developed by some of our older members of society, but through his exploration of the development of technical competence, Murnane also highlights what we take for granted in terms of developing technical literacy. This foray into more informal learning environments is followed by a chapter by Grimley, Nilsen, Kerr, Green, and Thompson in which they describe the use of computer games in science education. Grimley, Nilsen, Kerr, Green, and Thompson discuss the implications, of various massive multiplayer online role play games, and well known virtual environments, in terms of science education challenges and requirements. The idea of computer games in science education is explored further by Magnussen who also considers the use of computer games in science education, and describes a role play game. Magnussen's research shows how various facets that comprise and require multiple literacy (for example, complexity of game design, resources in the classroom, teachers, and other students) influence the enactment of representational inquiry competences.

So, are we just talking about literacy in science education when using contemporary tools of communication?

Susan Rodrigues
University of Dundee, Scotland

Acknowledgment

I would like to thank the authors in this book for meeting my various deadlines and for ensuring that the book made steady progress through the development stages. I would also like to thank my dad, Ferdinand Rodrigues. He read and commented on the penultimate draft. In addition, I also want to thank him, because our talks about competences required when using computer based technologies planted the seed for this book.

Susan Rodrigues
University of Dundee, Scotland

Chapter 1
A Conversation Between Colleagues:
Defining Multiple Literacy in Science Education

Colette Fortuna
University of Dundee, Scotland

Sheila Henderson
University of Dundee, Scotland

Joseph McLuckie
University of Dundee, Scotland

Susan Rodrigues
University of Dundee, Scotland

Lorraine Syme-Smith
University of Dundee, Scotland

Neil Taylor
University of Dundee, Scotland

Graham Williamson
University of Dundee, Scotland

ABSTRACT

At the University of Dundee, in the School of Education, Social Work and Community Education, there is a Science, Mathematics and ICT Education research group. The transcript that follows represents a conversation on the topic of literacy implications of technology driven practice. The authors hope the transcript will help situate their rationale for the inclusion of multiple literacy, rather than digital literacy, in the title of this book. In essence they suggest that multiple literacy in science education is the ability to identify, understand, interpret, create, apply, communicate, compute and use technology based resources associated within various science education contexts to achieve goals, develop knowledge, skills and understandings relevant to science information and ideas.

DOI: 10.4018/978-1-60566-690-2.ch001

Joe: *What is literacy all about? Is it to do with digital literacy?*

Susan: *Yes, digital literacy. It's because my understanding at the moment is that we are using the word digital literacy to mean lots of different things and if we don't have a shared meaning about it then someone else reading the book might be coming from a completely different place ...*

Colette: *I have looked at how I would develop digital literacy with the students in fourth year using the stimulus of a newspaper headline claiming that some breakfast cereals contain more sugar than chocolate bars. I don't know if that is being too specific? The students have to decide how to determine whether the newspaper report is accurate using the statistical facts presented by the manufacturers (on breakfast cereals and chocolate bars). The students then have to use mathematics as a "truth tool" to analyse the statistical data. They then use persuasive language to advertise their decision on the healthiest option.*

Susan: *What exactly is a truth tool?*

Colette: *In this case - it is deciding whether there is truth behind the headline and behind the advertising. They have to decide what constitutes a "healthy cereal" and use their statistical and numerical skills to decide which cereal they consider to be healthiest from a selection of cereals.*

Joe: *So that is more in line with what you mean by numeracy rather than digital literacy?*

Colette: *At that point – yes as they have to make sense of the statistical data (presented in a variety of ways) and decide whether the newspaper report is valid. They have to uncover the information that is presented (in a variety of formats) by advertisers (using persuasive language) to decide whether a cereal is healthy. Having selected their healthiest cereal, they have to advertise their selected cereal using a range of digital formats: power point, website, radio broadcast, rap, jingle, digital movie, pod cast or other digital format of their choice.*

Susan: *Colette, why is that not scientific literacy?*

Sheila: *I was just about to say that. Before we go anywhere we need to decide what our terms mean. What do you mean by numeracy, what do you mean by literacy, scientific literacy? So before we do anything we need to be sure we know what we are talking about.*

Colette: *I think that in answer to your question as to whether it is scientific literacy or not, I think it becomes numerical if they are using numbers to make their informed decision.*

Sheila: *But they are not just using numbers and things like that, because on packs of cereal there will be pie charts with information so they are actually using graphics as well, so it's more to do with mathematical literacy, as well as numerical.*

Susan: *So why is that then literacy? On your graphs you have got mainly language so they would have to understand what the scale is or what the title is. So why is that not literacy and why is that numeracy?*

Colette: *It's a combination of numerical and graphical.*

Joe: *We are introducing four distinct terms. There is digital literacy, there is numeracy, there is literacy and there is also this idea of scientific literacy. In relation to these four literacies, they are all going to sort of intertwine with each other. As far as digital literacy is concerned I tend to think of it as more a set of competences, a set of generic competences, if you want, that some of our students will have when they are entering* *undergraduate courses, some of the students will have when entering postgraduate courses in which case you would expect the competences to be greater. The same with school kids, there is an expectation that when they transfer from primary to secondary they have a certain amount of digital literacy already up their sleeve. So there are different levels of 'literacies'*

Susan: *Is that skill when you talk about competence?*

Joe: *Well it is a mixture of both skills and knowledge. And competences are a matter of intertwining the two. How do you use a database, you have the knowledge of how to use a database but can you actually put it into practice? If you are putting it into practice then that knowledge is becoming a competence.*

Lorraine: *I question whether there is one definition for digital literacy which is constant in time and across different societies. There are new technologies coming out all the time and hence digital literacy twenty/thirty years ago would have meant something quite different from what we might think of today. I also wonder whether there is a requirement for the phrase digital literacy, why do we have the word digital there? Are we not just talking about literacy using the contemporary tools of communication?*

Susan: *Would you say that Graham? Digital Literacy being a contemporary tool?*

Graham: *There are a couple of things that have been said that are central to the discussion, like the ones about demands of society and change and what our society sees as constituting literacy. The goal posts keep being moved. To be literate now a young person has to do much more in respect of any kind of literacy. Digital literacy now is like a technical aspect of literacy as a competence. Technology makes digital media more and more accessible in education. Up until a few years ago young people had to be digitally literate in terms of making electronic text, which would have needed special electronic equipment. But it has moved on from there. Literacy is a range of forms of making text and getting meaning from text. So I see literacy as more 'generic' and that digital literacy or other particular types of literacy do not much matter. Literacy is about getting meaning from text and making meaning with text. The way we use the word' text' here is very open. It now accommodates a big range of meanings. This may help us start to envelop or bracket some of the things people have been saying so far.*

Susan: *Yes, I think it has sort of taking on board from where Colette started from, because that newspaper clip that you held up was really in terms of how people make sense of text that is there with regard to numerical data that is entered.*

Colette: *The students had to identify the persuasive techniques of the advertiser and check whether the supplied numerical data supported the claim made by the newspaper or the manufacturers. The students then have to select five different numerical facts to use to advertise their chosen "healthiest cereal" using a range of digital formats: power point, blog, website, radio broadcast, rap, jingle, digital movie, pod cast. They had to identify the important mathematical and numerical facts for that cereal and present it in some sort of persuasive way. Finally the students were given an "unhealthy" cereal and they had to advertise it (truthfully and accurately) using a different digital format. This process should helps pupils understand how advertisers try to persuade consumers to buy their product.*

This should help them learn how to become informed citizens.

Joe : *That fits in kind of nicely with what Graham was saying. Graham was talking about this idea* *of literacy being the understanding of textual information and the actual application of textual information i.e. being able to use text to convey meaning or to convey knowledge but, in relation to the context that Colette has painted there, it's not just an understanding of textual information, it's an understanding of numerical information and, in the bad example or scenario, is also an understanding and conveying of both numerical information as well. So I would say if you think about digital literacy as an add on to that, digital literacy has got lots of different formats too. It's the understanding of text, it's the understanding of numerical data, it's the understanding of graphical information. It's the understanding of things like a scherzos versus scherzos communication. It's much wider than just literacy.*

Susan: *Is it an add on or is it just what you are talking about as society of culture.*

Colette: *I would say it's a response.*

Sheila: *I prefer the phrase mathematical literacy because it's wider. If you accept numeracy as just number, because the word almost says number to you, but if you accept mathematical literacy then you are thinking of more than number.*

Joe: *So we have now got mathematical literacy?*

Sheila: *I wonder about the narrow definition because you need something wider. It needs to encompass things like looking at graphs and looking at these things they put on cereal packets with the percentages of different calories, fat, sugar, all the rest of it, so even although they are very visual I would say anything visual like that is graphical literacy but to me mathematical literacy would encompass that.*

Joe: *So you have got mathematical literacy, scientific literacy and digital literacy – I am just beginning to wonder if the word literacy at the end of these three pairs is actually the right word because Graham's thinking in terms of literacy in relation to English language and communication skills, we are thinking about it in a broader context.*

Graham: *When I was using the word 'text', it was to signify a way of making meaning. It's not just meaning with words. It could be a combination of words, sounds, images, music - that is all text here and ways of making meaning.*

Joe: *So text itself could involve the media.*

Sheila: *Literacy is the encompassing term and within that you branch into the different literacies.*

Graham: *Yes but what you were saying is in the respect of mathematical literacy - ways to make and understand the different meanings of what I am calling text which is using the wider concept.*

Susan: *It probably makes sense in terms of the book originally started off being called digital literacy in informal and formal science and got changed to multiple literacy because of that notion of these perspectives.*

Joe: *So in the book the word literacy has got a much broader meaning rather than just the language context.*

Susan: *And this is why in terms of setting the scene for the other chapters to follow, it's having this kind of discussion with other people to see that we are not looking at literacy in its narrowest sense. One of the bits that we have not talked about is the fact that there is certain amount of emotional literacy involved like a self-efficacy thing, and confidence in being able to interpret the mathematical literacy or the digital or electronic. You might be really comfortable with them in maths but not very* *comfortable with the technology or you might be very comfortable with reading a newspaper and say they have not provided me with this and that evidence but not happy in communicating it further. I kind of wonder about that balance that we are asking pupils to have when we are talking about multiple literacies in the use of ICT.*

Joe: *But this book that you are talking about in which we will have a chapter is not just orientated towards school kids though, it's much more generic than that, is it not? Do you want our chapters to sort of reflect the idea within teacher education or pupil education or is it to be wider?*

Susan: *Wider. There is a chapter for example where they are talking about 90-year-olds and their use of technology. What does it mean to them in terms of their confidence, their access to information? People who are confined to a particular environment - are their abilities restricted? Do they still have a say in what is going on in the outside world since they can still see what is happening in their world because of the technology made available to them, providing they have the confidence and competence to use that technology.*

Colette: *My 77-year-old mother-in-law surfs the net, uses online shopping and internet banking*

Joe: *The implications behind that is that digital literacy is probably changing and the emphasis nowadays, is on communication. You are talking about these silver surfers buying laptops but are they buying laptops to use the spreadsheets and desktop publishing packages, to look at the web and e-mail? So the digital literacy is kind of going back to being a kind of communication tool. They are delivering photographs which are another form of communication via Facebook, etc.*

Colette: *They are using phones to download photographs to computers.*

Susan: *It is back to what Graham was talking about in terms of society. Is it the case that all we have done is become more high tech about the way we communicate or the way we use information?*

Joe: *So is digital literacy consuming all literacy and if you are digitally literate and competent in relation to using all these tools in digital scenarios does that mean you are automatically linguistically literate and mathematically literate and scientific?*

Colette: *I would say it is like sound bite – it catches people very quickly so it is going towards the visual one and the very short sound bite of advertising.*

Sheila: *The difference here is that many in society view literacy as whether or not you can spell and punctuate and numeracy on whether you can count, but in actual fact what I understand and what we are talking about is much, much wider. We do have this wider understanding and many ways of communicating but yet 'society' does not recognise that and in many cases it wants our citizens to have only the narrower competences of reading, writing and 'rithemtic.*

Graham: *Yes I think it is an important point that Sheila is making there and part of the problem is the social and political aspect. Politicians and the media create a difficulty here as when you ask what they mean by literacy, they tend to say something pretty mundane, pretty unreflective and pretty useless, such as "to be able to read, write and spell", as was required years ago, and basically that does not help in terms of what society needs to be from a citizen to be literate. So there is a real line to be drawn there between what the opinion formers in society describe as literacy and what is needed in terms of our view, as educators, of an appropriate education for the citizens of tomorrow.*

Susan: *So if we were to come up with a different definition then, if we were talking about somebody being able to use these various forms of literacies what do we mean when we talk about them?*

Sheila: *We mean the much bigger and broader picture. If you are numerate or mathematically literate then you are able to process mathematical information and communicate it to other people or to process it yourself and communicate it to yourself – any sort of data.*

Susan: *Ok, so are we saying it definitely needs the communication angle.*

Sheila: *Yes but it can be to yourself.*

Joe: *When you talk about being mathematically literate you are probably talking about the various levels of mathematical literacy. You are talking about kids being mathematically literate in relation to what they were like at the end of primary school, secondary school or at the end of a degree with honours. What I am getting at is in relation to this idea of mathematical literacy there is a whole series of graded points to mathematical literacy. It's not just you are either mathematically literate or you are not, you can't just say you are IT literate or you're not because IT literacy involves mathematics and a great deal of different levels of skills .*

Sheila: *You are able to deal with the data, whatever data it is and is being presented to you at the given time, so for a child in primary school what was being presented to that child is going to be different to what is presented to somebody in their twenties or thirties. I am thinking about things on government statistics, insurance and that sort of stuff and the important thing is that in order to be mathematically literate you have to be able to process and communicate even to yourself just and by communicating to yourself I mean the ability to inner process it at whatever level you meet it.*

Joe: *So it's making sense of the world as you see it?*

Susan: *Kind of informed judgements.*

Joe: *For instance when you are talking about scientific literacy using that analogy and making sense of say physics as you see it, if you are making sense say of a piece of metal you just see it as a piece of metal or you go down to the micro level and you are talking about that piece of metal being made up of molecules so now kids in primary school won't have any conception of what a molecule is but they will understand what a metal is.*

Susan: *I think it is interesting that you say that because at the primary school I was visiting, I was looking at kids talking about liquids and gases and these kids said its like Mickey Mouse and I thought what do you mean? And this boy says water is like Mickey Mouse and he drew it. And he drew an H20 molecule and it looks like a Mickey Mouse head, two ears for the hydrogen and the big oxygen between them. So they do know. But they might not use the language that we would use, but they have thought about it. I was like, where the heck did you get Mickey Mouse from but when he drew his model of H20 and it looks like a head with big ears I could see what he meant, and he was representing his view of the molecular level. And my understanding of scientific literacy is more in line with Colette's notion of mathematical literacy because it is about taking public information and making judgements about the capacity of the science that is there.*

Joe: *It's the application of knowledge and the interpretation of knowledge.*

Susan: *But having enough background science to be able to say no they have not given me all the bits of information to make judgements on this or yes they are using a mother and a daughter* *rather than the scientist here because they want my emotional vote, it's being able to judge how they are presenting that information as much as what that information is. I have always thought of IT literacy as having the skills to use the technology and I am not sure if that is my naivety.*

Joe: *That is probably historical now. I would say IT literacy itself has moved on and it's the skills to use the technology to communicate effectively so in a way it is becoming like ordinary literacy used to be.*

Colette: *But you have to almost teach various skills discretely, like doing a podcast or whatever and it's like projects like this that pull it all together where they have got to use one of their ways in presenting so they will have been taught the skill of doing a podcast initially and then they have a choice of podcast, digital film, PowerPoint, jingle, rap – they can select from those to do their advert or whatever.*

Susan: *Could you have multiple literacy without the skill, without the IT skill?*

Joe: *Could you have ordinary literacy without the application – yes you could. You need an understanding to apply it – is that what you are asking? I would say you can have some IT literacy skills, you can understand what you see on the web page, you can produce a web page.*

Susan: *If you go back to the silver surfers…*

Joe: *You want to represent that information in a different format as you have got that information on the web, how do you take meaning of that yourself, how do you take ownership of it, how do you actually change it – it's that idea of going from understanding to actually changing your own perception significantly .*

Colette: *What Susan says is do I need to. The silver surfers do not need to unless they particularly wanted to.*

Joe: *One of my golfers is 78 now and he is a silver surfer and the reality is he is applying it all the time because he is now not out all that much to apply the technology more than we would. One of the members of the male choir is 95 and he uses his web connections to create poetry which he publishes to all his friends so that is application as well. I don't think you can get away from the idea of multiple literacy being divorced from the actual application of knowledge once it has been collected. It's not just a static one way transmission process.*

Susan: *Again I think that is back to what Lorraine was saying in terms of the way society has evolved, it's changing the way we are. Like when I was on sabbatical, people said to me if you had taken sabbatical ten years ago you could have gone to a nice little cabin some place and just got on with stuff whereas now they can track you down and e-mail is there. You cannot simply jump ship, they have ways of finding you and so your ways of working have to change. Previously on sabbatical you may not have continued to supervise students because you were away whereas now the technology is available that allows you to carry on doing so. Is it just a question of not so much being able to apply it but needing to have the skills and the competence to do it.*

Sheila: *This is one of the arguments taking place within maths because it used to be the case you learnt maths for the sake of doing the maths and if you were able to do the exercise afterwards then it's fine but of course now what we are saying is that, especially in primary school, if you can't make that maths relevant and put it in a context relevant to children then there is no point in doing it. But there is another body of information that says actually by doing that we are diminishing* *the whole subject and we are not letting children actually study it in its own right because it actually develops your skills of logic.*

Colette: *I would say it's desirable to make it relevant but not necessary.*

Sheila: *One of the examples was to set up a travel agent in a primary school yet are children likely to book holidays? So in actual fact that context is not relevant to children. It's relevant to the teacher because she might be interested in it but it is not relevant to the children.*

Susan : *I actually take that whole notion one stage further in terms of relevance because I think you can teach maths in terms of the joy of the discipline, in terms of the subject if you actually had a conversation with the kids by talking to them, that is what makes it personally relevant by creating artificial contexts does not make it relevant.*

Sheila: *You are right that is the other scenario. You set up a shop with plastic money but children don't go in and buy tins of beans or things like that - they buy sweets.*

Lorraine: *I have seen that in Further education as well with beauty therapy students getting assignments about a building site, with questions like How many bricks? and How much mortar? Barriers were being presented by the subject before the students even thought about what the calculations they were going to do. Once the context of the questions was changed to being subject relevant the students were much more engaged and ready to answer the questions.*

Susan: *That is interesting because in Further education sector what would people want to see in terms of having competence, in terms of literacy, whether its numeracy, a word text or ICT competence or science? What's the push in the FE sector? What do they focus on? Are they expecting students to come in with all those?*

Lorraine: *Within group awards there are core skill requirements in numeracy, communication, ICT, problem solving and working with people. In general terms there is increasing emphasis on employability and the skills/literacies involved with that.*

Joe: *Graham, a wee while back, mentioned this idea of the social impact with regards to literacy but I don't think he was talking about digital literacy at that point, I am just wondering if that is another slant we should be taking. Has digital literacy had a profound impact in the way society now functions? I am thinking about the way solicitors and other professionals do their business now. They very seldom meet and most of their business is done via e-mail and the web, as a generic work place, probably has had some sort of impact in relation to the way you operate. Susan was talking about this idea that if you are away on sabbatical you are not out of touch. Now you are expected to keep in touch so it's had a profound impact in the way we work as educationalists, so maybe that is another slant we should take on it. The impact of digital literacy has had on culture in general.*

Graham: *Sorry I did not quite get all of that Joe but I am thinking back to the previous discussion about the context and I think it is interesting to take that idea further. You have to make it interesting and motivating in some kind of meaningful context and offer something they can relate to which is relevant. I think there is a difficulty with digital literacy and what you are kind of suggesting to us is something that is moving all the time – is that what you were saying?*

Joe: *Something like that. It's not just digital literacy we feel is moving all the time, the application of digital literacy in a sense in a social context is* *changing all the time and you can argue is that more significant or less significant and what impact is it having on the rest of society like being able to watch the Australian Open at 9 in the morning?*

Graham: *One of the points in what you are saying, which I see as very important, and that I want to pick up on and try to draw into the area of discussion is how it involves informal learning. What I mean by that is young people develop digital literacy and it has nothing to do with their school learning and they are able to go on and do something constructive making texts for social purposes, using messenger, using mobile phones, accessing social networking sites. All these are making texts using digital literacy for real purposes of their own. They have not been taught much of it, have learned it informally and in a sense are much more literate and using more technology than old guys like me. So that is something you may want to draw in.*

Joe: *Is the informal learning having greater impact than the formal learning?*

Graham: *There is a transforming aspect and that is what we have to try to get a hold of. Joe, you and I already talked about students saying "I don't look at Blackboard unless I have to; I don't use my Uni email account; I do everything through Bebo". They are all there for different purposes in other words. We have to find ways of making what we do in line with what they already do. There is a need to pick this up in higher education, to work with their informal learning, and take ideas from that into school education*

Neil: *If the discussion is to go forwards we need to have shared understanding of phrases such as 'digital literacy', 'numeracy' and 'literacy' (in terms of English language). For numeracy and literacy, the Curriculum for Excellence website will give us 'Scottish' definitions but for 'digital literacy' we need to look wider. I 'Googled' the phrase 'digital literacy' and one of the first links was the EDLC. As this is a recognised qualification then it may be the skill set which EDLC provides is a baseline of digital literacy. I also liked the link from Wikipedia. Digital literacy helps people communicate and keep up with societal trends, it also prevents us from believing* hoaxes *that are spread Online or are the result of* photo manipulation. *Digital literacy is the ability to locate, organize, understand, evaluate, and create information using digital technology. It involves a working knowledge of current high-technology, and an understanding of how it can be used. Digitally literate people can communicate and work more efficiently, digital literacy comprises of multiple aspects.*

Susan: *Ok, so perhaps instead of using digital literacy we use multiple literacy, would that convey the multiple aspects that comprise ....*

Neil: *Is it worth aligning it to a recognised definition?*

Susan: *Like what?*

Neil: *Are there any definitions within the Curriculum for Excellence, the web or recognised bodies like the OECD (Organisation for Economic Co-operation and Development/Programme for International Student Assessment)?*

Susan: *Do you know of any?*

Neil: *UNESCO? OECD? Learning and Teaching Scotland?*

Susan: *So what do they have in common?*

Neil: *Is literacy a skills set, for example, UNESCO suggests 'Literacy' is the ability to identify, understand, interpret, create, communicate, compute and use printed and written materials associ-*

ated with varying contexts. Literacy involves a continuum of learning to enable an individual to achieve his or her goals, to develop his or her knowledge and potential, and to participate fully in the wider society.' It maybe that all these things are just a basic skills set for communicating ideas, what Graham suggested earlier. So literacy is the ability to use, interpret, apply and communicate information. I'm sure CfE (Curriculum for Excellence) suggests something similar, they see literacy as ' the set of skills which allows an individual to engage fully in society and in learning, through the different forms of language, and the range of texts, which society values and finds useful.' where 'A text is the medium through which ideas, experiences, opinions and information can be communicated. Texts include those presented in traditional written or print form, as well as those presented orally, electronically or on film. Texts can be in continuous form, including traditional formal prose, or non-continuous text, such as a chart, graph or webpage. It will be important to provide opportunities for using various and appropriate kinds of texts.'

Susan: *So in essence you are saying that all of this is a part of 'literacy'? Where text isn't just words, but anything that is used to share or communicate understanding?*

Neil: *Yes! Like me, short and simple!*

Susan: *Hmmm! Let's not go there. So, in terms of people reading this book, what are we saying multiple literacy is?*

Neil: *What we are trying to do is try and have a shared understanding of literacy and text?*

Susan: *Yes.*

Neil: *So, based on Graham's suggestions, literacy is a set of skills which allows people to communicate their understanding, and text is any medium, graph, aural, digital.*

Susan: *Ok, so multiple literacy in science education with regard to the use of technology involves having skills to use, analyse, interpret and communicate science ideas and understanding through a variety of 'texts'. As Graham said, 'Literacy is about getting meaning from text and making meaning with text.' And you and Graham are saying that text is any medium.*

Neil: *Yes.*

Susan: *So, taking what's been said so far, we could say that multiliteracies in science education focuses on multimodal ways of conveying understanding, and the range of available information communication technologies afford multimodality that allow students to generate, communicate and demonstrate their science understanding and skills.*

Neil: *Yes.*

Chapter 2
Empowering Students to be Scientifically Literate through Digital Literacy

Wan Ng
La Trobe University, Australia

ABSTRACT

The objective of this chapter is to discuss the relationships between the three literacies that are mentioned above: digital literacy, science literacy and multiple literacies. The chapter will define digital literacy and scientific literacy and argue that being digitally literate would enhance the development of scientific literacy. It will look at the similarities in skills required for the two literacies (i.e., skills derived from learning science and learning to use digital technology). These are skills at both the operational and conceptual levels. The chapter will draw on these similarities to discuss how being digitally literate could better support the independent and personalized learning of science in the development of individuals who are scientifically literate. The use of multiple literacies pedagogy, the multimodal means of learning and communicating, as bridging the two literacies will be made. Specific examples will be used to illustrate the objective of this chapter. The chapter will conclude with a conceptual framework for the development of digital literacy in empowering students to become scientifically literate.

INTRODUCTION

We live in a society driven by science and technology. As more science and technology issues dominate public debates at national and international levels, it is important that we have global citizens who are scientifically literate. Science and technology based issues include those that concern personal and community safety, have impact on the environment or are ethically-based. People who are scientifically literate have sufficient knowledge and understanding of science to enable them to think critically in order to make sensible decisions about science related matters that affect their own lives. Since science is a mandated key learning area for primary and junior secondary students in most schools around the world, it is appropriate to teach students to be scientifically literate at these

DOI: 10.4018/978-1-60566-690-2.ch002

levels in order to prepare them for a society where science and technology are integral to their everyday living.

A similar argument to being scientifically literate is that of being digitally literate. Digital technologies are tools that are becoming more central to the individual's learning and social wellbeing. They are also central to the economic development and advancement of businesses and corporations. On the importance of digital literacy in terms of the economy and employment rates, Maria Wynne and Lane Cooper (2007, p. 4) had stated that:

At a national level, a growing number of experts predict that a lack of digital literacy will have a dampening impact on economic prospects. Consider that in the next eight years, according to Monthly Labor Review Online (November 2005 p.6) six out of every 10 new jobs will be in professional and service-related occupations requiring, at a minimum, a basic level of proficiency in computers……. Economic advantage and competitiveness will rest heavily on our ability to equip the 21st century workforce with competitive digital literacy skills

At the level of the individual, Berson and Berson (2003) noted that the youth of today are accessing a vast amount of information through the various media outlets and are simultaneously creating and disseminating their own messages and creative products through digital technologies. Due to the difficulties in parental or institutional control of young people's access to information from these outlets (for example, television and the Internet), educating them to be digitally aware is the most effective way of safeguarding them from being exposed to harm that being in these environments could bring. Berson and Berson (2003) also assert that effective citizenship is derived from a digitally literate population. This is a similar argument to a scientifically literate citizenship.

Equipping students with scientific and digital literacies and skills in order to prepare them for the 21st century workforce and citizenship are not separate educational processes. Multiple literacies,, a concept similar to 'multiliteracies' that was first proposed by The New London Group (1996), is based at one level on the influence of communication technologies on meaning making in education. A multiple literacies pedagogy focuses on multimodal ways of conveying understanding and digital technology offers multimodality that enables students to learn as well as demonstrate understanding in science.

DIGITAL TECHNOLOGIES

In the context of this chapter, digital technologies refer to a subset of electronic technologies that include hardware and software and which are used by children and adolescents for educational, social and/or entertainment purposes in the school and at home. These technologies would include desktops, mobile devices (laptops, tablet PCs, ultramobiles, mobile phones, smartphones, PDAs, games consoles), resources on the Internet (information, multimedia and communication resources), digital recording devices (cameras, voice and video recorders), data logging equipment and the myriad of software for different devices that are either commercial or free on the World Wide Web (WWW).

Digital Literacy

Accompanying the rapid growth of computer-based technologies and associated resources has been the increase in a range of terms related to its literacy. As a result there are many terms similar in definition to 'digital literacy'. Included in the array of computer-related literacies are ICT literacy, information technology literacy, technology literacy, media literacy, information literacy, net literacy, online literacy and digital literacy. Attempts at making distinctions between them have been made, for example, by Markauskaite (2006), Ng (2006) and Oliver and Tomie (2000).

Aviram and Eshet-Alkalai (2006) and Eshet-Alkalai (2004) have reviewed the numerous definitions for digital literacy in the literature, stating that the broadness of the definitions covers, singly or in combination, meanings that are technical, cognitive, psychological and/or sociological. An example of a broad, encompassing definition of digital literacy is one formulated by the *European Information Society* which states:

Digital Literacy is the awareness, attitude and ability of individuals to appropriately use digital tools and facilities to identify, access, manage, integrate, evaluate, analyse and synthesize digital resources, construct new knowledge, create media expressions, and communicate with others, in the context of specific life situations, in order to enable constructive social action; and to reflect upon this process. Martin (2005, p. 135)

At a more specific level, and in attempting to draw the different perspectives of digital literacy together by proposing a conceptual framework for the term, Eshet-Alkalai (2004) suggested that there are five types of literacies that are incorporated within the term 'digital literacy': (i) photo-visual literacy; (ii) reproduction literacy; (iii) information literacy; (iv) branching literacy and (v) socio-emotional literacy. The conceptual framework is aimed at enhancing the understanding of 'how users perform with tasks that require the utilisation of different types of digital skills'. (p. 94).

In this chapter, the definition of digital literacy will draw together common features of the different computer-associated literacies mentioned above and discuss the technical, cognitive and social dimensions of digital literacy. Most definitions of computer-associated literacies have not included skills required for learning and socializing in digitally-connected social networks such as Web 2.0 technologies. A discussion of these skills as part of being digitally literate will be made in the chapter. A major focus of the chapter will be on discussing digital literacy skills that will assist with the acquisition of science knowledge and the communication of science understandings.

The next three sections will discuss digital literacy in terms of (i) the technical and cognitive ability to use software (ii) search and assess information on the WWW and (iii) learning and socializing in networked communities. Based on these discussions, a definition for multiple literacy for this chapter will be made.

The Technical and Cognitive Ability to Use Software

Digitally literate individuals should be able to carry out basic computer-based operations and access resources for everyday use. They should be able to select and use the most appropriate technological tool(s) and features within the tool(s) to solve problems, synthesis new information or to create products that best demonstrate their understanding.

Technology tools include hardware and software. In education, hardware tools are the physical devices - desktops, laptops, handheld and mobile devices, printers, interactive whiteboards and other more specific subject-based tools such as datalogging equipment and probes. Handheld and mobile devices include PDAs, mobile phones, tablet PCs, games consoles, iPods, MP3 players, smartphones, digital cameras and video recording devices. At the most basic level, a digital literate person should be able to connect together a functional computer system for his/her own personal use, for example a desktop to a printer. The ability to read manuals to conduct basic technical activities is part of being digitally literate.

Another aspect of being digitally literate is the ability to use *application software*, tools that enable the end users to perform and accomplish specific tasks. The common types of educational software that are commercial and/or are open sourced and free are:

- Productivity software, such as tools for: word processing (e.g. Word) and authoring (e.g. Macromedia Dreamweaver, Microsoft FrontPage and Flash Animation Maker);
- Constructing spreadsheets (e.g. Excel), creating databases (e.g. FileMaker), concept maps (e.g. Inspiration), crossword puzzles and other worksheets/tests (e.g. Hot Potatoes);
- Drill and practice, games, tutorial and problem-solving software;
- Demonstration and presentation software, for example, PowerPoint and PhotoShop;
- Reference software e.g. Grolier Multimedia or Britannica Encyclopedia; dictionaries;
- Simulation software. There is a range of this, for example, Sims series, Crocodile series and many in the form of java applets for science;
- Data analysis software, for example, statistical packages (such as SPSS) and qualitative data analysis packages (such as NVivo); and
- Other shareware, demoware and freeware from the WWW.

There are both technical and cognitive skills involved in using these software packages effectively. Technical skills involve downloading programmes (if it is from the WWW), extracting (such as unzipping) and installing them into the computer or mobile devices. Other technical skills involve operating the software and understanding common features/instructions of software to navigate through them, for example when to click, double click on a function, or use highlight, drag, expand or drop features. Cognitive skills associated with using software include the ability to evaluate and select appropriate software programmes to construct new meanings. At the level of demonstrating understanding of knowledge acquired, selecting the most appropriate software to do this requires a good understanding of the capabilities of the software and how it can be used to convey the understandings.

Science teachers have expressed concerns for software programmes that have complicated and non-user friendly interfaces as it often deterred students (and themselves) from staying focused on tasks at the computer and spending too much time on the technical aspect of the software (Ng & Gunstone, 2003). Examples that these science teachers cited that are difficult to use technically include datalogging software and Flash Animation software. As part of being digitally literate in using software programmes, it is necessary to invest time in exploring this software, including reading the manual and help tools. Skills to evaluate the effectiveness of the software and to determine whether it is worthwhile using the software are necessary aspects of being digitally literate. Using technology for the sake of using it serves no advantages in helping a student increase his/her knowledge in any field of studies.

Search and Assess Information on the WWW

Information that is available digitally could be found online via the WWW or offline, usually from CDs or DVDs. This section will only focus on the accessing and using of information from the Web as the enormous amount of freely available information at no costs to learners from this source has implications on digital literacy skills. A digitally literate person should be able to search, locate and assess Web-based information effectively. Assessing information involves the ability to critique through analysing and evaluating digital content for accuracy, currency, reliability and level of difficulty.

The information on the WWW is presented in varied styles for learning and entertainment. Visual and audio presentations include text, images, graphics, diagrams, flow charts and multimedia formatted materials such as Flash animations, applets and other simulations. A Web-based learning environment also support communicative resources such as emailing, chat rooms, blogs, Skype and MSN, that enable communicating and sharing of ideas across distance easier and quicker. These features provide opportunities for flexible learning and for learners to carry out independent research and learning anytime and from any place where there is access to the Internet, including through the use of mobile phones and handheld computers. The variety of material also means that a commensurate sophistication is required in the accessing, comparing and evaluating of the content of the resources.

Digital literacy skills that are necessary for researching and using Web-based information include being able to:

- handle the WWW appropriately, such as being able to distinguish between the different search engines and using the more 'suitable' search engine for a particular purpose
- narrow down the search using appropriate keywords to maximise precision and to reduce the number of pages that the learner has to read, for example, using multiple (3-4) keywords in the search would yield better results than a single keyword
- critique the contents of Webpages in terms of accuracy, currency, reliability and level of difficulty. In addition, studies by Wallace, Kupperman, Krajcik and Soloway (2000) and Hoffman, Wu, Krajcik, and Soloway (2003) have shown that the use of 'search and assess' inquiry skills where information at one site is critically analysed before doing another search, is necessary for effective understanding of the content from Web-based resources (Ng, 2006)
- understand ethical and moral issues associated with writing that uses Web-based resources, for example copyrights and plagiarism
- synthesise new understandings using appropriate technology tools that will convey the meanings in the best manner:

These are higher order thinking skills, described by McKenzie (1998) as 'infotective' skills that are necessary for today's students to develop. McKenzie describes these students as 'free-range students' who have unlimited opportunities to graze the Net freely. He described students using the WWW as 'infotectives' who should be able to direct their own learning individually or through collaboration to solve problems by applying search and sort skills, and skills to evaluate their way through the piles of often fragmented information and to rearrange them in some meaningful format. Citing from his article *Grazing the Net: Raising a Generation of Free Range Students* at: http://www.fno.org/text/grazing.html, he indicated that:

We must also give students the tools to overcome the weaknesses of the new information sources.... The extensive information resources to be found in cyberspace are both a blessing and a curse. Unless students possess a toolkit of thinking and problem-solving skills to manage the inadequacies of the information landfills, yard sales, gift shoppes and repositories so prevalent on the "free Internet," they may emerge from their shopping expeditions and research efforts bloated with techno-garbage, information junk food or info-fat.

McKenzie further argued that the decision making skills in carrying out infotective actions are the same sorts of skills that students will use in deciding about important issues that affect their own lives. He added that to be successful with the promoting of the digital literacy skills mentioned above, there needed to be an emphasis on the development of questioning skills, and that topical research needed to be replaced with projects requiring original thinking.

Learning and Socializing in Networked Communities

Millions of young people 'meet' online to chat, exchange ideas, communicate socially and collaborate on project. Web 2.0 technologies such as Wikis, Flickr, MySpace, Blogs, Facebook and YouTube enable individuals to contribute to networked communities for learning or for socialising. As participating in online environments is fast becoming a lifestyle of the younger generations, the issue of cyber safety and potential risks in participating online is increasing (Conroy, 2007; Hanewald, 2008). Cyber-safety is about keeping young people safe online. The potential risks that children face when online include being bullied, stalked and harassed and exposed to identity fraud and inappropriate materials such as pornographic, illegal materials, spam and computer viruses. Hanewald (2008), in her literature review on the research into cyber safety cited work conducted in the US (Finkelhor, Mitchell & Wolak, 2000; Wolak, Mitchell & Finkelhor, 2006), in the UK (Smith, Mahdavi, Carvalho & Tippett, 2006) and in Canada (Li, 2004) to show an increase in primary school-aged children and adolescents' experiences with cyber-bullying and harassment. For example, the US studies indicated a rise in cyber bullying from 28% in 2000 to 48% in 2006. The various studies also indicated that text messaging, emails and chatroom postings are the most common means of cyber bullying. Unlike physical bullying where the perpetrator is known, the cyber perpetrator could be difficult to identify. The resulting impact of cyber bullying and harassment on children and young adolescents could lead to both short and long-term psychological harm which is manifested in depression and anti-social behaviour. This in turn would have negative implications on family lives and schooling.

As part of being digitally literate is an awareness of cyber safety (Becta, 2006) and the need to keep personal information as private as possible. A digitally literate individual should be able to recognise when (s)he is being threatened and know how to deal with it, for example whether to ignore, report or respond to the threat. The fostering of respect and responsibility when communicating through digital technologies and the understanding of ethical and cultural issues associated with digitally-based communities should be a part of the digital literacy education of students (Berson & Berson, 2003).

Digital Literacy and Related Skills

In light of the above discussions that are related to using digital technologies in educational settings, being digitally literate requires the development of a set of key skills that are both technical and cognitive. A digitally literate person should be able to:

1. carry out basic computer-based operations and access resources for everyday use
2. search, locate and assess information effectively. Assessing information involves the ability to critique through analysis and evaluation digital content for accuracy, currency, reliability and level of difficulty
3. select and use the most appropriate technological tool/features to solve problems or to create products that best demonstrate new understanding
4. behave appropriately in social network communities and protect oneself from harm in digitally enhanced environments.

Teaching these skills in context and providing opportunities to practice them in ways that demonstrate their significance in making choices is essential and invaluable to both the personal and academic development of the students as digitally literate learners.

In the next section, the concept of 'scientific literacy' will be discussed. This will be followed by a discussion on 'multiple literacies', a concept that links digital literacy and science literacy. An argument that being digitally literate will foster the development of scientific literacy will be made.

SCIENTIFIC LITERACY: WHAT IS IT AND WHY THE NEED FOR IT

Broadly, there are two arguments for the purpose of science education in the compulsory years of schooling in most countries across the globe. From a socio-economic perspective, and one that governments will argue for, is that science knowledge will advance research and development, bringing health to the nation's people and wealth to the nation's economy. Hence creating a 'clever' country where the training of future scientists and engineers begins in the school classroom is a worthwhile investment. Most educators, however, would argue for a more inclusive science education, that is, to teach students to become scientifically literate citizens as they enter the workforce in their adulthood and to become independent thinkers. The OECD/PISA[1] (2003, p.133) defines scientific literacy as:

The capacity to use scientific knowledge, to identify questions and to draw evidence-based conclusions in order to understand and help make decisions about the natural world and the changes made to it through human activity.

The definition emphasises that individuals should be able to use functional knowledge and skills in science to be active participants of society. Similar discourses about being scientifically literate are also made by Goodrum, Hackling and Rennie (2001), the National Research Council (1996), Whittle and Maharjan (2000) and the American Association for the Advancement of Science (1993) in Project 2061. For instance, Hackling, Goodrum and Rennie (2001, p. 7) asserts that scientific literacy should be a high priority for all citizens as it helps them (i) to be interested in, and understand the world around them (ii) to engage in discourses of and about science (iii) to be sceptical and question claims made about scientific matters (iv) to be able to identify questions, investigate and draw evidence-based conclusions and (v) to make informed decisions about the environment and their own health and well-being.

Shen (1975) however, breaks the broad definition of scientific literacy into three components: practical, cultural and civic. The practical aspect of Shen's scientific literacy is based on science being an integral part of everyone's life, and that science knowledge and skills provide practical assistance in helping people make informed decisions and choices of the way of life that are best suited for them. The cultural aspect of Shen's scientific literacy is about the importance of learning science in its social and human context while the civic aspect enables citizens to be able to read and understand reports and articles in the media, to engage in public conversations about the ethical, moral and social issues of scientific discoveries and to have influence over governments' decision making. Examples of scientific issues that are reported in the media and the subject of public debates include areas in cloning, nanomaterials, climate change, global warming, the entire genome sequenced, genetically modified food, space exploration, new design drugs, nuclear and other forms of alternative energy.

Hazen (2002) adds an aesthetic dimension of being scientifically literate as involving a deeper appreciation of everyday activities and living through the understanding of science. Apart from the aesthetic dimension, he also argues for the importance of scientific literacy from the civics and intellectual coherence perspectives. From the civics perspective, he argues that the general welfare of a nation is stronger if its citizens are scientifically informed. From the intellectual coherence perspective, he argues that because our society is so inextricably linked to science discoveries that they often play an important role in setting the intellectual climate of periods in human history. For example, nano-science and technology, hailed by some as the fifth industrial revolution (Treder, 2004) is currently one of the dominating topics of research and discussions.

Hazen (2002) summarises scientific literacy as simply a mix of concepts, history, and philosophy that help an individual understand the scientific issues of his/her time. He stated that a scientifically literate person has a broad understanding of the most general principles of science and possesses within his/her knowledge sufficient facts and vocabulary to understand scientific issues in print and electronic media with similar ease as understanding news in politics, sports or the arts. Hazen also noted that scientists with expertise in a particular field often lack scientifically literacy characteristics. He asserts that it is not uncommon to find physicists being uninformed in areas outside their field of expertise, such as lacking in the broad knowledge of biological sciences. The converse of biologists lacking in the broad knowledge of the principles of the physical sciences is also true.

In addition to the ability to use scientific knowledge and processes for living a better life and for making sensible social and personal decisions, Enger and Yager (2001) added that a scientifically literate person should also be able to understand that there are strengths and limitations associated with science and its endeavours.

SCIENTIFIC LITERACY AND RELATED SKILLS

Scientific literacy can have diverse and important roles in people's lives, and its values and skills, inherent in the knowledge and processes of science, provides a strong justification for science education in schools. Based on the views presented above, the skills associated with being scientifically literate are the ability to:

1. develop an understanding of the general principles of science and the process of how scientific knowledge is generated;
2. demonstrate understanding of scientific principles and processes learned, including the strengths and limitations of science theories, and questioning the integrity of evidence shown and any ethical or social concerns that may be associated with it;
3. read and critique media reports and if necessary to undertake further research and analysis of the topic;
4. collaborate and communicate with peers, experts (e.g. via emails or blogs) and the general public (e.g. through newspaper, journals or magazines) about science-related issues; and
5. have an awareness that technology, including digital technology, plays a major role in the advancement of science either for discoveries or for communication and collaborative purposes.

MULTIPLE LITERACIES, WHERE DIGITAL TECHNOLOGY INFLUENCES MEANING MAKING IN SCIENCE

There is growing recognition that science learning entails multiple representations of concepts under study as well as the ability to conceptually link the different modes of representations (multimodal) in order to demonstrate their reasoning processes and understandings of the concepts (Ainsworth, 1999; Prain & Waldrip, 2006; Waldrip, Prain & Carolan, 2006). Multiple representations in the study of science mean the re-representing of the same concept in different ways, for example, in a practical report or PowerPoint slide or through role play. Multimodal representations refer to the simultaneous use of two or more modes of representations e.g. a table of results also represented in graphical form and using text or audio recording

Table 1. Modes of representation in science enabled by digital technology and software

Mode of representation	Demonstrated in	Digital technology/software (examples of)
Written	Text	Word
	Worksheets; puzzles	Word; Hot Potatoes
	Assignments; projects	Word; Web search for information
Verbal	Oral presentations	Audio/video recording; podcasts
Visual	Drawings/diagrams	Drawing software; digital camera to capture diagrams; scanner
	Concept maps	Inspiration; Kidspiration
	Tables and graphs	Excel
	Animations/simulations	Flash, java applets; simulated experimental work
	Presentations	PowerPoint
Embodied	Role play; science drama	Digital. video recorder; digital camera
	Experimental work & report	Word for text; Excel for graphs; digital camera to capture results; video recording a process
	Modelling (hands-on)	Digital camera and or video recorder
3-dimensional modes	Models (visual)	Modelling software e.g. RasMol for molecular structures

to further describe and explain the graphical representation. Russell and McGuigan (2001) assert that learners need to be provided with opportunities to construct a variety of representations of a concept where they refine their representations in different modes as they make more explicit their conceptual understandings.

The influence of digital technologies on multimodal meaning-making is one aspect of 'multiple literacies', a term first proposed by the New London Group (1996). The term is intended to highlight two arguments: (i) as today's society becomes more linguistically and culturally diverse, meaning making differs according to cultural, social and professional contexts and (ii) media and communication technologies enable meaning making to be increasingly multimodal where the written-linguistic modes are integral of visual, audio, gestural and spatial patterns of meaning. In science the modalities of representations afforded by digital technology are shown in Table 1. These include written, verbal, visual, embodied and 3-dimensional mode. The table indicates that the different modes could be presented or captured with digital technology.

An example of multimodal demonstration of 'gravity' by year 6 students using handheld computers and assisted by their teacher, is shown in Figure 1. The first representation is in text mode only, defining gravity as the 'pull by the centre of the earth'. The student provided further evidence to say that he is being pulled by gravity since he is not floating. The second representation is presented in two modes – drawing and text (labeling). The downward arrow indicates gravitational force pulling the person 'downwards' while the balloon is not affected by gravity and is moving in an upward direction. In addition to the multiple and multimodal representations of the first two diagrams, the third representation is an animation (website http://www.physicsclassroom.com/mmedia/newtlaws/efff.cfm) of the impact of gravity on an elephant and a feather in the absence of air resistance and (website http://www.physicsclassroom.com/mmedia/newtlaws/efar.cfm) in the presence of air resistance. The animations are each accompanied by a text description (which is

Figure 1. Multiple and multimodal representations of 'gravity'

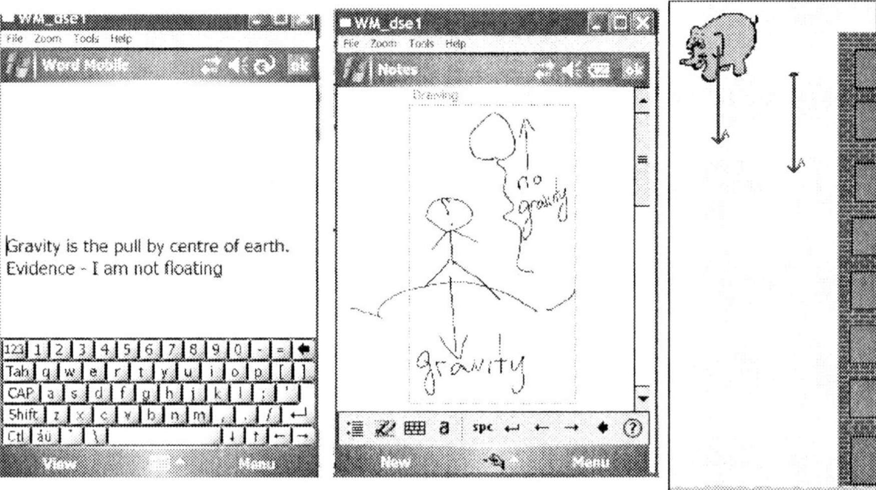

complex for year 6 students) for the free falls in the absence and presence of air resistance. Similar to the previous two representations, this last representation also demonstrates gravity pulling the elephant and feather, with contrasting weights, toward the earth.

The additional dimension of forces shown by these animations is the concept of 'air resistance' exerting an upward force. The pictures in Figure 1 depict different modes of representing the students' initial view of 'gravity' and each could be probed deeper in terms of the accuracy of the representations and meaning of the concept followed by re-representing its meaning in similar or other modes. Hence teaching students to represent understanding in different modes is fostering science literacy and being digitally literate helps with selecting the appropriate application software to demonstrate them. The latter point will be elaborated in the next section.

ENHANCING SCIENTIFIC LITERACY THROUGH DIGITAL LITERACY

Multiple literacies and multimodality emphasise the many ways and contexts in which we communicate and develop understanding. The argument in this chapter is that the more digitally literate an individual is, the better equipped (s)he is to learn science and to communicate that understanding. The influence of technology on many aspect of students' learning of science is highlighted below. The skills required to learn science independently to develop scientific literacy with the help of technology will also be presented.

Use of Technology in Science Learning

Some of the ways students learn science, via the different modes, are:

- Digital devices have made it possible to learn science through audio means, for example listening to podcasts for instructions or lectures in MP3 players or iPods. This allows students to move at their own pace with the learning.
- Information in text books are static and contain text and visual representations. Most text books these days come with CDs which contain interactive modes of representations, including links to websites.

- Students record experimental work by taking digital images or video recordings of procedures and results that could be incorporated into their reports. They use Excel to plot graphs or create tables. They use drawing software to create diagrams of experimental results.
- Some students may share their data from practical work with a wider audience by posting their video recordings on YouTube while others may describe experiences on blogs.
- There are numerous software packages that assist students to learn. These include visualisation software that helps students to understand abstract concepts better or be engaged with virtual laboratory experiments, for example, performing titration or building an electrical circuit.
- There is a huge amount of information for almost every science topic on the Web. This and other Web-based resources such as animations, videos and interactive simulations enable students to take their learning as deep as they wish.
- Through online learning communities such as listserves, virtual conferences and open access encyclopedia such as Wikipedia, students can further increase their knowledge in science.

Whether students listen to podcasts, read electronic text books or access information on the Web, learning science would require them to be able to evaluate the information, identify keywords and their meanings and make appropriate connections from one concept to another. The student may be required to conduct further research if understanding of concepts under study is not forthcoming. Further research could take the form of questioning the teacher or another adult, consulting reference materials online or from the library or posting questions in online networked learning communities. A digitally literate student would possess the necessary skills, as discussed in the section 'Search and assess information on the WWW' above to effectively support his/her research for more science-related information pertaining to the topic under study. Being digitally literate when working with information on the Web also means an understanding of the types of resources that are available. For example, in approaching online communities and forums to help clarify thinking about a problem, the student would search for appropriate websites where scientific Q & A forums are available. For example, for higher education students, it would be appropriate to visit and post questions at the *Ask the Expert at Scientific American* website (http://www.sciam.com/askexpert_directory.cfm), and for more school-based questions and discussions, it would be appropriate to visit the *Ask Science Questions* website at http://www.asksciencequestions.com/, the *Science Club* website at http://scienceclub.org/kidquest.html or the Science and Mathematics section of Yahoo site at http://uk.answers.yahoo.com/dir/. Understanding the 'culture' of virtual learning communities such as these and other virtual forums and having the necessary academic and social skills to communicate in these environments are important aspects of developing students' scientific literacy. As these skills are skills that digitally literate students learning in digital environments should possess, being equipped with these digital literacy skills will assist students to learn science better and quicker to develop as scientific literate citizens as well as life-long learners.

Demonstrating and Communicating Science Understanding through Technology

Another aspect of being scientifically literate is the ability to (i) demonstrate understanding of scientific principles and processes learned and (ii) question the integrity of evidence shown and any ethical or social concerns that may be associated with it.

Digital tools provide effective means of constructing science understanding and communicating it. Digitally literate students who are well-versed with an understanding of the nature and operation of the different types of available software can selectively use these tools to demonstrate science understanding and use them for the synthesis of new ideas. At the operational level, it is essential that students have mastery of the features that are offered by particular software and the technical skills to operate them. For example, to construct an interactive piece of work on the lifecycle of a butterfly for primary students does not need special application software since the standard Word programme is able to do it, as shown in Figure 2. In constructing the life cycle in Word, the technical and cognitive skills required include: (i) the use of text boxes for writing text and moving images around the document; (ii) providing background colours to highlight differences; (iii) inserting pictures and arrows in the document; (iv) finding copyright-free images and animations on the Web or from books, and if it is from the latter, the images will have to be scanned in and saved; (v) deciding on whether to save images as jpg, bitmap or gif files and whether it is necessary to compromise file size with resolution and (vi) use of hyperlinks. Hyperlinking is a powerful feature that allows readers to go into more depth into information related to the topic. Links could be made to Web pages showing animation (as shown in Figure 2) or to images, videos and other text files, all of which could be online and/or offline.

In doing presentations to convey understanding, students should be able to decide if a PowerPoint presentation, a Web page or an alternate technology is the better way to get across the ideas that (s)he would like to convey. For example, if the student wanted to show the number of, and the different food chains in a food web, then using animated food chains where one chain at a time shows up on the screen would be one way of presenting the information. This could be done

Figure 2. An interactive page on the lifecycle of a butterfly using Word features

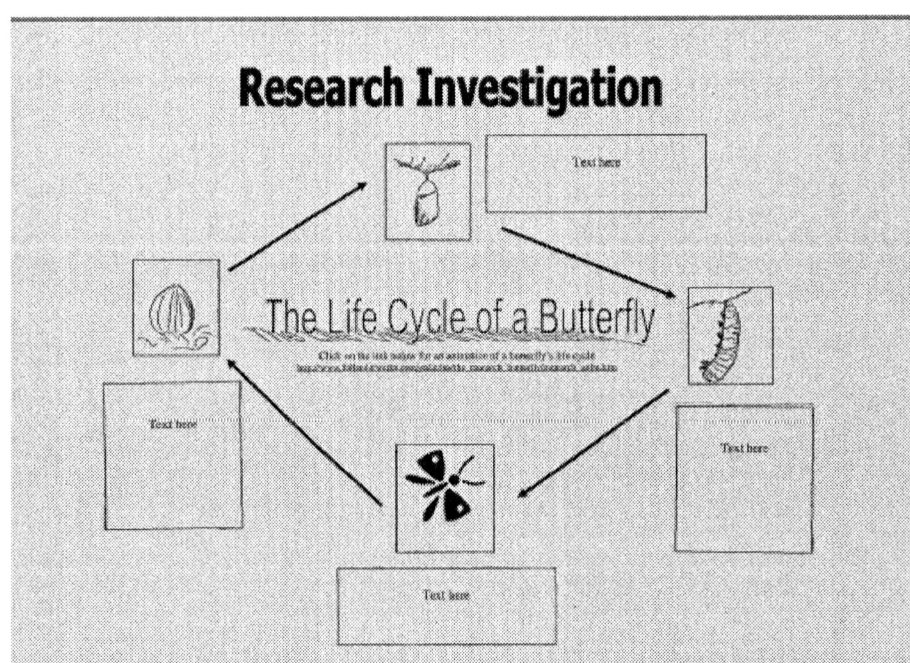

using PowerPoint or Flash animations. Presenting the student's understanding in this manner may demonstrate clarity of thinking in his/her mind of what a food chain is and how they are connected to form a food web. In other PowerPoint presentations where there is a need to show a table that cannot fit into a slide of the presentation, a hyperlink from the PowerPoint presentation to a html page that shows the table would be a better way of presenting the table as the student could scroll up and down the page to show a complete table to the audience.

Digitally literate students are also able to decide on the manner of presentation for project work. As an example, the essay on *Ethics in Space Exploration* in the Appendix is a learning outcome from a group of second year tertiary students studying about Ethics in virtual teams (Ng, 2008a). Accompanying the essay or as an alternative presentation to the essay could be a visual presentation (see Figure 3) constructed with the concept mapping software *Inspiration.* The concept map provides an 'at-a-glance' graphical organisation of key concepts (keywords) and their inter-relationship via linking phrases. Apart from using key words to represent key concepts or key ideas in the nodes, the students could insert images, multimedia or video files to represent the ideas. The software also has auditory functions for learners to voice record short statements or explanations in the nodes. To show the relationships between these multimodal ways of representing key ideas, labeled links are created between them. The concept map in Figure 3 shows visually the three main ethical issues of (i) the high costs associated with space exploration, (ii) potential risks experienced by space travelers, including a disparity between the rich and poor and (iii) issues related to animal experimentation during space exploration. The linking nodes provide details to key issues or outcomes surrounding these ethical concerns.

Figure 3. Visual representation of Ethics in Space Exploration

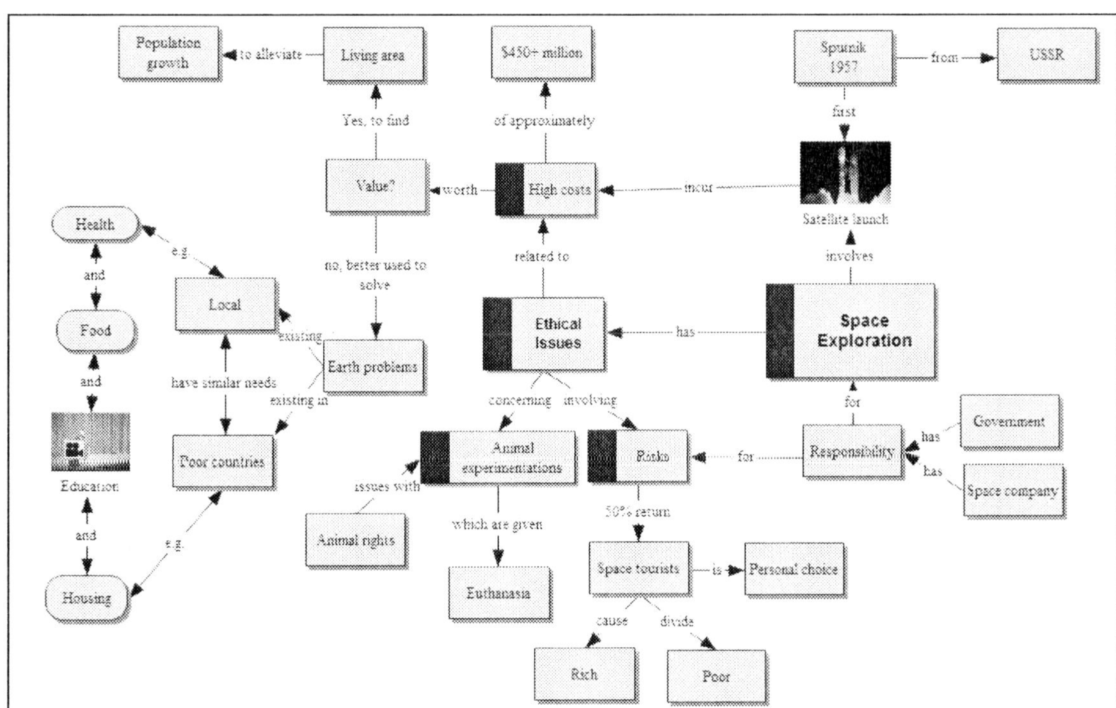

Concept mapping tools provide flexible means for organizing conceptual understanding, allowing students to organize ideas in a logical but not rigid manner, with the option to build on the maps as they progress through their learning. Understanding these options as digitally literate students would assist them to reflect and re-organise thinking in developing their science literacy skills and content knowledge.

Developing Science Literacy through Collaborative Work in Online Environment: WebQuests and Web 2.0 technologies

Developing skills to work collaboratively in teams to produce meaningful learning outcomes is an aspect of being scientifically literate. Collaborating through WebQuests projects and the use of Web 2.0 technologies as collaborative and display tools are ways of fostering the collaborative skills of science literacy. A digitally literate student would be able to contribute effectively to teamwork undertaking WebQuest projects or working with Web 2.0 technology-based projects.

WebQuests

WebQuests are inquiry oriented, web-based activities requiring students to work in small teams to explore online a large body of information in a content area and complete a research task (Dodge 1997). WebQuests are created by teachers according to a standard format (Dodge, 1997) and are usually uploaded to the school's intranet or server. A WebQuest poses an authentic problem and provides relevant websites to direct students' learning. The six components in a standard WebQuest design (Dodge, 1997) are: (i) *Introduction* where background information is provided (ii) *Task* where the finished product that is expected from students is described (iii) *Resources* where primarily web-based resources are listed (iv) *Process* where the WebQuest process is broken up into steps providing a framework for students to follow (v) *Evaluation* where the criteria for the evaluation of the product, usually in the form of an evaluation rubric is outlined and (vi) *Conclusion* which brings closure to the quest and students reflect on what they have learnt and propose future directions where appropriate.

WebQuests are often multidisciplinary in its approach and promotes collaborative learning where authentic, real-world problems have to be solved. A well-written WebQuest requires students to analyse a variety of resources and use their creativity and critical-thinking skills to derive genuine and reasonable solutions to a real-world problem (Yoder, 1999). For example, in a group work task, students take on such different roles as that of a biologist, a chemist, an environmentalist and an economist to investigate environmental, health and cost issues associated with acid rain. Technology-based interdisciplinary learning of this type relates closely to students' daily experiences (Raizen, Sellwood, Todd, & Vickers, 1995) and fosters higher order thinking skills when they have to apply their various skills in the different discipline areas to analyse, evaluate, synthesise and communicate information in solving the problem given in the task. The communication aspect of WebQuest tasks is similar to that described in the section above where decisions need to be made on the selection of appropriate software and format of presentation of synthesised meanings. Students who are digitally literate would focus their time on the cognitive rather than technical aspects of the tasks.

Collaborative Learning Using Web 2.0 Technologies

Another option of working collaboratively in digital environment to construct meaningful learning outcomes in science is the use of Web 2.0 technologies. The WWW has traditionally been a source of information for learners. In recent years, changes in the way software developers and end-users make use of the Web has led to the development of hosted services where online communities collaborate to create and share information on the Web. These hosted services, known as Web 2.0 technologies, include wikis, blogs, video sharing sites (YouTube) and social-networking sites such as the popular MySpace and FaceBook sites. In a case study (Ng, manuscript under review) with primary pre-service teachers studying how to teach science, a task was to work collaboratively in small virtual teams to produce science learning outcomes that reflected science knowledge and activities that could be used in the classrooms. Of the 19 virtual groups, eight groups produced wikis and blogspots. Examples are *The science of ice cream* at the wikispaces website http://onlinelearninggroup6.wikispaces.com/ and *Kitchen science* at

http://onlinesciencecollaboration.blogspot.com/2008/03/introduction.html.

For these students, their digital literacy was at a level that provided them with sufficient confidence to focus on the science learning aspect of the task and not the technical aspect of using these technologies. They were able to search for and evaluate relevant information and videos and include them in their wikispace and blogspot sites, producing comprehensive materials with science information and activities that could be shared with their peers for the primary classroom.

LEARNING THEORIES THAT UNDERPIN THE DEVELOPMENT OF SCIENTIFIC LITERACY IN DIGITALLY ENHANCE LEARNING ENVIRONMENT

Within the context of this chapter, the learning theories that support the development of scientific literacy in digitally enhanced environments are educational constructivism and situated learning theories. Educational constructivism draws on the cognitive and social theories of Piaget (1955, 1972) and Vygotsky (1962, 1978) respectively. It posits that the learner is an active participant in the construction of his/her own knowledge and that prior knowledge and a socially interactive environment influence this learning. In learning science in a technologically mediated environment where the interaction between learners and material that is displayed on screen is open and non-linear, learners self-direct his/her own learning by actively analysing, evaluating and making decisions while manipulating the material at hand in order to construct new knowledge or solve a problem. They will constantly have to compare their own prior knowledge of a body of information with that presented in the learning environment and seek means of re-confirming their prior knowledge or to de-construct and re-construct new meanings. In situated learning (Lave & Wenger, 1991) students learn by doing and learning in context. The varying degrees of complexity and the different styles ICT are capable of offering, for any one topic, means that students could actively construct meanings and learn in context at a pace and style that suits them.

Blended into constructivist learning theory is the theory of constructionism. In his books *Mindstorms* (1980) and *The Children's Machine: Rethinking School in the Age of the Computer* (1993), Papert linked constructivism to technology. According to Papert, students are more motivated and engaged in learning when constructing a public artefact that others will see, critique and use. The artefact could be a sand castle, a programmes for a game, a PowerPoint or a theory of the universe. An example of constructionism in the development of science literacy in a digital environment is the production of a wiki on a science topic. Team collaboration fosters social interactions, with each team member bringing with them their prior knowledge on the topic that is shared to begin the construction of the wiki. At the level of the individual team member, (s)he is actively constructing new meanings through (i) active research and assessing of new information (ii)

Figure 4. Conceptual framework of digital literacy empowering the development of science literacy

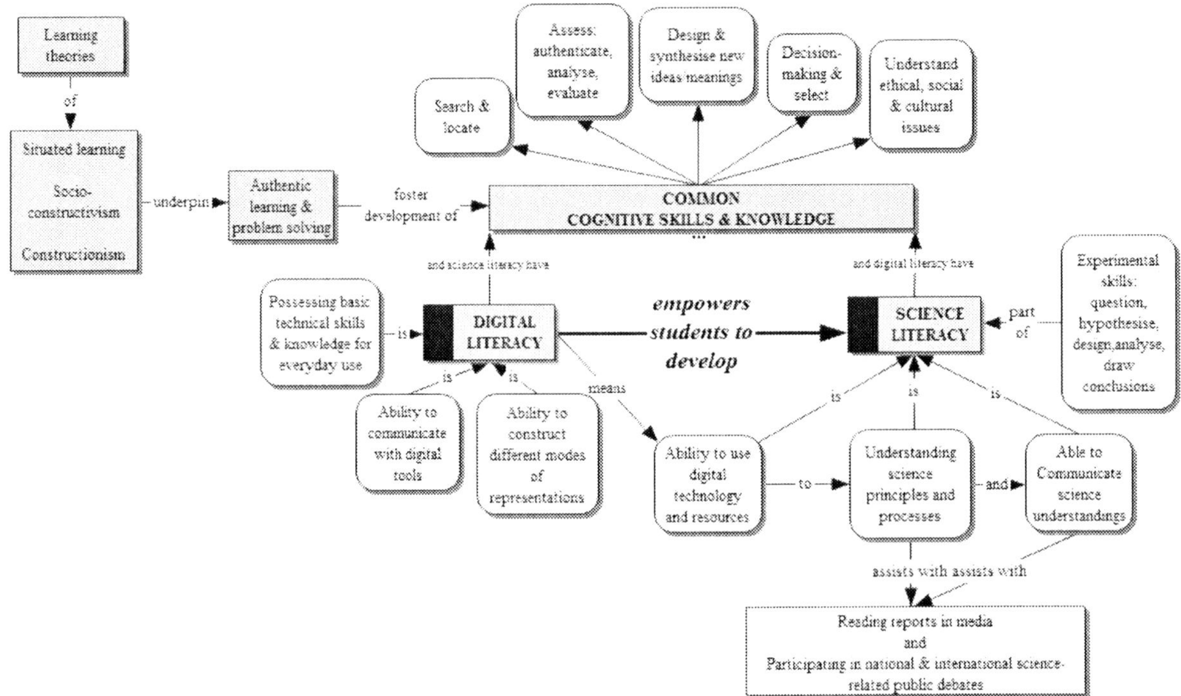

contributing and responding to discussions from other team members and (iii) selecting relevant bits to build the wiki.

CONCLUSION

This chapter has put forward an argument that students can be empowered to develop science literacy if equipped with good digital literacy skills and knowledge. The points argued in the chapter are summarised in Figure 4, which is a framework for enhancing science literacy through digital literacy. The framework

- broadly captures the skills required to develop digital and science literacies. These have been discussed above in the sections for these literacies;
- shows the relationship between the two literacies in that (i) there is a set of skills that are common to the development of both literacies. These are critical thinking and research-based skills that include being able to search, locate, decide, assess (analyse and evaluate) and select for appropriate information; design and synthesise new ideas and meanings and understand cultural, ethical and social issues pertaining to the topic under study and (ii) the ability to use digital technology and resources to learn science and to communicate understandings through multimodal representations empowers students to better develop science literacy
- indicates the need to provide authentic learning opportunities for students to use digital technology to foster development of their digital literacy skills which in turn would help develop their science literacy.

The motivational impact of technology on students' learning has been well documented in the literature (for example, Dywer, 1994; Pedretti, Mayer-Smith & Woodrow, 1998; Mistler-Jackson & Songer, 2000; Ng, 2008b; Ng & Gunstone, 2002, Wallace 2002; Pittard, Bannister & Dunn, 2003). Students are motivated to learn with technology as they have ownership and control over their learning in terms of pace and choice of content. Educators should be capitalising on this aspect of technology to motivate students to learn more effectively. However a concern that teachers have with the use of technology for science learning is the time spent on the technical aspects of the technology which distracts the student from focussing on the science learning itself. For example in undertaking activities like WebQuests, students could lose their focus on learning the content due to their inability to connect information from different Web pages together or because they become distracted with the technical skills of using technology in researching or creating a presentation. Providing time and opportunities to develop digital literacy skills will alleviate these concerns when using technology for learning science.

REFERENCES

Ainsworth, S. (1999). The functions of multiple representations. *Computers & Education, 33*, 131–152. doi:10.1016/S0360-1315(99)00029-9

American Association for the Advancement of Science. (1993). *Science for all Americans: Project 2061*. New York: Oxford.

Aviram, R., & Eshet-Alkalai, Y. (2006). Towards a theory of digital literacy: Three scenarios for the next steps. *European Journal of Open Distance E-Learning*. Retrieved December 29, 2008, from http://www.eurodl.org/materials/contrib/2006/Aharon_Aviram.htm

BECTA. (2006). *Safeguarding children in a digital world: Developing a strategic approach to e-safety*. Retrieved January 5, 2009, from http://publications.becta.org.uk/display.cfm?resID=25933

Berson, I. R., & Berson, M. J. (2003). Digital literacy for effective citizenship. *Social Education, 67*(3), 164–167.

Conroy, S. (2007). *Labor's plan for cyber-safety (fact sheet)*. Retrieved January 5, 2009, from http://www.alp.org.au/download/now/labors_plan_for_cyber_safety.pdf

Dodge, B. (1997). *Some thoughts about WebQuests*. Retrieved January 5, 2009, from http://webque.sdsu.edu/about_webquests.html

Dwyer, D. (1994). Apple classrooms of tomorrow: What we've learned. *Educational Leadership, 51*(7), 4–10.

Enger, S., & Yager, R. (2001). *Assessing student understanding in science: a standards-based K-12 handbook*. Thousand Oaks, CA: Corwin Press.

Eshet-Alkalai, Y. (2004). Digital literacy: A conceptual framework for survival skills in the digital era. *Journal of Educational Multimedia and Hypermedia, 13*(1), 93–106.

Finkelhor, D., Mitchell, K., & Wolak, J. (2000). *Online Victimization: A report on the Nation's Youth*. National Center for Missing and Exploited Children Bulletin, Alexandria, VA, US Department of Justice, Washington, DC.

Goodrum, D., Hackling, M., & Rennie, L. (2001). *The Status and Quality of Teaching and Learning of Science in Australian Schools*. Canberra, Australia: Department of Education, Training and Youth Affairs.

Hanewald, R. (2008). Confronting the Pedagogical Challenge of Cyber Safety. *Australian Journal of Teacher Education, 33*(3), 1–16.

Hazen, R. (2002). *Why Should You Be Scientifically Literate?* Retrieved December 29, 2008, from http://www.actionbioscience.org/newfrontiers/hazen.html

Hoffman, J. L., Wu, H. K., Krajcik, J. S., & Soloway, E. (2003). The nature of middle schools learners' science content understandings with the use of on-line resources. *Journal of Research in Science Teaching*, *40*(3), 323–346. doi:10.1002/tea.10079

Lave, J., & Wenger, E. (1991). *Situated Learning. Legitimate peripheral participation*. Cambridge, UK: University of Cambridge Press.

Li, Q. (2004). *Cyber-bullying in school: Nature and extent of adolescents' experience.* Retrieved January 5, 2009, from http://www.ucalgary.ca/~qinli/publication/cyberbully_aera05%20.html

Markauskaite, L. (2006). Towards an integrated analytical framework of information and communications technology literacy: from intended to implemented and achieved dimensions. *Information Research, 11*(3). Retrieved December 29, 2008, from http://InformationR.net/ir/113/paper252.html

Martin, A. (2005). DigEuLit – a European Framework for Digital Literacy: a Progress Report. *Journal of eLiteracy, 2,* 130-136.

McKenzie, J. (1998). Grazing the Net. [from http://www.fno.org/text/grazing.html]. *Phi Delta Kappan*, *79*(September), 26–31. Retrieved January 5, 2009.

Mistler-Jackson, M., & Songer, N. B. (2000). Student motivation and Internet technology: Are students empowered to learn science? *Journal of Research in Science Teaching*, *37*(5), 459–479. doi:10.1002/(SICI)1098-2736(200005)37:5<459::AID-TEA5>3.0.CO;2-C

National Research Council. (1996). *National Science Education Standards*. Washington, DC: National Academy Press.

New London Group. (1996). A pedagogy of multiliteracies: Designing social futures. *Harvard Educational Review*, *66*(1), 60–92.

Ng, W. (2006). Web-based Technologies, Technology Literacy and Learning. In L. W. H. Tan & R. Subramaniam (Eds.), *Handbook of Research on Literacy in Technology at the K1-2 Level* (pp. 94-117). Hershey, PA: Idea Group Publishing.

Ng, W. (2008a). Virtual teamwork: students learning about ethics in an online environment. *Journal of Research in Science & Technological Education*, *26*(1), 13–29. doi:10.1080/02635140701847421

Ng, W. (2008b). Self-directed learning with web-based sites: How well do students' perceptions and thinking match with their teachers? *Teaching Science*, *54*(2), 26–30.

Ng, W., & Gunstone, R. (2002). Students' perceptions of the effectiveness of the World Wide Web as a research and teaching tool in science learning. *Research in Science Education*, *32*(4), 489–510. doi:10.1023/A:1022429900836

Ng, W., & Gunstone, R. (2003). Science and computer-based technologies in Victorian government schools: Attitudes of secondary science teachers. *Journal of Research in Science and Technology Education*, *21*(2), 243–264. doi:10.1080/0263514032000127266

OECD. (2003). *The PISA 2003 Assessment Framework – Mathematics, Reading, Science and Problem Solving Knowledge and Skills*. Paris: OECD.

Oliver, R., & Tomei, L. (2000). Information and communications technology literacy: getting serious about IT. In *Proceedings of ED-MEDIA 2000. World Conference on Educational Multimedia, Hypermedia and Telecommunications*, VA. Retrieved January 5, 2004, from http://elrond.scam.ecu.edu.au/oliver/2000/emict.pdf

Papert, S. (1980). *Mindstorms.* New York: Basic Books.

Papert, S. (1993). *The children's machine: Rethinking school in the age of the computer.* New York: Basic Books.

Pedretti, E., Mayer-Smith, J., & Woodrow, J. (1998). Technology, text and talk: Students' learning in a technology enhanced secondary science classroom. *Science Education, 82*, 569–589. doi:10.1002/(SICI)1098-237X(199809)82:5<569::AID-SCE3>3.0.CO;2-7

Piaget, J. (1955). *The construction of reality in the child.* Routledge & Keegan Paul: London.

Piaget, J. (1972). *Psychology and epistemology: Towards a theory of knowledge.* London: Penguin University Books.

Pittard, V., Bannister, P., & Dunn, J. (2003). *The big pICTure: The impact of ICT on attainment, motivation and learning.* Nottinghamshire,UK: DfES publishing Department for Education and Skills. Retrieved January 5, 2009, from http://www.dcsf.gov.uk/research/data/uploadfiles/ThebigpICTure.pdf

Prain, V., & Waldrip, B. (2006). An exploratory study of teachers' and students' use of multi-modal representation of concepts in primary science. *International Journal of Science Education, 28*(15), 1843–1866. doi:10.1080/09500690600718294

Raizen, S., Sellwood, P., Todd, R., & Vickers, M. (1995). *Technology education in the classroom.* San Francisco, CA: Jossey-Bass Publishers.

Russell, T., & McGuigan, L. (2001). Promoting understanding through representational redescription: an illustration referring to young pupils' ideas about gravity. In D. Psillos, P. Kariotoglou, V. Tselfes, G. Bisdikian, G. Fassoulopoulos, E. Hatzikraniotis, & E. Kallery (Eds.), *Science education research in the knowledge-based society. Proceedings of the Third International Conference of the ESERA* (pp. 600-602). Thessaloniki, Greece: Aristotle University of Thessaloniki.

Shen, S. P. (1975). Science Literacy and the Public Understanding of Science. In S. B. Day (Ed.), *Communication of Scientific Information* (pp. 44-52). Basel, Switzerland: Karger.

Smith, P., Mahdavi, J., Carvalho, M., & Tippet, N. (2006). *An investigation into cyber bullying, its forms, awareness and impact, and the relationship between age and gender in cyber bullying* (A Report to the anti-Bullying Alliance). London. UK. Retrieved January 5, 2009, from http://www.anti-bullyingalliance.org/

Treder, M. (2004). *Jolt to the system: the transformative impact of nanotechnology.* Retrieved December 29, 2008, from http://www.crnano.org/Speech%20%20Troy%20%20Mike%20Treder,%20CRN.ppt

Vygotsky, L. (1978). *Mind in society: The development of higher psychological processes.* Cambridge, MA: Harvard University Press.

Vygotsky, L. S. (1962). *Thought and language.* Cambridge, MA: MIT Press.

Waldrip, B., Prain, V., & Carolan, J. (2006). Learning junior secondary science through multi-modal representations. *Electronic Journal of Science Education, 11*(1), 87–107.

Wallace, R., Kupperman, J., Krajcik, J., & Soloway, E. (2000). Science on the Web: Students on-line in a sixth grade classroom. *Journal of the Learning Sciences, 9*(1), 75–104. doi:10.1207/s15327809jls0901_5

Wallace, R. M. (2002). The Internet as a site for changing practice: the case of Ms Owens. *Research in Science Education*, *32*(4), 465–487. doi:10.1023/A:1022477832695

Whittle, P., & Maharjan, D. (2000). Scientific and technological literacy for sustainable development in the 21st Century. In *Proceedings to the International symposium BioEd 2000: The challenge of the next century*. Paris. Retrieved January 5, 2009, from http://www.iubs.org/cbe/pdf/whittle.pdf

Wolak, J., Mitchell, K., & Finkelhor, D. (2006). *Online victimization of youth: five years later* (National Center for Missing & Exploited Children bulletin - #07-06-025). Alexandria, VA.

Wynne, M. E., & Cooper, W. L. (2007). *Power up: The campaign for digital inclusion*. Retrieved December 28, 2008, from http://www.digitalaccess.org/pdf/White_Paper.pdf

Yoder, M. B. (1999). A productive and thought-provoking use of the Internet. *Learning and Leading with Technology*, *26*(7), 6–11.

ENDNOTES

[i] Organisation for Economic Co-operation and Development/Programmesme for International Student Assessment

[i] www.seasky.org/spaceexp/sky5d.html

[ii] http://portal.unesco.org/shs/en/files/8460/11223752131RationalesSpaceExplor.pdf/RationalesSpaceExplor.pdf

[iii] http://www.nasa.gov/centers/kennedy/about/information/shuttle_faq.html#10

[iv] www.engin.umich.edu

APPENDIX

Space Exploration: The Ethics

Team 2.

Unit: Science & Technology in Contemporary Society

In terms of ethical debates and issues, the arguments underlying space exploration are not very often raised in the media, unlike those commonly discussed, such as human cloning and abortion for example. However once the statement "Should so much money be spent on space exploration with so many problems on our own planet?" is prompted, a very heated debate between opposing sides arises.

The ethical debate surrounding space exploration started at the dawn of the 'space race' in the late 1950'swhen the former USSR successfully launched Sputnik, a satellite into the outer atmosphere of earth.[i] From that instant, governments and organisations have hurled things into space while the moral and ethical concerns have been intensively debated. The initial space flights sparked debates about animal rights where numerous animals such as dogs and primates were sent into space to predict how humans would respond to this new and strange environment. These animals however never returned to earth but were euthanized near the end of their 'mission'.

The development of space exploration has led to the possibility of tourists in space. As a result the ethical considerations of such a possibility need to be considered. The probability of not returning from a space expedition (to the moon) is 50% [ii]. Is it ethically viable to allow tourists to enter into space with such a high probability of not returning? Would it be up to the company running the space expedition, the government or the tourist to decide whether they were going to accept this risk? Would the desire for money of the company governing the space tours outweigh their desire to make tourists fully aware of the potential hazards of such an expedition? Additionally, the cost of being a tourist on a space expedition is undoubtedly going to be rather expensive as launching a space shuttle costs approximately $450 million[iii]. Thus, space tourism will only be available to those who are quite affluent, consequently creating yet a further divide between the rich and the poor. So, is it ethically just that only the rich have access to, and the opportunity to obtain all of the benefits that might arise from, tourism in space?

The risk to, and loss of human life is just the tip of the iceberg when it comes to the moralities surrounding space exploration. Such topics include unnecessarily excessive funds being diverted to space programmes in many countries when that money could be better spent on healthcare, food and housing for the poor or education. Other people are concerned about such things as storage of nuclear waste in space, isn't it enough that we fill our own planet to the brim with highly toxic rubbish that we now have to resort to shooting it into space? [iv]

The possibilities that space exploration could unfold may be enormous, however the ethics associated, and fundamentally those to do with economics, could have a colossal influence on the extent to which such exploration progresses.

Chapter 3
Achieving Multiple Literacy in Science Education: A Classroom Teacher's Perspective

Sinclair Mackenzie
Science Teacher (Physics), Scotland

ABSTRACT

This chapter provides a teacher's view of the role and influence of multiple literacy in secondary school science. Multiple literacy from the author's perspective, (who is a secondary classroom science teacher) is the concept of doing things that teachers have always done in their lessons, but achieving them by incorporating new, engaging, ICT-rich strategies.

"Do you know how to find information?
do you know how to validate it?
do you know how to synthesise it?
do you know how to leverage it?
do you know how to communicate it?
do you know how to collaborate with it?
do you know how to problem-solve with it?
That's the new 21st Century set of literacies"
(Kay, 2008)

INTRODUCTION

Although not specifically aimed at science education, Kay (2008) provides a framework upon which to hang a definition of multiple literacy within the science classroom. How is this possible despite the absence of "digital" in the statement? Multiple literacy from the science teacher's perspective is the concept of doing things that science teachers and educators have always done in their lessons, but achieving them by assimilating new, engaging, ICT-rich strategies, styles and approaches. Although these interpretations of good pedagogy are new, they sit comfortably within existing constructivist models of learning and teaching, such as that described by Brooks & Brooks (1999):

"The teacher's responsibility is to create educational environments that permit students to assume the responsibility that is rightfully and naturally theirs. Teachers do this by encouraging self-initiated

DOI: 10.4018/978-1-60566-690-2.ch003

inquiry, providing the materials and supplies appropriate for the learning tasks, and sensitively mediating teacher/student and student/student interactions" (Brooks & Brooks, 1999, p49)

The shift towards a digital perspective on literacy has been, and continues to be, driven by forces outside of school itself. Beyond the classroom, from an early age, pupils are bombarded with highly stimulating material, including animation, games consoles, instant messaging and streaming video. The ability to bring these familiar modes of communication into the classroom as a tool for learning has become an essential skill. My own discussions with learners in the classroom indicate that, given the option, pupils prefer learning via interactive simulations and animations in preference to traditional text and practical delivery methods. Pupils have singled out the accessibility of animations and simulations as the main reason for their choice. Pupils can, and often do, respond well when they see that their teacher has made an effort to adapt to the preferred learning style of 'digital natives' (Prensky, 2001).

The rate at which learners have embraced the digital age has been highlighted by the 'Did You Know' (Fisch, 2006) and 'Did You Know 2.0' (Fisch & McLeod, 2007) presentations. Yet the idea of 21st Century (or digital) skills and literacy remain contentious (Matthews, 2009) and some science educators have made a call for the return of more basic skills (Doyle, 2009).

Often those arguing against the inclusion of a 21st Century skills set in the classroom suggest that a focus on ICT competency is achieved at the expense of course content, resulting in a lack of knowledge and understanding. The truth is that the two need not be mutually exclusive. Drexler (2008) illustrates the potential benefits of embracing digital literacy in the classroom with the 'Networked Student' presentation. Drexler's vision of the networked student shows that digital literacy entails the harnessing of technology to meet and redefine outcomes. While the objectives themselves may change little over time, the methodology and skills employed to meet objectives embrace new technology.

The shift in thinking presented by Drexler (2008) offers an opportunity for progressive reskilling of learners, who have not always benefited from the slow pace of change in our education system. A graphic illustration of this lack of change in pedagogical methods is provided by Picardo (2009) who argues that we, as educators, cling to traditional approaches to teaching by continuing to teach in the same way we ourselves were taught. In the process, we do little to meet the changing needs of pupils.

These differing opinions are one example of the digital divides that exist within school communities. Perhaps the most obvious of the digital divides results from lack of inclusion due to socio-economic factors (Selwyn & Facer, 2007). The perception of pupils as digital natives compounds the problem. Many teachers are reluctant to trial new ICT ideas in a classroom filled with 'expert users'. Hence there is potential for a section of school staff being left behind as new ICT strategies are adopted by more confident colleagues. A US-based survey (RIT, 2008) reported that many teachers continue to be uncomfortable supervising pupils while they worked on computers as they regard pupils' knowledge to be superior to their own. Respondents cited a lack of CPD in this area as the principal reason for their lack of confidence. Against this background, it is easy to see why many teachers see those colleagues who are embedding digital literacies in their classroom practice as risk takers, not innovators (RIT, 2008).

USING ICT IN SCIENCE EDUCATION

In science education there are many opportunities to embed the use of ICT into learning and teaching. While the current trend is for greater use of online tools and the social internet, often referred to as *Web2.0*, providing internet access to all members of a class for the duration of a lesson can present difficulties for schools. Access to the facilities can generate a barrier in itself as pre-booking of an ICT suite several weeks in advance does not lend itself to digital and ICT resources being embedded seamlessly in the learning process. Fortunately, many valuable day-to-day ICT-based resources and strategies may be implemented off-line.

The use of data loggers in science has increased in recent years. There are several identifiable benefits of introducing data loggers into science lessons. One of the key advantages is the immediacy of results. Before data loggers became available, any quantitative study of motion relied heavily on the ticker timer and an analysis of ticker tape. It can be argued that the kinaesthetics of working with ticker timer charts can enhance pupils' understanding of speed-time graphs (Nuffield Foundation, 2004). However, the very analysis of these paper tapes may act as a barrier to pupils who lack the necessary numeracy skills to understand the construction and interpretation of the charts. Removing the need to draw a graph from a table of data, where choice of scale and plotting of points can be obstacles to success, allows more pupils to achieve the object of the lesson, which is rarely 'draw a graph'. By making a visual representation of the results available to all pupils, the teacher can level the playing field and make the opportunity of achieving an outcome available to all.

Regardless of individual abilities or learning styles, there are circumstances where the tedious nature of a task, such as recording data for a cooling curve, can lead to difficulties keeping students on task. Using the data logger allows students to start the experiment and then work on another task while the experimental data is collected.

When interviewed, pupils are quick to highlight positive experiences with data logging; citing the sensitivity of the probes, fast sampling and the automated graphing of data. Criticism from pupils suggests a boredom factor as the data logger is perceived as 'doing' the experiment on their behalf, a notion of deskilling whereby pupils feel their own practical expertise is no longer developed to the same extent as with traditional practical sessions and a shift in group dynamic from those who are competent at science towards those with stronger IT skills (Ng & Yeung, 2000).

These negative aspects of pupil feedback are a warning that simply using technology for the sake of having technology in the lesson plan does not enhance the learning experience. Rather than simply replacing thermometers or voltmeters with a sensor in a 'straight swap', the way in which practical work is performed and assessed must be adapted.

Consider an investigation of cooling curves. It would be a poor use of technology if pupils stood passively watching an emerging graph for the duration of the lesson. When taking temperature readings manually, pupils at least have the opportunity to practice their observation and recording skills. So how might the activity be structured to take advantage of the technology? Data logging tasks such as these can be set up at the beginning of a lesson to give time for completion, then left running as a background activity. As the experiment progresses, pupils can work on a critical thinking task, such as predicting the shape of the temperature/time graph, explaining the reasons for their hypothesis to others in the group, discussing possible methods to speed up or slow down the cooling process. Comparing these ideas to the results displayed by the logging software and evaluating the validity of their own, and each other's, predictions is a powerful use of self-assessment.

By incorporating the use of the data logger in this way, when practical work is complete, the teacher is free to explore pupils' understanding of what the graph shows in terms of underlying science, rather than spending learning time checking a rote skill has been used, i.e. that each pupil can plot the raw data. This is an opportunity that may otherwise be overlooked. Significant teacher time is required to check the plotting and shape of each pupil's graph, which may not allow time for discussion of understanding, i.e. what graphs reveal about the topic under investigation. Ensuring that pupils can use a source of information, such as a graph, extract the information it displays, use this to inform understanding and pass that learning on to others is a key communications skill, as suggested by Kay (2008).

Opportunities for differentiation in this area include the testing of ideas to alter the rate of cooling, a comparison of cooling curves for different materials or examining the effect of sampling rate on the shape of the curve obtained. By automating the act of taking many measurements, recording them and producing a graph, especially one in which the result is a curve rather than a straight line, we allow all pupils to achieve the initial outcome and choose the direction in which to extend their work with the opportunity to create, test and verify or amend their own hypothesis.

An important, and often overlooked, advantage of data logging is that it enables pupils to succeed in specific practical activities that have been logistically difficult to realise from a traditional perspective. Often there are experiments:

requiring longer than a single science lesson, e.g. annual variation on soil temperature (Turner, 2005)

based on fast measurements, e.g. investigating impulse (Data Harvest, 2008)

requiring very large data sets for valid statistical analysis, e.g. determining radioactive nature of K-40 isotope (Jamieson & Steele, 2008).

Encouraging the use of computers and data loggers within the science classroom can therefore enrich practical activities by providing a greater degree of autonomy for pupils as they work towards learning outcomes and provide better opportunities for personalisation, reflection and self-evaluation of experimental work.

A key aspect of any practical activity in the science classroom is the recording and presentation of the results obtained. Rather than communicating discoveries to the public, the role of this writing in the context of the science classroom has traditionally been to communicate pupils' understanding of a concept to their teacher. The chance to encourage pupils to reflect on their own learning by expression in writing is rarely taken. As a form of assessment, the use of reports may change in nature from formative in earlier years to summative in senior classes. Pupils can find the writing of science reports difficult at many levels due to the challenges presented by unfamiliar vocabulary and constraints on the style of writing (e.g. choice of tense or use of third person where the first person may seem a more natural voice for them when they did, after all, do the experiment themselves).

While teachers can help by modeling good report writing and scaffolding pupils' work with templates or writing frames, learners' self-confidence can be damaged by the marking process. When marking, teachers can forget the sense of ownership a pupil feels toward their own writing. One method for encouraging pupils in their writing is to set aside classroom time for word processing of reports. In doing so, the teacher acts to remove any embarrassment that may be caused by poor handwriting or spelling and the use of a common document template can ensure that each pupil has the correct layout.

Word processing allows rapid sharing of projects and reports for commentary from either the teacher or a peer group Assignments such as reports can be submitted electronically, shared across a network, transferred on to a flash "memory stick" or submitted by email. A further advantage of word processed tasks arises from pupils' different perceptions of word processed and handwritten work. Classroom conversations during the review of written work reveal that pupils treat electronic and handwritten feedback differently. Having access to an electronic copy, instead of a printout, allows the teacher to provide feedback using the commenting option inside the software. Many pupils see this form of feedback as more positive in nature than traditional red pen on a handwritten copy. Electronic feedback is perceived as personalised for the submitted item of work without becoming personal in nature. Learners do not have the same bond with typed work as existed previously with handwritten documents.

The ability to cut and paste means pupils are not left with the task of rewriting by hand an almost perfect page of work for the sake of a single line and do not have to laboriously redraw diagrams or tables on the same page. Using the technology allows more time for the real learning as opposed to time spent copying out sections and redrawing diagrams because of errors elsewhere in the handwritten copy. These are highly motivating reasons for pupils to accept constructive feedback that may have been ignored previously due to the effort required to make changes. The same cut and paste functionality may tempt some students to incorporate the work of others into their own assignment. This is an area where the individual class teacher must be alert to the possibility of plagiarism and, while online detection algorithms exist (e.g. Klug, 2002), education and knowledge of a particular pupil's style of working is often a sufficient solution.

In the same way as word processing can help with structure and presentation of a piece of work, spreadsheet packages lend themselves well to the manipulation and display of experimental data. The ability to display data effectively can have a significant impact on the success of any report in communicating to the reader. Teachers of science often discuss with a class the reasoning for selecting a line graph rather than a bar chart. However, the correct choice of format with an inappropriate scale, lack of a key, inaccuracies in plotting, missing labels/unit or the need of a second y-axis can prevent clear communication of information.

The use of a spreadsheet package allows pupils to focus on how data is presented to the reader, rather than struggle through the counting of squares and selection of a best-fit line or curve. By moving the process away from traditional pencil and paper methods towards the highlighting of data ranges and a sequence of mouse clicks, the teacher empowers pupils to test, and evaluate for themselves, the value of various chart formats before selecting an appropriate style for the display of data that will support and improve the overall communication of information. The associated problem-solving and justification, whether internalised or presented to a group, is an in-depth critical thinking process.

In science classrooms, educators often see pupils express a preference for a kinaesthetic style of learning. However, there are situations where pupils are unable to make connections between the practical activity and the more abstract scientific concept. Within physics, pupils often have difficulty in achieving success in activities where they are required to build and observe the behaviour of a circuit. This is seldom due to a lack of reasoning ability in analysing circuit behaviour and more often due to an inability to see a circuit drawing in terms of the physical components, a problem identified by Driver et al (1994);

"circuit diagrams can be seen as either pictures or as abstractions but it is clear that pupils often find it hard to recognise the circuits in the practical situations of real equipment" (Driver et al, 1994, p124)

Simulation software can be a useful strategy to assist pupils who have difficulty moving between the abstract concept of the circuit diagram and the physical reality of wired components forming a tangible circuit. Software packages such as Yenka Electricity (Crocodile Clips, 2008) and the PhET Circuit Construction Kit (University of

Colorado, 2008) are useful classroom aids which allow pupils to perform the experimental work set for the group without leaving the abstraction of the diagram.

Pupils respond well to simulation software and it can be a valuable method of allowing all pupils to access activities set for a lesson, since it removes barriers to learning and facilitates a more inclusive classroom. In my experience, girls often express a preference for working through tasks concerning electronics on simulation software rather than connecting up physical components. Pupils with poor fine motor skills or dyspraxia can also have difficulty working with small components such as resistors and light emitting diodes (LEDs). For such pupils, the problems become greater as the complexity of the circuit is increased. This might include using parallel, rather than series, circuits, or the introduction of orientation specific components, such as electrolytic capacitors, diodes and transistors. The use of simulation software again allows opportunities to challenge pupils, extend their work beyond limitations of equipment availability, practical ability or real or perceived barriers.

While school resources may limit the opportunities for all pupils to simultaneously sit down in front of a PC to use these, and similar, applications, the teacher should be encouraged to manage resources such that all pupils can gain experience of simulation software and not just those who find assembling of practical circuits difficult. Additionally, certain packages, including those mentioned above, provide web-based versions of their simulators, or offer a more basic version for student and teacher use at home, subject to an end user license agreement for home use only. These solutions can prove to be beneficial, since pupils could continue to have access outside of the classroom for extension work, revision and homework.

ENGAGEMENT STRATEGIES

One of the difficulties common to both practical electronics work and ICT-based simulations is that pupils do not always stop to reflect on the outcome of their activities. Pupils press switches, record their answer in a workbook and move on to the next section of the given task. This experience is echoed in a review of research into the use of computers in education which showed that, while pupils' understanding may increase due to the ability to perform a greater number of simulations than would be possible with real-life experiments:

'students often did not understand that they should be reflecting upon what they found' (Harlen, 1999, p24)

To prevent superficial engagement in similarly abstract or simulated activities, it is good practice to employ formative assessment strategies to enable pupils to identify what it is they are setting out to achieve when participating in the lesson. The use of process success criteria (Clarke, 2005) can be particularly helpful in this regard. Providing process success criteria to pupils can be as simple as telling pupils in advance of an activity what steps will be required of them to fulfill the learning objectives of the lesson. At first glance, this might seem to be a form of spoon-feeding but the aim of sharing success criteria is

'not to give them a simple fix-it list, but rather to remind them of those aspects of the task on which they most need to focus' (Clarke, 2005, p34)

A further advantage of sharing process success criteria is that pupils have a clearer understanding of what is (and is not) important and are better informed for the process of peer and self-assessment as a result, since learning outcomes are made explicit. For example, pupils may be engaged in an activity where a specimen question and answer are discussed, firstly in pairs and then as a class. The objective is for the pupils to produce a mind map while developing a set of criteria to help in producing a "good answer". These criteria are then used as rubric for subsequent lessons whenever the operation of a circuit is described. As a result of this activity, the teacher can direct pupils to the mind map, reminding them of the things the class themselves thought would be needed to improve their answer. This identification of the gap between current achievement and the self-defined criteria for success is particularly useful in that feedback to the pupils can be made specific and relevant to the learning outcome of providing a description of the circuit behaviour.

Teachers can themselves model the operation of a circuit using a single computer and digital projector from the front of the classroom although this should be made more active for the learners, perhaps with an interactive whiteboard (IWB) and a series of group tasks with peer review. The IWB is often perceived as more of a "teacher toy" than a teacher tool (Association for ICT in Education, 2001) despite the large financial resources being allocated to IWB provision in our schools. A major cause of ineffective IWB use in our classrooms is the lack of training beyond initial use of the board accessories and an introduction to the proprietary notebook or flipchart software. There are occasions when it is the software that the classroom teacher finds useful for a valuable teaching point (Mackenzie, 2008a) rather than the hardware of the IWB itself. To obtain true value from the investment in these tools, individual classroom teachers must seek out their own continuing professional development (CPD) opportunities rather than rely on school or local authority provision. Many of the best CPD opportunities are to be found on the internet, with a growing number of IWB experts sharing their knowledge through audio and video podcasts (Hazzard & Badger, 2006; Promethean Inc., 2006) supported by episode specific resource downloads to allow the listener to practice on their own IWB during or after the show. Advanced classroom practitioners also share their IWB skills through weblogs (Picardo, 2008a) and collaborative online documents (Barrett, 2008).

Moving the electronic device from the front of the classroom and directly into the hands of the learner is, in many ways, a more effective strategy than investing in IWBs. The growth of so-called 'handheld learning' has been fueled by the emergence of a variety of consumer electronic devices coupled with an innovative approach to learning by a number of educators seeking to bridge the gap between formal and informal learning (Smith, 1999).

USING RECREATIONAL TECHNOLOGY IN FORMAL ENVIRONMENTS

Much has been made of the Nintendo DS as a platform for handheld learning (Collins, 2009; Evans, 2009; Robertson, 2007). Projects evaluating Nintendo DS games-based strategies for pupil motivation and attainment have reported successful outcomes (e.g. Bray, 2009; Miller & Robertson, 2009). A significant proportion of the focus to date has been at the primary school level with numeracy-related activities taking a lead role.

Handheld devices for learning can also be implemented within the secondary school environment. For the strategy to succeed, the teacher must ensure that the presentation of resources, whether commercial off-the-shelf (COTS) packages or developed in-house, is pitched at an appropriate level to avoid patronising the teenage end user. There are three main methods available for publishing of material on the DS platform; text documents, graphics files and the xhtml mark-up language. Current developments on the Nintendo DS for secondary pupils are centred on the use of plain text files rather than graphics to enable use of the system as a form of eBook reader (Bray, 2008). Alternative methods of enabling the device to be used as an eBook reader are available. One method, supported by The Moon Book Project (Mullins, 2006), provides DS-specific applications to its members, who are then able to download not only games, but also classic literature, films and comics. Similar homebrew' (Wikipedia, 2009) information sites are available, offering different styles of reader software and downloadable content.

These applications offer much more than an opportunity to consume content in a new format, they have changed the role of the pupil from consumer to creator of digital content. When used in conjunction with a basic text editor, pupils can produce their own, self-contained science notes in a language and structure that is more familiar to the teenage brain. Notes can be swapped and shared, almost virally, among classmates as a form of peer review and peer support. Classroom-wide use of the DS therefore offers pupils a chance to take responsibility for their learning by researching a topic and exercising higher-order thinking to re-present the information in a more appropriate language and format for the use of themselves and their peer group. The idea becomes self-differentiating, since pupils can adapt the presentation of descriptions, definitions or procedures until they themselves feel they have achieved clarity.

The ability of the Nintendo DS handset to display information written in xhtml code remains largely untapped by educators in any curricular area at present. This is, in part, due to the difficulty of working within a stricter set of rules than traditional html used to produce web pages. However, xhtml will provide a powerful opportunity for future learning when applications to produce files via a 'what you see is what you get' (WYSIWYG) interface are commonly available. The advantage of the xhtml format is the flexibility it brings by accommodating specification of the document layout and embedding of objects, such as images, within the text of any note. The open source software community has made available a suitable reader application (Haleblian, 2007) although a straightforward content creation and conversation tool has yet to be produced. In the interim, more digitally aware pupils and teachers may develop workaround strategies to create their own resources.

Handheld learning strategies, such as those outlined above, break down the barriers erected by traditional models of education. Rather than learning at a set time of day, in a fixed location, with a certain group of people, the learning can take place at a time and place of the learner's own choosing. Thus learning may occur on the way to or from school, in the home, when traveling or while absent from school through illness. What is important is that students are empowered with the choice to learn when they themselves feel ready to learn, rather than at a time or place that is convenient to the education system. The opportunity presented by handheld learning is best summarised by Heppell

'every turned off device is potentially a turned off child' (Heppel, 2008)

While the Nintendo DS offers opportunities in this informal learning zone, it is the mobile phone that has achieved the greatest penetration of the secondary school cohort.

There are, of course, difficulties with the use of mobile phones since their use is often banned during the school day. However, the advent of the built-in camera in mobile phones has resulted in many handsets having a reasonable image browser or gallery application provided by the manufacturer as a standard feature. The software used is capable of displaying any file in the JPG picture format, whether a genuine photograph or a computer-generated graphics file. Recently, it has been proposed to exploit this facility by moving slideshows and quizzes away from the PowerPoint and individual pupil response handset strategy towards taking these tests on the pupils' own mobile phones (Soon, 2008).

The simplicity of the graphics file allows the use of free applications to generate the quiz slide. Often a similar software package is available on school library or home computers and so the quizzes can be created in any location equipped with a computer. Simplicity is again advantageous in that it lowers both the age and ability barriers for pupils to generate their own content rather than consume resources made by others.

The distribution of image-based quiz material is flexible in nature. Slides may be downloaded by usb direct from the computer, downloaded from the internet or, more actively, files may be shared virally by establishing peer-to-peer bluetooth connections. Completed quiz materials can be used in class either individually or in groups as a stimulating formative assessment activity. Pupils may also choose to try teacher-generated quiz materials alongside those created by their peers for diagnostic revision purposes outside the classroom, taking their work into the informal learning environment.

The text-based opportunities of the Nintendo DS and image slide quiz possibilities of mobile phone handsets are both offered by the Apple iPod family. Basic quiz functionality is obtained through the iPod photo album while text notes are stored in a default notes folder on the iPod that can be accessed using the host computer's file manager instead of the proprietary iTunes application (Vincent, 2008). Unlike mobile phone handsets, homogeneity across the iPod range has allowed third parties to develop quiz software (Aspyr Studios, 2007) that may be used in place of the photo gallery.

Beyond text-based information, any definition of digital literacy should include the ability to access, use, create and edit both audio and video forms of communication. The use or creation of audio materials can be difficult in a classroom setting due to background noises. However, the thought processes required for even the shortest of audio clips, in terms of devising an idea, developing the format, scripting the recording and the subsequent editing process ensure that the learning experience is active and meaningful. Many teachers see audio as a consumable resource rather than an opportunity for students to develop their own multimedia portfolio of work, citing the need for hardware or software as obstacles. In truth, usb microphones are relatively inexpensive and there is often suitable basic software, such as Microsoft Voice Recorder or Apple Quicktime already available on the computer.

Simple activities, such as the creation of a 'nanostory', a short recording of perhaps only two or three sentences, can be used as an introduction to audio work. Such short pieces present a relatively low barrier to achievement and can be produced with minimal scripting and editing. Nanostories can be self-supporting pieces of work. In science, a pupil could use a nanostory to;

- summarise the structure of an atom
- inform the listener of the parts of a plant or animal cell
- describe how to test for the presence of oxygen gas.

There is opportunity to split discussion of a larger topic or scheme of work into smaller nanostory compatible sections, such as an introduction to the different forms of energy, the Periodic table or photosynthesis. This breaking down of a larger topic provides a collaborative learning activity where each member of the class can point to their own individual success in achieving an outcome. All pupils can experience success as each individual nanostory can be produced with or without support in the classroom or in a resourced base.

While the nanostory idea lends itself well to small aspects of a topic or an executive summary of an entire unit of work, it is less appropriate for detailed work or capturing a record of discussion. When nanostories have been "outgrown" as a methodology, audio may take a more structured, scripted approach more suited to the concept of podcasting. A podcast is a longer, often scripted, piece of audio work that is made available online for the listening of others. The audience subscribe to a website that syndicates the material and pushes it out for consumption once it has been published (Richardson, 2006).

Producing a podcast, individually or as a group, is more demanding in terms of scripting and post-production editing. Scripting sessions can make excellent collaborative exercises, often stimulating genuine topical debate between pupils that would be difficult to initiate and sustain in teacher-led, whole class discussion. The debating phase is valuable since it gives pupils an opportunity to research an area of science, exercising higher-order skills to develop an informed opinion based on the information available and linking this to arguments put forward by others in the group before producing their own audio content as evidence of their learning. Producing a podcast can give pupils a context and meaning for group discussion on topics such as nuclear power or climate change. There is scope for peer assessment in both the scripting and editing phases. Additionally, the involvement of fellow classmates can be a powerful motivator for other pupils who then try to raise the standard of their work, since peer assessment is more direct, and often more valued, than feedback from the teacher.

There are several software packages available for audio editing. The two most popular applications, Garageband (Apple Inc, 2009) and Audacity (Audacity, 2009), are free. The Garageband application is specific to Apple hardware, while Audacity is a cross-platform Open Source application that can be used on several different operating systems.

A further motivating factor for pupils is the ability to publish the final audio file containing the work to the internet, making it accessible for direct listening on a web site or downloadable via an aggregation service such as Apple's iTunes, a free application for PC and Mac. Placing the audio online provides motivation by bringing pupil's work to a real audience outside the classroom. The value of podcasting lies partly in this sharing of work. Classroom walls are removed as subscribers listen to the pieces and pupils see the download statistics as evidence for people taking the time to listen to their pieces.

Podcasting can be taken beyond the use of audio alone. Enhanced podcasts (Wikipedia, 2008) are a means of displaying a sequence of still images synchronised to an audio piece. The image is changed at trigger points, or chapters, set by the creator during editing. Pupils can discuss the selection of an image and justify how it supports the background narrative of the podcast. Issues such as intellectual property and copyright may be introduced at this stage, since it is important that pupils can identify whether or not permission is required to use an image they have selected.

Enhanced podcasts offer an excellent medium for additional support needs. Whether produced by pupils or a member of staff, the accompanying images can greatly assist pupils to compare, contrast and organise information for themselves. Having images with supporting audio is good practice for all learners, since

'understanding occurs when learners are able to build meaningful connections between visual and verbal representations' (Mayer, 2005, p5).

The creation of enhanced podcasts is a straightforward process on Apple computers (Apple, 2008) but the effect can be difficult to achieve on other platforms, although advice is available (e.g. Ludington, 2005). One solution is to create video podcasts, sometimes referred to as vodcasts. Most computers have the ability to perform basic editing of video footage using software applications provided as standard by the manufacturer, e.g. Windows Movie Maker or Apple iMovie. The video content can take the form of live action recorded by a camcorder, digital camera or mobile phone. Alternatively, a sequence of slides from presentation software such as PowerPoint may be inserted at identified points in the timeline. This latter technique essentially replicates the format of the enhanced podcast format, publishing it as a video file.

When filming within the classroom, the capture and publication must be carefully controlled to ensure compliance with child protection measures in place at both school and local authority levels. Footage of pupils should not be placed on the internet without consent but may be used within the classroom or school as a whole. Students are often keen to rehearse a particular experiment for the camera. Once a topic has been selected, pupils are competitive in sharing individual ideas on aspects that must be included. From personal experience, pupils may require some direction from the teacher to avoid multiple interpretations of popular experiments such as the water-sodium reaction being filmed (Mackenzie, 2008b). In previous filming sessions, pupils have insisted on the inclusion of close-up scenes showing the use of personal protective equipment such as safety goggles or a perspex screen. While this may demonstrate bravado to some, it also displays a willingness to work safely and remind the viewers that risks have been assessed prior to the task. Examples of good practice, highlighting what can be achieved with video in the classroom, are available (e.g. Bradley, 2008).

The sequencing of scenes for filming demonstrates an understanding of the order in which the steps of an experiment must be performed, e.g. placing a leaf in boiling water before removing chlorophyll in alcohol. Transitional effects in the editing software may enhance the video presentation if used sparingly. It is important to ensure that continuity is not compromised as a result. This may be achieved through narration or, if this is not possible, a running text-based commentary added to the video. As with the enhanced podcast, the use of additional materials such as images or background music can stimulate discussion around copyright, creative commons (Creative Commons, 2009) and podsafe music (e.g. Podshow, 2005). Mainstream artists are now beginning to make some of their work available under a licence for educational use (Moby, 2007).

A particularly powerful use of video is the screencast (Friedman & Lennartz, 2008). In screencasting, the image displayed on a computer screen is recorded during the completion of a task. This allows a teacher to model the approach to any particular piece of work in advance, making the video available for viewing at a later point in time. Recognising the enhanced learning potential when meaningful audio is presented alongside video (Mayer, 2005), the best screencasts are accompanied by a narrative explaining each step of the process as it is replayed in the clip. Depending on the software application, a narrative may be recorded simultaneously or added subsequently. The latter method is more time intensive but has the advantage of separating audio and video tracks for independent editing. Splitting video and audio into different 'takes' also allows the creator to focus on one element of the presentation at a time. This is especially desirable if the video capture device is an IWB, since a trailing microphone lead can be hazardous.

Screencasting can serve many purposes. The technique may be used to review topics already covered in class or support learners at regular intervals as they progress through a unit of work (Basler, 2009). They are also useful as a means of providing solutions to exercises attempted in class or at home (Mackenzie, 2008c). Feedback from pupils indicates that the introduction of screencasts is a welcome innovation that aids understanding more than a sheet of paper showing the answers;

'It's like you're helping me, standing here going over bits I can't do.' (feedback received from S5 girl)

Learners are free to pause the presentation at any point to review their progress and replay short sections, or the entire clip, as necessary. Thus students may make as much, or as little, use of the screencast as they require, using it as personalised support for their learning.

MANAGING DISTRIBUTION, ACCESS AND OWNERSHIP

When multimedia files (nanostories, podcasts, screencasts) are created, a distribution mechanism to facilitate resource sharing must be established. The most flexible method is to use a weblog (or blog) as the vehicle for information sharing. Media files, additional notes for extended reading or revision and useful links to other web pages may be embedded within a classroom blog to create an extensive resource bank for pupils to tap into as required. Correctly configured, a blog can syndicate new content on podcast directory services such as Apple's iTunes (Mackenzie, 2008d). Learners can subscribe to the podcast and automate the downloading of new content on to their home computer, giving an opportunity for informal learning on the computer or a portable device such as an iPod.

Creating a classroom blog as an online resource centre for pupils is a worthwhile activity. However, the true value of the blog can only be leveraged when the exchange becomes a two-way process. Persuading learners to make the transition to participation is a difficult process. To blog is to expose thoughts and opinions in a public arena that lies outside the social networking sites that form a young person's comfort zone. The roll of a classroom blog is not to enter that recreational space, the creepy treehouse phenomenon (Krutsch, 2007), but to offer a chance to engage in a conversation of learning in a virtual extension of the classroom. Encouraging pupils to leave comments as feedback on the usefulness of resources is one strategy to initiate the dialogue and several educators have gone further, developing the 'become a better blogger in 30 days' theme (e.g. Waters, 2007; Dembo, 2008).

Ownership of a personal learning space may be achieved if pupils are allocated individual blogs. The opportunity to encourage dialogue and write reflectively about the learning journey represents the full power and educational value of blogging. This development of literacy, reading and writing skills can be difficult to implement in a traditional science classroom but

'blogging places literacy within the context of today's information age' (Warlick, 2005, p110).

Producing written work in the context of the blog may also expose misconceptions that can act as a focus for continuing the conversation (Crosby, 2008). Blogging provides an excellent opportunity for the practice of realistic dialogue involving scientific vocabulary and the chance to debate different viewpoints according to the interpretation each has placed on the source material.

The appearance of microblogging services is a recent development. Posts to a microblogging site are restricted to a maximum number of characters, 140 in the case of Twitter.com (LeFever, 2008). Such small entries make microblogging a very accessible application and short posts charting each step of learning can provide motivation for less-able pupils. While the low character limit might, at first, appear to be an impediment to finding educational value in microblogging, short messages could be used in the science classroom to;

- exchange properties or uses for an element
- create a word equation to represent the process of aerobic respiration
- describe alpha, beta and gamma radiation
- record and share experimental results in real time
- conduct a quick poll
- share learning outcomes or success criteria

A perceived weakness of the Twitter system is that an audience for messages cannot be selected, a member's posts are either wholly public or viewable only by confirmed followers (Murphy, 2008). This can be turned into a strength for classroom microblogging by preventing access to outsiders. With the option for pupils to register their mobile phone number, group members can receive updates as SMS text messages without having to disclose their number to another member. This combination of web and SMS alerts on Twitter.com has been employed in a secondary physics class for posting of assignments, key date reminders and help requests (Willard, 2008). A similar, closed system developed specifically for education called Edmodo.com has been demonstrated as a viable alternative microblogging service for secondary schools (Picardo, 2008b).

Schools are often concerned that social networking platforms, including blogs and microblogs, can provide a new, fertile territory for antisocial behaviours such as bullying. It is important that schools develop policies and put training in place to support pupils and staff by ensuring their understanding of the acceptable use of ICT facilities and the internet in particular. While online bullying may attract more attention, the most common breach of e-safety in schools remains the viewing of unsuitable online content (BECTA, 2007).

Teachers can take practical steps to minimise these breaches of school e-safety guidelines and reinforce expectations of acceptable use. The chosen classroom blogging software should have the facility to moderate, or approve, comments before they are published for general viewing. This allows inappropriate content to be filtered before publication. Classroom activities for different age groups, designed to model safe use of the internet and demonstrate the potential risks and consequences of inappropriate actions, are distributed by the Child Exploitation and Online Protection Centre (CEOP, 2007). Comprehensive advice is also available from groups such as Respect Me, an anti-bullying organisation (Scottish Association for Mental Health, 2008). While some teachers prefer to operate classroom blogs within a "walled garden", allowing only authorised individuals to view their sites, others are keen to promote pupils' work to as wide an audience as possible. Each approach has advantages (Edtechroundup, 2008) and teachers should consider the intended outcome of the project before determining the most appropriate solution for their situation.

When choosing an application for blogging or microblogging in the classroom, it is wise to consider aspects such as the ongoing maintenance of the site. While some schools are well resourced and capable of hosting their own web presence, others may find themselves relying on their service provider to keep a backup copy of information. With regular launches of new web2.0 tools, it is a fact of digital life that many will have a transient existence. Since many web2.0 services run on the web server rather than as installed applications on the user's computer, the software may cease to be available if the company fails. Innovative practitioners often sign up and test services before moving on to the next new application to appear. These early adopters are therefore less likely to be affected by the withdrawal of an application than the average user and we should be mindful there is no "money back guarantee" with a free online programmes.

CONCLUSION

With the progressive adoption of ICT in learning and teaching, education is moving into a digital space that allows others to look inside the virtual classroom. The persona of the networked teacher is now projected out to a larger community than that associated with the conventional notion of a school campus. Educators must consider whether there is a need to separate the digital persona from the personal persona, e.g. does the web-based interaction "at arms length" give the impression of an uncaring professional? Teachers can support each other to make the transition to an online environment by forming a Personal Learning Network (PLN) (Couros, 2008). The PLN can act as a powerful community providing professional development and a personal support network. Directories of already networked teachers have been established to allow educators working in similar curricular areas to make contact (e.g. Hartman, 2008). It is common for these directories to leverage the power of microblogging by indexing members according to their username rather than an email or web address.

For the first time, the potential exists for the science classroom to act as a creative centre, providing a stimulating and challenging environment for learning. The proficient use of ICT can make a real difference to the effectiveness of science education. By adopting technology, and embedding digital literacy at the heart of education, science is opened up beyond the constraints of traditional pedagogy, creating genuine opportunities for all students to progress, through discussion, communication and reflection, towards true understanding.

REFERENCES

Apple Inc. (2008). *GarageBand.* Retrieved February 15, 2009, from http://docs.info.apple.com/article.html?path=GarageBand/4.0/en/6624.html

Apple Inc. (2009). *iLife.* Retrieved February 15, 2009, from http://www.apple.com/ilife/

Aspyr Studios. (2007). *iQuiz Maker.* Retrieved February 15, 2009, from http://www.iquizmaker.com

Association for ICT in Education. (2001). *Teaching ICT volume 1 issue 2.* Retrieved February 15, 2009, from http://acitt.digitalbrain.com/acitt/web/resources/pubs/Journal%2002/whiteboards.htm

Audacity. (2009). Retrieved 15th February 15, 2009, from http://audacity.sourceforge.net

Barrett, T. (2008). *Interesting ways to use your interactive whiteboard.* Retrieved February 15, 2009, from http://docs.google.com/Presentation?docid=dhn2vcv5_106c9fm8j&hl=en_GB

Basler, D. (2009). *Baslercast Physics.* Retrieved February 15, 2009, from http://www.aasd.k12.wi.us/staff/baslerdale/Ppodcasts.asp

BECTA. (2007). *Signposts to safety: teaching e-safety at Key Stages 3 and 4,* Retrieved February 15, 2009, from http://publications.becta.org.uk/display.cfm?resID=32424&page=1835

Bradley, J. (2009). *DNA spooling experiment.* Retrieved February 15, 2009, from http://scienceguyinatie.blogspot.com/2009/02/videocast-dna-spooling-experiment.html

Bray, O. (2008). *Nintendo DS in Education - Harper Collins 100 Classic Book Collection.* Retrieved February 15, 2009, from http://olliebray.typepad.com/olliebraycom/2008/12/nintendo-ds-in-education-harper-collins-100-classic-book-collection.html

Bray, O. (2009). *Computer Games Based Learning: Sharing good practice between Musselburgh, East Lothian and Sydney, Australia.* Retrieved February 15, 2009, from http://olliebray.typepad.com/olliebraycom/2009/01/computer-games-based-learning-sharing-good-practice-between-musselburgh-east-lothian-and-sydney-australia.html

Brooks, J. G., & Brooks, M. G. (1999). *In search of Understanding: The Case for Constructivist Classrooms.* Alexandria, VA: Association for Supervision and Curriculum Development.

CEOP. (2007). *ThinkUKnow - Teachers and Trainers Area.* Retrieved February 15, 2009, from http://www.thinkuknow.co.uk/teachers/

Clarke, S. (2005). *Formative Assessment in the Secondary Classroom.* London: Hodder Murray

Collins, T. (2009). *Maths DS Lite Project Adventures from Campie Primary 6b*. Retrieved February 15, 2009, from http://edubuzz.org/blogs/campiep6b/2009/01/23/maths-ds-lite-project/

Couros, A. (2008). *What is a PLN? Or, PLE vs. PLN?* Retrieved February 15, 2009, from http://educationaltechnology.ca/couros/1156

Creative Commons. (2009). Retrieved February 15, 2009, from http://creativecommons.org

Crocodile Clips. (2008). *Yenka Electricity.* Retrieved February 15, 2009, from http://www.yenka.com/en/Yenka_Electricity/

Crosby, B. (2008). *One reason we blog - finding and clarifying misconceptions* Retrieved February 15, 2009, from http://learningismessy.com/blog/?p=550

Data Harvest Ltd. (2008). Retrieved February 15, 2009, from http://www.dataharvest.co.uk/distributor/files/documents/sensors/3143/3143_ds038_2_force.pdf

Dembo, S. (2008). *Be a better blogger in just 30 days*. Retrieved February 15, 2009, from http://www.teach42.com/2008/10/23/be-a-better-blogger-in-just-30-days/

Doyle. (2009). *Our Kids are being Jobbed.* Retrieved February 15, 2009, from http://doyle-scienceteach.blogspot.com/2009/01/are-kids-are-being-jobbed.html

Drexler, W. (2008). *Networked Student.* Retrieved February 15, 2009, from http://www.youtube.com/watch?v=XwM4ieFOotA

Driver, R., Squires, A., Rushworth, P., & Wood-Robinson, V. (1994). *Making Sense of Secondary Science: Research into Children's Ideas.* London: Routledge Falmer.

Edtechroundup. (2008). *Edtechroundup podcast episode 4.* Retrieved February 15, 2009, from http://www.edtechroundup.com/2008/05/05/edtechroundup-podcast-episode-4/

Evans, A. (2008). *Nintendo DS in the class - What's good beginning to look like?* Redbridge Primary ICT Consultant. Retrieved February 15, 2009, from http://redbridgeprimaryit.blogspot.com/2008/05/nintendo-ds-in-class-whats-good.html

Fisch, K. (2006). *Did You Know?* Retrieved February 15, 2009, from http://thefischbowl.blogspot.com/2006/08/did-you-know.html

Fisch, K., & McLeod, S. (2007). *Did You Know 2.0?* Retrieved 15th February 15, 2009, from http://thefischbowl.blogspot.com/2007/06/did-you-know-20.html

Friedman, V., & Lennartz, S. (2008). *Screencasting: How to start, tools and guidelines Smashing Magazine*. Retrieved February 15, 2009, from http://www.smashingmagazine.com/2008/08/19/screencasting-how-to-start/

Haleblian, R. (2007). *DsLibris - an ebook reader for the Nintendo DS.* Retrieved February 15, 2009, from http://rhaleblian.wordpress.com/dslibris-an-ebook-reader-for-the-nintendo-ds/

Harlen. (1999). *Effective Teaching of Science: A Review of Research. SCRE*. Retrieved February 15, 2009, from http://www.scre.ac.uk/pdf/science.pdf

Hartman, G. (2008). *Twitter4Teachers.* Retrieved February 15, 2009, from http://twitter4teachers.pbwiki.com

Hazzard, B., & Badger, J. (n.d.). *SMARTboard Lessons Podcast.* Retrieved February 15, 2009, from http://pdtogo.com/smart/

Heppell, S. (2008). *Learning to Change/Changing to Learn.* Pearson Foundation. Retrieved February 15, 2009, from http://www.pearsonfoundation.org/pg5.6.html

Jamieson, J., & Steele, G. (2008). *Radioactivity Experiments.* Dunfermline, UK: Scottish Schools Equipment Research Centre.

Kay, K. (2008). *Learning to Change/Changing to Learn.* Pearson Foundation. Retrieved February 15, 2009, from http://www.pearsonfoundation.org/pg5.6.html

Klug, B. (2002). *The Plagiarism Checker.* Retrieved February 15, 2009, from http://www.dustball.com/cs/plagiarism.checker/

Krutsch, J. (2007). *Are you building a creepy treehouse?* Retrieved February 15, 2009, from http://technagogy.learningfield.org/2007/11/19/are-you-building-a-creepy-treehouse/

LeFever, L. (2008). *Twitter in Plain English.* Retrieved February 15, 2009, from http://www.commoncraft.com/twitter

Ludington, J. (2005). *Windows media enanced podcast.* Retrieved February 15, 2009, from http://www.jakeludington.com/project_studio/20051004_windows_media_enhanced_podcast.html

Mackenzie, S. (2008a). *Smart scale diagrams.* Retrieved February 15, 2009, from http://blog.mrmackenzie.co.uk/2008/06/16/smart-scale-diagrams/

Mackenzie, S. (2008b). *Is there really "dead time" in the school year?* Retrieved February 15, 2009, from http://blog.mrmackenzie.co.uk/2008/04/07/is-there-really-dead-time-in-the-school-year/

Mackenzie, S. (2008c). *Youtube solutions to latest higher hw.* Retrieved February 15, 2009, from http://mrmackenzie.co.uk/2008/12/16/youtube-solutions-to-latest-higher-hw/

Mackenzie, S. (2008d). *iPod my Physics.* Retrieved February 15, 2009, from http://blog.mrmackenzie.co.uk/2008/11/05/ipod-my-physics/

Matthews. (2009). The Latest Doomed Pedagogical Fad: 21st-Century Skills. *Washington Post*. Retrieved February 15, 2009, from http://www.washingtonpost.com/wp-dyn/content/article/2009/01/04/AR2009010401532.html

Mayer, R. E. (2005). *The Cambridge Handbook of Multimedia Learning*. New York: Cambridge University Press.

Miller, D. J., & Robertson, D. P. (in press). Using a games console in theprimary classroom: Effects of 'Brain Training' programmesme oncomputation and self-esteem. *British Journal of Educational Technology*.

Moby. (2007). *Film music.* Retrieved February 15, 2009, from http://mobygratis.com/film-music.html

Mullins, B. (2006). *The Moon Books Project.* Retrieved February 15, 2009, from http://www.moonbooks.info

Murphy, J. (2008). *Micro-blogging for science & technology libraries.* Retrieved February 15, 2009, from http://eprints.rclis.org/12752/2/Murphy_Microblogging.pdf

Ng, P., & Yeung, Y. (2000). Implications of Data-logging on A.L. Physics Experiments: A Preliminary Study. *Asia-Pacific Forum on Science Learning and Teaching, 1*(2), Article 5. Retrieved February 15, 2009, from http://www.ied.edu.hk/apfslt/issue_2/phng/phng3.htm

Nuffield Foundation. (2004). *Ticker-timers for investigating speed.* Retrieved February 15, 2009, from http://www.practicalphysics.org/go/Experiment_266.html

Picardo, J. (2008a). *Interactive Whiteboard.* Retrieved February 15, 2009, from http://www.boxoftricks.net/?cat=8

Picardo, J. (2008b). *Edmodo: What students think.* Retrieved February 15, 2009, from http://www.boxoftricks.net/?p=432

Picardo, J. (2009). *I teach, therefore you learn... or do you?* Retrieved February 15, 2009, from http://www.boxoftricks.net/?p=863

Podshow. (2005). *Podsafe Music Network*. Retrieved February 15, 2009, from http://music.podshow.com

Prensky, M. (2001). *Digital Natives, Digital Immigrants - A New Way to Look at Ourselves and Our Kids*. Retrieved March 15, 2006, from http://www.marcprensky.com/writing/Prensky%20-%20Digital%20Natives,%20Digital%20Immigrants%20-%20Part1.pdf

Promethean Inc. (2006). *Promethean Planet - Teacher Feature*. Retrieved February 15, 2009, from http://feeds.feedburner.com/PrometheanPlanetTeacherFeature

Richardson, W. (2006). *Blogs, Wikis, Podcasts and other powerful web tools for classrooms*. Thousand Oaks, CA: Corwin Press

RIT Center for Multidisciplinary Studies. (2008). *Survey of internet and at-risk behaviors*. Retrieved February 15, 2009, from http://www.rrcsei.org/RIT%20Cyber%20Survey%20Final%20Report.pdf

Robertson, D. (2007). *Dr. Kawashima's Brain Training - Outcomes and Evaluation*. Dundee, Scotland: Learning and Teaching. Retrieved February 15, 2009, from http://www.ltscotland.org.uk/ictineducation/gamesbasedlearning/sharingpractice/braintraining/outcomes.asp

Scottish Association for Mental Health. (2008). *Cyberbullying*. Retrieved February 15, 2009, from http://www.respectme.org.uk//cyberbullying/cyberbullying/cyberbullying.html

Selwyn, N., & Facer, K. (2007). *Beyond the digital divide: Rethinking digital inclusion for the 21st century*. Futurelab. Retrieved February 15, 2009, from http://www.futurelab.org.uk/resources/documents/opening_education/Digital_Divide.pdf

Smith, M. K. (1999). Informal Learning. In *Encyclopedia of Informal Education pub*. London: InFed.org. Retrieved February 15, 2009, from http://www.infed.org/biblio/inf-lrn.htm

Soon, L. (2008). *Creating quizzes for the phone*. Retrieved February 15, 2009, from http://video.google.com/videoplay?docid=-2479360146328027324&ei=_qFrSdmTM5-QiQLPhvyqBQ&q=2479360146328027324

Turner, S. (2005). *Answering the Temperature regulation problem*. Retrieved February 15, 2009, from http://www.esf.edu/EFB/turner/termite/The%20temperature%20problem%202.html

University of Colorado. (2008). *PhET Circuit Construction Ki.t* Retrieved February 15, 2009, from http://phet.colorado.edu/simulations/sims.php?sim=circuit_construction_kit_dc_only

Vincent, T. (2008). *iPods in Education* (Learning in Hand Podcast). Retrieved February 15, 2009, from http://learninginhand.com/ipod/clickwheel.html

Warlick, D. F. (2005). *Classroom Blogging: A teacher's guide to the blogosphere*. Raleigh, NC: The Landmark Project.

Waters, S. (2007). *31 Day Project*. Retrieved February 15, 2009, from http://aquaculturepda.wikispaces.com/blogs1

Wikipedia. (2008). *Enhanced podcast*. Retrieved February 15, 2009, from http://en.wikipedia.org/wiki/Enhanced_podcast

Wikipedia. (2009). *Nintendo DS Homebrew*. Retrieved February 15, 2009, from http://en.wikipedia.org/wiki/Nintendo_DS_homebrew

Willard, R. (2008). *The Tech Teachers show #83*. Retrieved February 15, 2009, from http://thetechteachers.blogspot.com/2008/08/tech-teachers-show-83.html

Chapter 4
Using Multimedia Learning Aids from the Internet for Teaching Chemistry:
Not as Easy as it Seems?

Ingo Eilks
University of Bremen, Germany

Torsten Witteck
Engelbert-Kaempfer-Gymnasium, Lemgo, Germany

Verena Pietzner
University of Hildesheim, Germany

ABSTRACT

Large advances in technology in the last few years have made computers cheap and presentation technologies easily available in most secondary schools, at least in industrialised countries. Due to recent developments in software technology nearly anyone can create animations and visualisations. The Internet has helped to make the distribution of such graphic tools both wide and fast. Thus, using multimedia in science teaching is becoming more and more common. Today, integrating visualisations and animations from the Internet into the science classroom seems an obvious choice for enhancing science lessons. But are all of the animations offered on the Internet really helpful for promoting understanding? This chapter discusses what might occur while working with animations taken from the Internet and how these multimedia illustrations can potentially interact to reinforce rather than resolve students' misconceptions about chemical principles. Daniell's voltaic cell serves as a good example to illustrate the ways in which visual aids can be interpreted differently by experts and novices. The following discussion takes place in the form of an exaggerated example. It is meant to appear as a critical interjection making readers more aware of the myriad, often invisible, potential drawbacks which exist when first selecting promising-looking animated illustrations for classroom use.

DOI: 10.4018/978-1-60566-690-2.ch004

INTRODUCTION

The development of ICT was and remains a very fast phenomenon. It was only a little over a generation ago when computers were exclusively adopted for special industrial applications and research purposes. At that time, computers were prohibitively expensive and therefore generally not available to private users. Additionally, the handling of computers was a very demanding challenge, one only performed by experts. Nowadays computers are available everywhere. Working with them and their wide range of gainful applications no longer requires specialised competencies, *e.g.* skills in understanding computer-programming languages. Computers are used everywhere in our life for business or private purposes. The use of computer applications, like cognitive tools, became easier step by step in recent years, consequently those developing learning environments can no longer ignore this progress (Falvo, 2008).

One of the major steps in making computers easy and comfortable to use was the development of graphic desktops and WYSIWYG technology ("What you see is what you get"). Until the shift towards WYSIWYG occurred some 15 years ago, cognitive tools like text editors and graphic tools were unable to accurately depict materials on a computer monitor while simultaneously working with and altering them. Also, the picture quality of the representations was highly inferior to that which we take for granted today when working with electronic media. Frequently, this led to various surprises when printing out a hard copy of the material. Today, displaying correct fonts, page layouts or graphical elements on the computer screen while simultaneously editing the respective documents is no longer a problem. It has become possible to create texts, pictures, and animations in nearly any form, combination or context we prefer. Modern computer games beautifully demonstrate the current, cutting-edge breakthroughs in developing illustrations and animated programmes, up to and including artificial, "virtual" realities.

The creation of pictorial, animated materials no longer demands the professional expertise of specialised programmers. Cognitive tools like Macromedia Flash are quite cheap to purchase and have become easier to use. This has also had wide-reaching consequences in the educational domain. Curriculum developers and teachers are now able to create animated classroom aids using these tools, and actively do so. This leads to a constantly-increasing number of animated visual aids. The Internet has helped to make these visual aids on a variety of science topics available for everyone at very nearly no cost to the user.

Perhaps because of this rapid development of hard- and software, it seemed obvious and logical to enhance science lessons by integrating animated visualisations from the Internet into the science classroom. This was done because of the promising perspectives hoped for by the inclusion of multimedia applications (e.g. Pavio, 1986). It was also undertaken in order to keep pace with students' positive computer-based multimedia experiences in their private lives (Falvo, 2008). A lot of promises were made and educational research supported expectations that the integration of visualisation and animation via the use of computers could potentially support learning. One example is the thoughtful embedding of computer technology into teaching by binding it to other learning activities (e.g. Mayer, 2001; Falvo, 2008).

But the use of multimedia visualisations in science classes might seem to be easier than it truly is. We can derive such a conclusion from the dangers seen when working with potentially misleading illustrations in textbooks (e.g. Hill, 1988; Eilks, 2003). The question should be allowed, as to whether all animations used for illustrating scientific concepts are genuinely useful in supporting learning. In particular, we must reflect on whether any specific illustrations taken from the Internet actually provide us with visual stimuli that are feasible to support learning in the direction of scientific correctness and the intended learning objectives.

Drawing an analogy with the WYSWIG technology (Eilks, Witteck, & Pietzner, 2009), teaching using computer-based visualisations and animations can be reflected upon in terms of learning theory. We know from constructivist theories of learning (e. g., Bodner, 1986) that visual stimuli are not simply captured in learners' heads. They are filtered and then re-interpreted using the framework of the viewer's preconceptions. Students do not anchor into their minds exactly what they see in an easy analogy. In terms of the language of ICT: Learning with visual stimuli does not work through a "copy-paste mode". We do not get into the students' heads what we see, maybe not even what they see.

The failure of this "easy" mode of learning is of special importance for learning chemistry. Chemistry necessarily tries to explain macroscopic properties and phenomena on the sub-microscopic level. However, the inaccessibility of the sub-microscopic level to direct human senses also is one of the main problems (e.g. Johnstone, 1991). The sub-microscopic level is invisible to the human eye and, therefore, largely dependent on human imagination to depict and describe. New impressions gained by illustrations therefore are interpreted with reference to the learners' pre-knowledge. This process of transformation has to be taken into account when offering visualisations of new concepts "to be learned".

THE SPECIAL ROLE OF LEARNING BY VISUALISATION FOR TEACHING CHEMISTRY

Modern interpretations of the learning of science generally refer back to the theory of constructivism as described, for example by Bodner (1986). One of the major ideas of constructivism is that students will develop their knowledge and understanding through an active process of construction. They start with newly gained information, *e.g.* something accessed via a computer screen, in the foreground of the cognitive framework which they already have in their minds. This means that all information already existing in the mind of the learner will influence the interpretation of newly-acquired information. The newly-constructed framework in the learner's mind will be a conglomerate of their pre-knowledge and the new pieces of information gained.

Unfortunately, not all ideas brought by students into the classroom fit the accepted scientific understanding of science. Nor do they offer a fruitful basis upon which to build scientifically-reliable concepts. One of the major sources of students' pre-conceptions is their everyday-life experience. Personal experiences are frequently applied to explain scientific phenomena, although radically different concepts must actually be applied to scientific concepts and/or learning. One easy example given by Pfundt (1982) may illustrate this nicely: From our everyday experience, we know that candles shrink when they burn. It seems an obvious conclusion that the wax 'disappears' due to something inherent in the process of combustion. This idea is over-generalised by students. They often believe that objects are becoming 'lighter' and disappear during any process of combustion. So far so good. Indeed, after combustion has ended, the candle is 'no longer there' and the wax has obviously 'disappeared' (at least from the place where the candle originally was). The problem begins when students think that the same is true at the sub-microscopic level – the level of atoms and molecules - and that matter can simply 'disappear' on this level as well. This understanding is in direct contradiction to the scientific concepts of the Law of Conservation of Mass and the Law of Conservation of Atoms. In science, we use the concept that atoms and their constituting units never disappear during chemical changes. This means that their mass is conserved and that the sub-microscopic entities involved in the combustion process will not disappear, but only change in certain, specific fashions. The two aforementioned concepts are, therefore, contradictory to one another. But the more familiar explanation (the one gained earlier and on the macroscopic level in this case) will influence further learning about new patterns of explanation on the invisible, sub-microscopic level. This is why the theory of constructivist learning became one of the driving forces for intense empirical research into students' pre- and alternate beliefs about chemistry (Wandersee, Mintzes, & Novak, 1994). It is also why much evidence has been gained in the time since the early works of Pfundt (1975) or Novick and Nussbaum (1978) in many different fields (e. g., Garnett, Garnett, & Hackling, 1995).

Starting from constructivist learning theory, the aim of curriculum development is to create learning environments in which the alternative beliefs of students are taken into account. Different scholars have attempted to make such beliefs more explicit and to seriously discuss them, in order to replace them with less naïve and more scientifically-defined concepts which are more in line with current, reliable scientific theories. One of the most popular ideas for overcoming alternative conceptions is to provoke 'cognitive conflict' in the learner. This conflict can then be used to promote conceptual change, e. g. where naïve ideas can be falsified by the use of experiments (Posner, Strike, Hewson, & Gertzog, 1982).

One example using Daniell's voltaic cell may help to better illustrate the idea of cognitive conflict theory. Classroom experience frequently reports that the ions within an electrochemical cell are thought to be 'produced by' the cell by many students. They also state that such particles 'disappear' after being uncharged at the corresponding electrode. From this point of view, during the whole process the electrodes are not subjected to any change themselves. They remain unchanged and play a passive role, just like the glass beaker or the wires of the outer electric circuit do. This is true for some galvanic cells, e.g. when working with inert electrodes, but not for all. In the learners' minds, the whole process is principally disconnected from a possible gain (or loss) of mass by the two oppositely-charged electrodes. The students only take into account the role of the ions with respect to the macroscopic phenomenon of achieving a flow of electricity. But the students neglect the sub-microscopic character of the ions as particles which must be conserved. The two chemical Laws of Conservation of Mass and Conservation of Atoms are incorrectly applied in this case. By means of introducing a cognitive conflict into the experiment, such an interpretation can easily be challenged and disproved. If, for example, a Daniell voltaic cell is connected to an electronic device for a few hours, both electrodes can be weighed. The loss in mass of the zinc anode and the increase in mass of the copper cathode can be specifically measured. The result can be used to provoke cognitive dissidence and a discussion in those students who neglect or misinterpret the change in mass occurring to the electrodes.

Unfortunately, this scenario is limited to the phenomenological level in school education. But misinterpreting macroscopic phenomena is only one part of problem in chemistry learning. Chemistry also encompasses a theoretical side. The discipline itself is performed on different levels and, quite especially, includes the invisible level of sub-microscopic changes (Eilks, Möllering, & Valanides, 2007). Johnstone (1991) in his very famous contribution "Why is science difficult to learn?" explained that learning and thinking in chemistry always takes place in an area of conflict between the macroscopic, sub-microscopic, and symbolic level (Figure 1). If this area and the relationships between the levels are overlooked, misinterpretations are a foregone conclusion. Thus, if there is a mismatch in students' thinking concerning the observable, macroscopic level, we may use a certain phenomenon or experiment to bring the cognitive conflict to the floor. Unfortunately, we cannot do the same with sub-microscopic misinterpretations in most cases of

Figure 1. The 'Johnstone triangle'

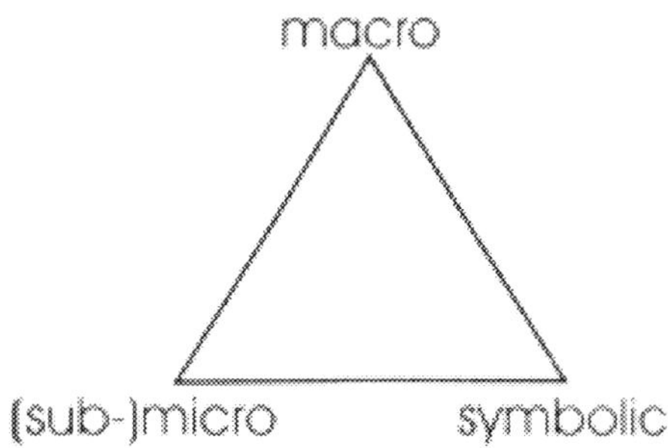

school-level chemistry. Therefore, we must remain keenly aware that dealing with sub-microscopic explanations also touches directly upon another field resulting in major problems in learning chemistry: the use of models and modelling. In chemistry, we often use models to help us better understand phenomena at the sub-microscopic level. But, dealing with different models and differentiating them correctly, however, is often not an easy task to accomplish (Justi, Gilbert, 2002a,b).

Because of their "invisible", non-tactile and, therefore, necessarily model-based nature, the sub-microscopic aspects dealt with in chemistry teaching and learning must be relegated to and discussed almost exclusively on a theoretical level. Such theoretical explanations can and should be supported by static visualisations (e.g., Brandt, Elen, Hellemans, Heerman, Couwenberg, Volckaert & Morisse, 2001) as well as animations (e.g. Williamson & Abraham, 1995). With the improvements in modern ICT, it became very common to plea for the use of computer-generated animations and simulations in learning situations. Such added detail could both aid and support an explanation of the sub-microscopic world far beyond the capability of static pictures alone, in order to further challenge learning and promote understanding. Such computer-animated illustrations have great potential for improving teaching and provide considerable advantages over static images (Mayer, 2003). They allow us to visually demonstrate the dynamic nature of the sub-microscopic world and can lead to a better understanding of the underlying chemistry concepts (Williamson & Abraham, 1995; Sanger & Greenbowe, 2000; Yang, Greenbowe, & Andre, 2004). One concise overview about relevant studies with relation to chemistry learning is given by Kozma and Russell (2005a/b).

In recent years, ICT has become more and more popular in science education. Starting with very simple animations and computer based data logging more than twenty years ago, nowadays a whole range of computer-based tools became available for science lessons: spreadsheets, presentation tools, learning software, WebQuests, molecular representations, animations or simulations, chemistry chat rooms, online test platforms, podcasts, Wikis, etc.. In addition, web 2.0 communities like Youtube can be a promising source for cartoons, animations or films showing risky experiments like the lightning of a barbecue with liquid oxygen. Most of these applications offer new advantages for all school subjects. But, as discussed above, the tools offer completely new perspectives to chemistry education where chemistry comes to the illustration of its theories. 3D-visualization tools for molecules or interactive calculations of the electrostatic surfaces of molecules or solid bodies can show things which cannot be illustrated dynamic by any other media. Animations and simulations can contribute to a better presentation of the dynamic nature, e.g., of reaction mechanisms, chemical equilibria, or technical applications.

In principle, animations and simulations can be considered as powerful tools for teaching and learning chemistry. They can foster students' understanding of three-dimensional structures (Williamson & Abraham, 1995) and even help developing learners' spatial ability (Barnea & Dori, 1999). They provide a resource which may help to reduce students' misconceptions about basic chemical principles (Kozma & Russel, 2005b; Sanger & Greenbowe, 2000; Yang, Greenbowe, &

Andre, 2004) and the use of animations has been shown to motivate students' learning of chemistry (Tsui & Treagust, 2004). On the other hand, common factors possibly impeding chemistry learning through animations and simulations include such negative aspects as: 1) limited learner attention spans when viewing animations, due to the high transfer rate of information (Ploetzner, Bodemer, & Neudert, 2008), 2) inadequate meta-cognitive competencies (especially in non-linear learning environments) (Schwartz, Andersen, Hong, Howard, & McGee, 2004; Azevedo, 2004), 3) a lack of prior knowledge (Shapiro, 1999) or 4) too little ability in recognizing or using proper spatial relations (Lee, 2007; Huk, 2007). However, this has also proven itself to be the case in every other field of learning when attempting to include the use of multimedia-based animations and simulations.

In general, we can say that promising potential for animations and simulations exists for teaching/learning chemistry. However, this potential is not self-evident. Schnotz and Bannert (2003) discussed the fact that picture-use in multimedia learning processes are not necessarily beneficial to learning in every case. The same considerations discussed above in the burning candle example hold true for learning with multimedia-based pictorial illustrations. Schnotz and Bannert (2003) summarized that only semantic processes will lead to an understanding of pictures and pictorial information and this is always related to the pre-knowledge of the learner. Therefore, if effective learning is to take place, illustrations and visual aids must be structured to take the learner's pre-knowledge of a given topic into account. Additionally, the relationship towards the scientifically accepted explanation and – if relevant – the model nature of the chosen explanation (Kelly, 2005, as discussed in Falvo, 2008) must also be taken into account.

In our opinion, two opposing ideas seem to be central with reference to the above outlined framework when attempting to structure, select or apply multimedia aids for supporting chemistry learning: 1) If the learner's preconceptions are scientifically reliable, illustrations should confirm, foster and strengthen them. 2) If the learner's preconceptions of a topic are scientifically unreliable, illustrations should induce a cognitive conflict which leads to overcoming the formerly-held ideas. In both cases it is necessary to use illustrations that are scientifically reliable and which do not demonstrate or call upon incorrect or conflicting explanations. Generally, we might be tempted to think that such conclusions can logically be taken for granted, however, even static illustrations in school textbooks do not always meet these two criteria (e. g., Hill, 1988; Eilks, 2003).

In our experience, a task which is described (see the next section) can be beneficial for chemistry teacher training, for example. It can sensitise teachers and learners alike to: 1) the role of alternative conceptions for learning, including 2) the potential and limitations of learning with multimedia visualisations.

A LEARNING TASK FOR 'LEARNING WITH MULTIMEDIA VISUALISATIONS' AS DERIVED FROM AN INTERNET SOURCE DISCUSSING DANIELL'S VOLTAIC CELL

The Daniell voltaic cell is the standard example for voltaic cells used in most chemistry textbooks and many science curricula. It is both easy-to-use and leads to experimentally reproducible results. Because the Daniell cell is an often discussed topic in chemical education, many visual aids are readily available on the Internet. Table 1 lists a small number of such resources from different countries.

Figure 2 shows example screenshots from an animation of the Daniell voltaic cell taken from a German website, which is explicitly thought to provide learning-support and self-study material for secondary and tertiary students. It will serve

Table 1. Animations of the Daniell voltaic cell taken from the Internet (last accessed October 14, 2009)

www.chemgapedia.de/vsengine/vlu/vsc/de/ch/13/vlu/echemie/galvanische_elemente/batterie.vlu/Page/vsc/de/ch/13/pc/echemie/galvanische_elemente/daniellapplet.vscml.html
www.chemgapedia.de/vsengine/vlu/vsc/de/ch/1/pc/pc_02/pc_02_01/pc_02_01_02.vlu/Page/vsc/de/ch/1/pc/pc_02/pc_02_01/pc_02_01_12.vscml.html
http://home.scarlet.be/at_home/pileflash.htm
www.ltam.lu/chimie/DaniellElementCD.html
www.mhhe.com/physsci/chemistry/essentialchemistry/flash/galvan5.swf
www.chem.iastate.edu/group/Greenbowe/sections/projectfolder/flashfiles/electroChem/volticCell.html
www.chem.iastate.edu/group/Greenbowe/sections/projectfolder/flashfiles/electroChem/voltaicCell20.html
www.edunet.tn/ressources/resdisc/physique/monastir/pile/epd.htm
www.chemie-interaktiv.net/html_flash/ff_galvanische_zelle.swf
www.chempage.de/theorie/galvanischeselement.html
www.youtube.com/watch?v=nNG5PMIHSoA
de.youtube.com/watch?v=A0VUsoeT9aM&NR=1
www.chembio.uoguelph.ca/educmat/chm19105/galvanic/galvanic1.htm
www.wainet.ne.jp/~yuasa/EngF2.htm
www.doitpoms.ac.uk/tlplib/batteries/basic_principles.php
iwan.chem.tu-berlin.de/~lehre/moses/Kapitel9%20Elektrochemie/Kapitel9.ppt (Slide 6)

as an example to show the inherent problems in 'seeing' animations which do not 'really' animate the science concept commonly accepted within the scientific community (Eilks, Witteck & Pietzner, 2009). The task we regularly give to our student teachers-in-training in our chemistry seminar is: 'Watch the animation and write down what you see!'

What do our students regularly see when they look at the animation mirrored in Figure 2? They describe an animation representing Daniell's voltaic cell. The formation of zinc ions stemming from the zinc electrode is recognised. The electric circuit itself is completed via a salt bridge, which connects two beakers holding solutions of electrolytes. Starting at the zinc electrode, zinc ions are solved by the aqueous solution around the anode. Connected to this process, two electrons are set free and conducted toward the copper electrode by the external electrical circuit. These electrons are therefore available to reduce copper ions (taken from the electrolyte solution) at the copper electrode. The copper ions move toward the copper cathode and accept the electrons, thereby forming copper atoms, which are physically deposited on the copper electrode. Overall, we see a flow of electric current, which would theoretically be able to power a small engine.

But is this what we actually see? Most of the things described above cannot, in fact, be seen in the animation. What is described above is an interpretation stimulated by the student teachers' pre-knowledge.

What, then, would we really see, if we view the animation through the eyes of an inexperienced secondary school student, who cannot even define a voltaic cell or how it functions? What we see is a red bar, which we interpret as being an electrode (an analogy from the macroscopic level). Taken from the symbolic level, we interpret 'Zn' as showing a zinc electrode, since this is the abbreviation (element symbol) for the metal, zinc. This electrode

Figure 2. An animation of the Daniell voltaic cell from the Internet

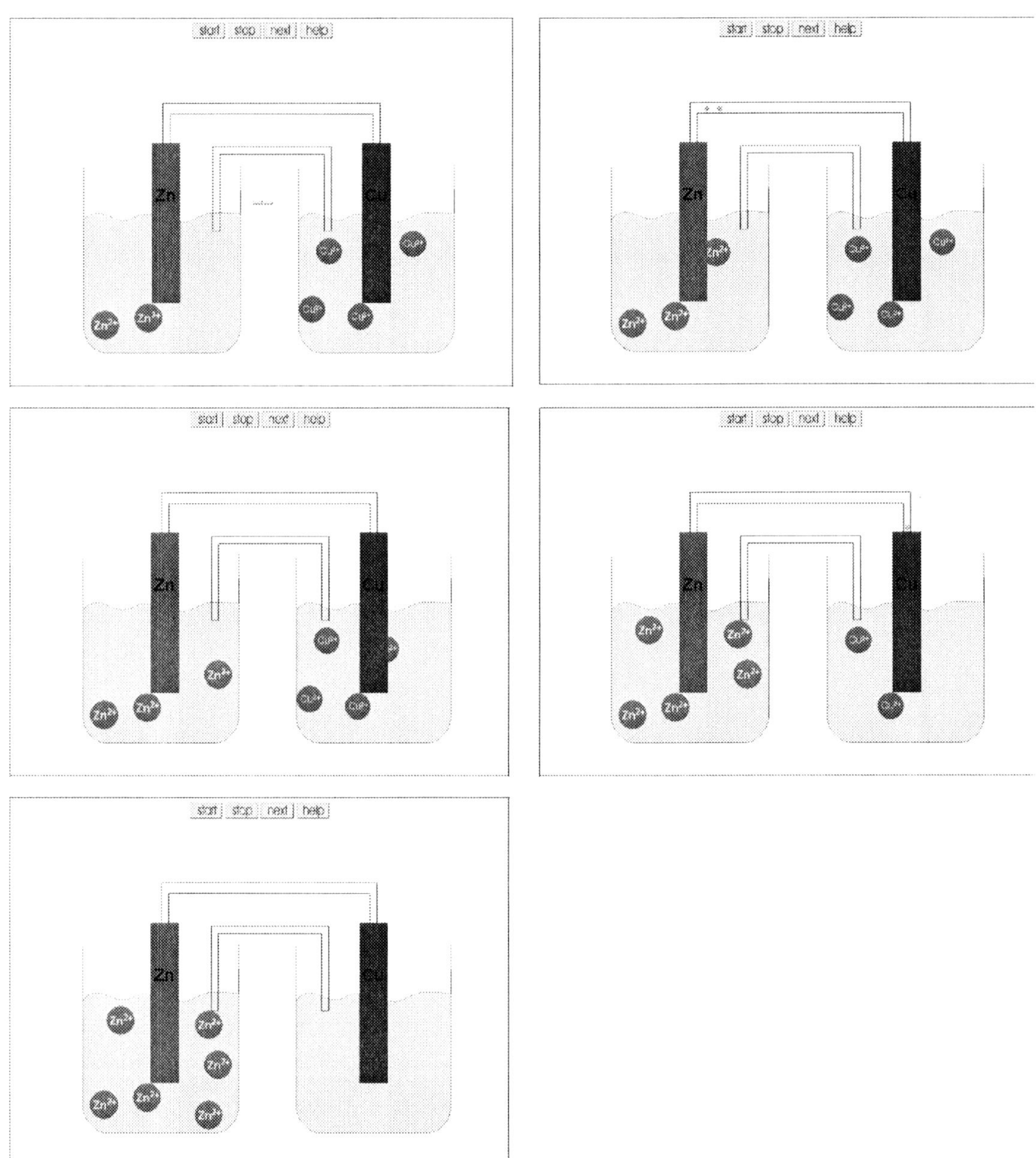

is represented as a continuum; it does not consist of particles, in this case individual atoms. (Coincidentally, if we take into consideration the fact that zinc atoms are larger than their corresponding ions, it is impossible for us to believe that the zinc electrode consists of zinc atoms. The electrode is thinner than any of the actual zinc atoms could possibly be, since the zinc ions shown in the animation are exactly as thick as the electrode!) We also see a blue bar. This should properly be identified as the model representation of a copper electrode. Both electrodes are dipped into a grey continuum containing various 'particles', which (obviously) should represent the respective zinc and copper ions in 'solution'. (Taken for granted, of course, that the learner clearly knows what a 'solution' is!)

This animation is made to give a model explanation of the dynamic processes occurring within a Daniell cell while the electric circuit is closed. It shows that, starting at the zinc electrode, zinc ions move into the solution. The ions either 'come out of' the electrode or 'appear' from behind it. These ions move into a uniform, grey 'continuum', which is not part of or involved in the entire process, during which there are no changes to the zinc electrode's size or form. (It does not get smaller.) Accompanying the appearance of zinc ions, two electrons are somehow released. These same two electrons move through an envisioned 'electron channel' toward the copper electrode. (By the way: This is a questionable construct of electrical conductivity in metals.)

The copper electrode is also represented as an unaffected continuum. A random copper ion from the solution moves towards the charged electrode, where (together with the disappearance of the two zinc-generated electrons) a copper atom is formed and disappears into or behind the copper electrode. Similarly, this second electrode does not physically change throughout the process. The zinc electrode does not appear to become smaller, nor does the copper electrode appear to get larger, either in terms of mass or in volume. Within this explanation, both electrodes are presented as having no more in common with the redox reaction than the external wire circuit or the salt bridge. The salt bridge exists, but is not directly involved in the dynamics of the entire process. We can assume that a grey continuum represents some kind of 'electrolytic solution'. However, the space within the salt bridge looks completely different. It is empty and looks like the external wire connection. Particles or charges are not shown to be transported via the salt bridge and the salt bridge seems to belong to the same category as the external wire circuit. In interpreting the external circuit ('the electron channel'), only electrons can be transported. The salt bridge looks the same, since ions seem to be too large for passage through the salt bridge. Even the transport of charges in electrolytes as a free-flow of electrons is an oft-documented misconception among students (Garnett & Treagust, 1992b).

But why then do we as experienced chemists 'see' what we believe we see? Our 'expert knowledge' leads us to perceive what we want to see: an animated illustration of the Daniell voltaic cell. Within seconds, we reconstruct our knowledge using the impulse of the animated picture in order to obtain the 'correct' view. The observed content is no longer of interest to us; unfortunately we can't expect the same for our inexperienced students.

Some of the animations in Table 1 are actually quite good models of explanation for the Daniell cell in terms of commonly-accepted scientific views. Others are questionable in a similar way. Perhaps the reader would like to analyse the other examples on their own.

CONSTRUCTIVIST LEARNING AND THE IMPORTANCE OF STUDENTS' ALTERNATIVE IDEAS REVISITED

One may criticise our interpretation of the animation in Figure 2 as being exaggerated. This is true. One could say that this is only a model or a model-based explanation (a 'teaching model' as discussed in Justi & Gilbert, 2002b). It is part and parcel of the nature of models that they always use shortcuts and simplifications to represent their targets (Van Driel & Verloop, 1999). Therefore, if a model is sufficiently discussed and reflected upon in the classroom, there may no longer be any misunderstandings. But can we be completely confident in this belief when using this kind of model explanations in secondary chemistry classes?

If we view model-based thinking as a serious task in scientific learning (Van Driel & Verloop, 1999), we have to be extremely careful that it is not the above-mentioned simulation that is represented as the scientific model. The simulation is merely an illustration of the scientific model (Justi & Gilbert, 2002b). To be more exact, it is just a teaching model which was derived by different steps of interpretation from the scientifically-accepted curriculum model (Taber, 2008). The scientific model (or scientific theory, as we may call it) is the set of ideas behind the whole: the scientific models of particles, atoms and atomic structures, or the model of electron-transfer. Teachers as experts will recognise rather quickly where illustrations and scientifically-accepted models depart from one another. Can this, however, be expected from our students in school? In most cases, students do not have a sufficiently-developed understanding of scientific models and modelling (Grosslight, Unger, Jay, & Smith, 1991). Unfortunately, this may also be true of some teachers (e.g. Van Driel & Verloop, 1999; Harrison, 2000; Justi & Gilbert, 2002a and b, Sprotte & Eilks, 2007). In the same vein, students lack a developed understanding of the processes occurring in the sub-microscopic world and its differentness from the macroscopic level.

In recent decades, empirical research has revealed a large body of evidence concerning the variety of alternative beliefs which students hold about matter and chemical change. These results consequently suggest many points which have to be kept in mind when structuring or evaluating simulations as a potential tool for learning. Table 2 gives some selected results which may be important for understanding visual representations of a Daniell cell as it was discussed above.

However, it is obviously difficult (or maybe even impossible) to create an accurate visual aid which shows 1) all of the processes within the Daniell voltaic cell, while 2) simultaneously recognizing all the consequences of students' alternative beliefs, and 3) being comprehensible and not too complex. Simplifications are necessary to reach some sort of clarity for the learner. But which kind of simplification is acceptable, and which will only serve to nurture students' alterna-

Table 2. Selected results from empirical research about scientifically-unreliable concepts from students, with relevance for understanding electrochemical cells (see also Garnett & Treagust, 1992a, 1992b; De Jong & Treagust, 2002)

The particulate nature of matter often is misinterpreted as an understanding of particles within a continuum (Novick & Nussbaum, 1978; Johnson, 1998). An understanding of the particulate nature of matter within a system is often limited to single substances. These are those substances which are the focus of the discussion (Ahtee & Varjola, 1998). These kinds of mixed interpretations sometimes are retained during the entire time-period of school educ ation (Nakleh, 1992).
Macroscopic changes are sometimes explained by similar changes in the submicroscopic particles (Lee, Eichinger, Anderson, Berkheimer, & Blakeslee, 1993). This can be true even for mass changes. If matter is no longer visible, it seems to disappear along with its particles (Stavy, 1990). This is not only important to keep the principle of conservation of mass in mind, but also to retain the principle of conservation of atoms (Gomez, Pozo, & Sanz, 1995).
Current flow often is over-generalised as a flow of electrons. This is a reliable idea for portraying conductivity in metals. But an understanding of electrolytic conductivity as a flow of free electrons through the electrolyte has sometimes been observed. The concept of the movement of ions is not applied (Grosslight, Unger, Jay, & Smith, 1991; Garnett & Treagust, 1992b; Ogude & Bradley, 1994, 1996).
Current flow is not necessarily connected to the concept of a circuit. There are misinterpretations in understanding electric current as something flowing from the source to the device (de Posada, 1997). The necessity of a salt bridge therefore is not understood. The external and internal electric circuits are not seen as one entity (Burger, 2000).
Students sometimes adhere to a concept where electrolytes do not contain ions, but instead salts. Ions are formed at the moment voltage is put across the solution or an electrolytic process started (Ogude & Bradley, 1994, 1996; Butts & Smith, 1987).

Figure 3. Draft of an animation on the Daniell voltaic cell

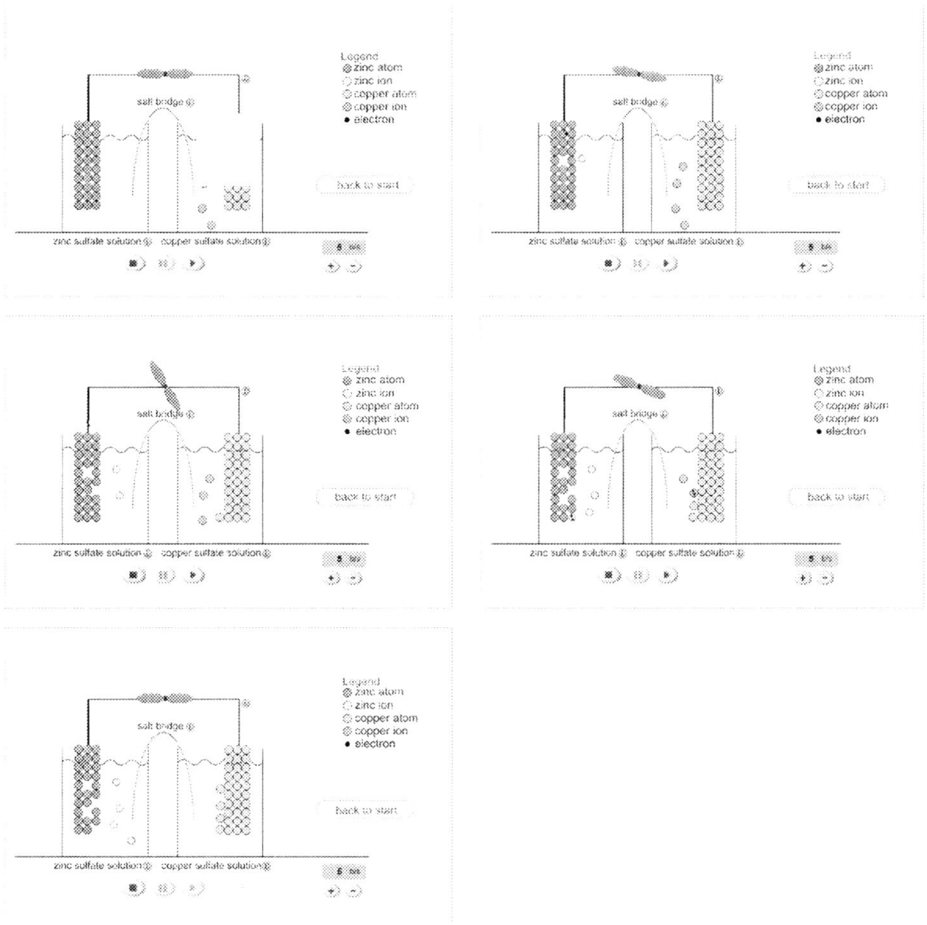

Table 3. Some ideas from empirical research with some compromises being made

Considered ideas	Compromises
Electrodes are composed of atoms of zinc and copper, respectively. During the reaction, zinc atoms are changed into ions. The zinc ions are dissolved into the solution. Solvated copper ions are changed into copper atoms and form new copper atoms upon the surface of the copper electrode.	The particles of the solvent are not shown. The level of the solution is sketched, but a grey or colourful sketch of a continuum is not shown. Additional information is available as a pop-up window, where a picture is available showing all particles within both half-cells (Figure 4).
No particles seem to spontaneously appear from nothing or disappear into nothing. The Laws of Conservation of Mass and the Conservation of Atoms are considered.	
Chemical change in particles leads to change in the electrodes' mass and structure. The zinc electrode becomes smaller in mass and volume, the copper electrode becomes larger in mass and volume.	
There is no visualisation of a flow of electric current only going from one half-cell to another.	Conductivity of the external electric circuit and the salt bridge is not explicitly shown in the animation. Additional information is available as a pop-up window, where both the processes of conductivity in metals and in electrolytes are explained.

Figure 4. Picture from a pop-up window showing all particles within the copper half-cell

Information about the copper sulfate solution

The copper electrode is submersed in 1 M copper sulfate solution (1 mol $CuSO_4$/liter). For better visibility the electrode is shown as a cutaway view. The water and sulfate ions are totally omitted because they don't undergo any changes during the reaction. If all of the particles were shown by the model in solution, the copper half-cell would look like this:

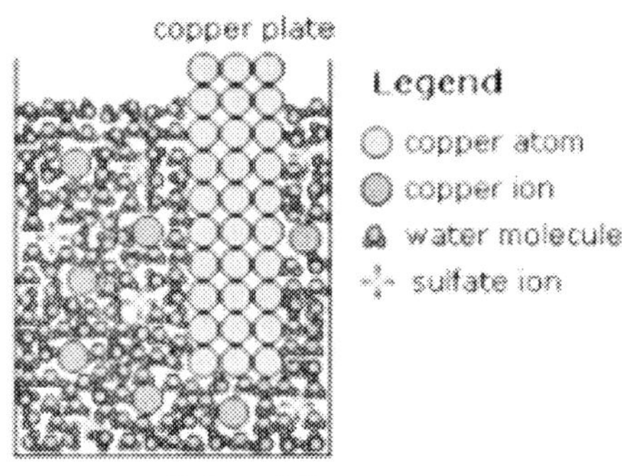

close window

tive beliefs? Of course, we don't have a definitive answer. But Figure 3 invites discussion, since it shows a draft visual aid, developed within an action research project to structure a student-active lesson plan to deal with electrochemistry based on cooperative learning (Markic & Eilks, 2006). Table 3 discusses some results from the empirical research which had been taken into consideration, but it also shows some compromises.

IMPLICATIONS

In all seriousness, if anyone would start criticising the animation in Figure 3 in the same exaggerated fashion with which we did the animation from Figure 2, even our own animation would fail the test. Even in this case, compromises were necessary. Our interest in writing this chapter was not to describe what a perfect Daniell voltaic cell simulation has to look like. This chapter was intended to serve as a critical warning. We must all be more contemplative when employing animated illustrations and critically reflect upon their potential effects upon novices and learning. And our discussion is just one case. Similar examples can be found for a broad range of other science topics, e.g. the states of matter and their changes (Figure 5).

Selected references for research on students' alternative conceptions:

- The particulate nature of matter often is misinterpreted as an understanding of particles within a continuum (Novick & Nussbaum, 1978), which we can see in the left-most animation. It is often difficult for students to accept that there is empty space between the particles. They often consider the particles of a substance as being 'dissolved in air or water' (Johnson, 1998). Within such an interpretation, students sometimes consider a substance like water itself to consist of 'water particles' within a continuum of liquid water (Lee et al., 1993). An interpretation like this can be seen in the animation on the left, where the particles are 'dissolved' in a grey continuum which obviously belongs to the water (in the solid and liquid state) but is not defined in more detail. The particles later on move into a continuum of air (or into completely empty space).

Figure 5. Animations of the state of matter and their changes (Capri, 2004)

Explaining the change in the state of matter of water

Changing Matter
Imagine a **block of ice in a beaker** with the temperature at **-30 degrees Celsius.**
Supposing the particles could be seen then they would be **gently vibrating.**
Increase Heat

Changing Matter
Eventually, as the temperature continues to increase the **particles break away from their fixed points** and **move around the container.** The solid has turned into a liquid. The particles **still stay in close contact** with each other.
Increase Heat

Changing Matter
The temperature continues to increase and the **particles of water continue to move faster and faster.**
Increase Heat

Explaining the liquid and gaseous state of water

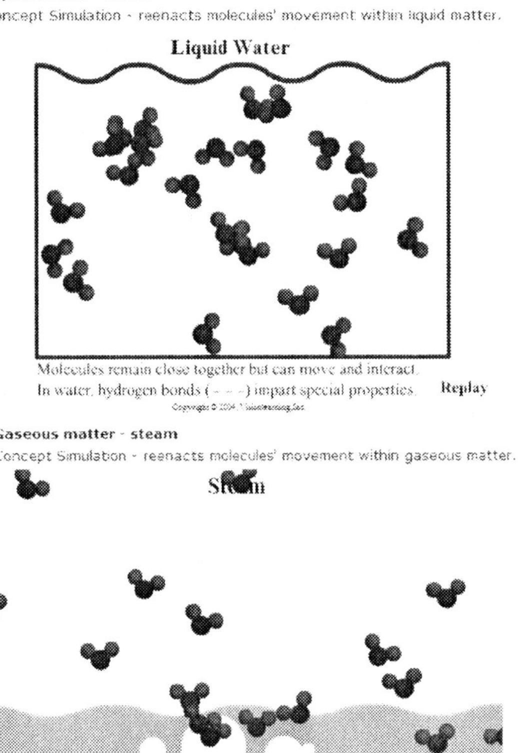

- Macroscopic changes are sometimes explained by similar changes in the sub-microscopic particles (Lee et al., 1993). If matter is no longer visible it seems to disappear along with its particles. This is especially important for processes of evaporation (Osborne, Bell & Gilbert, 1983; Stavy, 1990), as we can see in the animation on the left where the red particles disappear after having left 'the liquid'. But students sometimes think that new substances or particles are formed during the process of evaporation, which had not been part of the liquid in advance (Osborne et al., 1983), as can be seen in the right animation (lower picture) where the water molecules 'come out of' bubbles in the boiling water. This can also be seen in the left picture where continuous particles are leaving the grey continuum, without either a loss in the number of particles 'in the liquid' or any effect upon the grey continuum.

- Students often consider particles in the liquid state to have a considerable distance between themselves. The size is often emphasised as being somewhere in-between the distance the particles have within a solid or the gaseous state (Garnett, Garnett, & Hackling, 1995; Johnson, 1998). This can be seen in both the animations. This interpretation is in direct contradiction to the behaviour of liquids as relatively incompressible (just like solids, but opposite to gases) and having more-or-less the same volume per amount of substance as the solid, but far less than that compared to the volume of the same amount of substance in the gaseous state.

Starting with the point-of-view that every innovation in the chemistry classroom has to begin with the expertise of the teachers (Eilks, Ralle, Markic, Pilot, & Valanides, 2006), we think that discussions like the one here can help pinpoint specific aspects discussing the development of teachers' expertise.

In recent years it has become quite common to understand the teacher's knowledge base in terms of content knowledge (CK), pedagogical knowledge (PK) and pedagogical content knowledge (PCK). Pedagogical Content Knowledge (PCK) was originally described by Shulman (1987) as *"that special amalgam of content and pedagogy that is uniquely the province of teachers, their own special form of professional understanding."* This 'definition' was later expanded upon and more precisely described by different scholars. For example, Van Driel, Verloop, & De Vos (1998) considered PCK to be a specific form of craft knowledge and Loughran, Milroy, Berry, & Gunstone (2001) defined PCK as *"the knowledge that a teacher uses to provide teaching situations that help learners make sense of particular science content."* Following Van Driel, Verloop, & De Vos (1998) such a knowledge base incorporates: 1) knowledge of students' conceptions with respect to a domain or topic, understanding specific student difficulties in that area, 2) knowledge of representations of subject matter for teaching, and 3) knowledge of instructional strategies incorporating such representation. The last area of the definition by Van Driel et al. (1998) also encompasses the question of representations and instructional strategies. This obviously includes knowledge about multimedia representations and their application in teaching.

Therefore, the above discussion demands the development of teachers' PCK, specifically the availability and accuracy of respective animations for different topics to be taught in science. But the discussion also touches upon teachers' critical skills while reflecting upon and evaluating animations offered via the Internet. The discussion within educational areas dealing with ICT acknowledges that ICT may be not only one medium among many, but takes into account the very special aspects of learning with digital technology and computers. For example, Mishra & Koehler (2006) suggested picking up technology as a fourth, interacting dimension in the discussion of PCK. They suggested the construct of *Technological Pedagogical Content Knowledge* (TPCK) (see Figure 6). TPCK is thought to combine the PCK of teachers with their technological knowledge and skills. It also discusses the teachers' knowledge base of exactly how to integrate computers appropriately into domain-specific learning processes. This is not a completely new idea, but instead looks

Figure 6. The Framework of Technological Pedagogical Content Knowledge (TPCK) (Koehler, Mishra, & Yahya, 2007)

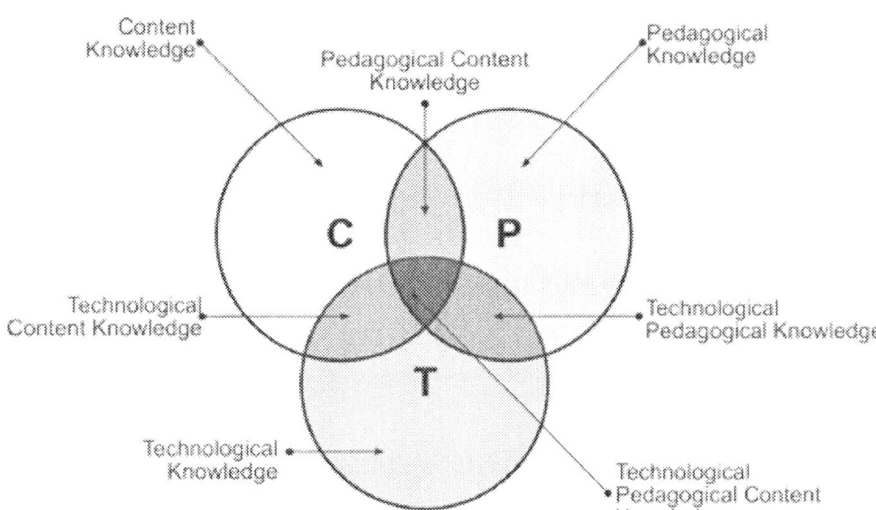

back into the 1980s, when the first papers dealing with professional technological knowledge of teachers were published (e.g. Lacina, 1984; Ellis & Kuerbis, 1985).

For Mishra and Koehler (2006), TPCK is the source of successful teaching and learning with technology and consists of:

- *understanding the representation of concepts using technologies;*
- *pedagogical techniques that use technologies in constructive ways to teach content;*
- *knowledge of what makes concepts difficult (or easy) to learn and how technology can help redress some of the problems which students face;*
- *knowledge of students' prior knowledge, theories and epistemology;*
- *knowledge of how technologies can be used to help build on existing knowledge and to develop new epistemologies or even strengthen old ones.*

In particular, TPCK discusses how difficult it is to create a potentially-helpful visual aid for chemistry students. Our above examples show that it is essential to have the results of empirical research about students' alternative conceptions and learning processes in mind in order to properly evaluate the benefits or dangers of a certain animation. Another lesson which we can derive from the above discussion is that it is often very risky to use information technologies or visual aids which have not been thoroughly tested and considered, if we are attempting to identify any potential problematic interpretations from the student's point-of-view. Animations and simulations seem all too often to have a larger potential for fostering misconceptions than they do for promoting scientific understanding, especially if they have been constructed without sufficient reflection upon the learners' perspective and foreknowledge.

Table 1 offers different examples for animations of Daniell's voltaic cell in various languages. We encounter many possible and highly-interesting interpretations of the data if we consider them from an inexperienced or 'naïve' point-of-view. This same viewpoint is potentially used by novice students, who know little about the chemical principles behind the visual aids, and who do not have a developed understanding of model usage. After the preliminary activity described in our teacher training seminars as described above, we follow up with a second activity. Our students are asked to write down what they really saw during the demonstration and to exaggerate their interpretation, as we have performed above. The act of writing down such simplistic, 'naïve as possible' explanations of pictures or animations, while simultaneously attempting to tie them to available research literature about alternative frameworks, has proven to be a fruitful exercise in teacher-training seminars. We have witnessed a great potential for 1) making students more critical of the use of multimedia animations, 2) promoting their own understanding of empirical research and its potential for changing students' alternative frameworks, and 3) making the students themselves more sensitive to the difference between expert and novice thinking. The same holds true for chemistry teachers, chemical educators, or software developers, after having themselves structured a multimedia learning tool.

ACKNOWLEDGMENT

We thank Bill Byers for his critical review on the discussion within an earlier version of the manuscript and his helpful remarks.

REFERENCES

Ahtee, M., & Varjola, I. (1998). Students' understanding of chemical reaction. *International Journal of Science Education*, *20*(3), 305–316. doi:10.1080/0950069980200304

Azevedo, R. (2004). Using hypermedia as a metacognitive tool for enhancing student learning? The role of self-regulated learning. *Educational Psychologist*, *40*(4), 199–209. doi:10.1207/s15326985ep4004_2

Barnea, N., & Dori, Y. (1999). High-school chemistry students' performance and gender differences in a computerised molecular modelling learning environment. *Journal of Science Education and Technology*, *8*(4), 257–271. doi:10.1023/A:1009436509753

Bodner, G. M. (1986). Constructivism – A theory of knowledge. *Journal of Chemical Education*, *63*(10), 873–878.

Brandt, L., Elen, J., Hellemans, J., Heerman, L., Couwenberg, I., Volckaert, L., & Morisse, H. (2001). The impact of concept mapping and visualization on the learning of secondary school chemistry students. *International Journal of Science Education*, *23*(12), 1303–1313. doi:10.1080/09500690110049088

Burger, N. (2000). *Vorstellungen von Schülern über Elektrochemie - eine Interviewstudie [Students' conceptions about electro-chemistry – an interview study]*. Unpublished doctoral dissertation, University of Dortmund.

Butts, B., & Smith, R. (1987). What do student perceive as difficult in H.S.C. chemistry? *Australian Science Teachers' . Journal*, *32*(4), 45–51.

Capri, A. (2004). Matter: States of matter. *Visionlearning, CHE, 3*(1). Retrieved November 1, 2008, from http://www.visionlearning.com/library/module_viewer.php?mid=120

De Jong, O., & Treagust, D. F. (2002). The teaching and learning of electrochemistry. In J. K. Gilbert, O. De Jong, R. Justi, D. F. Treagust, & J. H. Van Driel (Eds.), *Chemical Education: Towards research-based practice* (pp. 317-338). Dordrecht, The Netherlands: Kluwer.

De Posada, J. M. (1997). Conceptions of high school students concerning the internal structure of metals and their electric conduction: structure and evolution. *Science Education*, *81*(4), 445–367. doi:10.1002/(SICI)1098-237X(199707)81:4<445::AID-SCE5>3.0.CO;2-C

Eilks, I. (2003). Students' understanding of the particulate nature of matter and some misleading illustrations from textbooks. *Chemistry in Action, 69*, 35–40.

Eilks, I., Möllering, J., & Valanides, N. (2007). Seventh-grade students' understanding of chemical reactions – Reflections from an action research interview study. *Eurasia Journal of Mathematics . Science and Technology Education, 4*(3), 271–286.

Eilks, I., Ralle, B., Markic, S., Pilot, A., & Valanides, N. (2006). Ways towards research-based science teacher education. In I. Eilks & B. Ralle (Eds.), *Towards research-based science teacher education* (pp. 179-184), Aachen, Germany: Shaker.

Eilks, I., Witteck, T., & Pietzner, V. (2009). A critical discussion of the efficacy of using visual learning aids from the Internet to promote understanding, illustrated with examples explaining the Daniell voltaic cell. *Eurasia Journal of Mathematics . Science and Technology Education, 5*(2), 145–152.

Ellis, J. D., & Kuerbis, P. J. (1985). *Development and validation of essential computer literacy competencies for science teachers.* Paper presented at the Annual Meeting of NARST, French Lick Springs, IN.

Falvo, D. (2008). Animations and simulations for teaching and learning molecular chemistry. *International Journal of Technology in Teaching and Learning, 4*(1), 68–77.

Garnett, P. J., Garnett, P. J., & Hackling, M. W. (1995). Students' alternative conceptions in chemistry: a review of research and implications for teaching and learning. *Studies in Science Education, 25*, 69–95. doi:10.1080/03057269508560050

Garnett, P. J., & Treagust, D. F. (1992a). Conceptual difficulties by senior high school students of electrochemistry: electric circuits and oxidation-reduction equations. *Journal of Research in Science Teaching, 29*(2), 121–142. doi:10.1002/tea.3660290204

Garnett, P. J., & Treagust, D. F. (1992b). Conceptual difficulties experienced by senior high school students of electrochemistry: electrochemical (galvanic) and electrolytic cells. *Journal of Research in Science Teaching, 29*(10), 1079–1099. doi:10.1002/tea.3660291006

Gomez, M.-A., Pozo, J.-I., & Sanz, A. (1995). Students' ideas on conservation of matter: effects of expertise and context variables. *Science Education, 79*(1), 77–93. doi:10.1002/sce.3730790106

Grosslight, L., Unger, C., Jay, E., & Smith, C. (1991). Understanding models and their use in science: conceptions of middle and high school students and experts. *Journal of Research in Science Teaching, 28*(9), 799–822. doi:10.1002/tea.3660280907

Harrison, A. (2000). *How do teachers and textbook writers model scientific ideas for students?* Paper presented at the NARST Annual Meeting, New Orleans, LA.

Hill, D. (1988). Misleading illustrations. *Research in Science Education, 18*(1), 290–297. doi:10.1007/BF02356607

Huk, T. (2007). Who benefits from learning with 3D models? The case of spatial ability. *Journal of Computer Assisted Learning, 22*(6), 392–404. doi:10.1111/j.1365-2729.2006.00180.x

Johnson, P. (1998). Progression in children's understanding of a 'basic' particle theory: a longitudinal study. *International Journal of Science Education, 20*(4), 393–412. doi:10.1080/0950069980200402

Johnstone, A. H. (1991). Why is science difficult to learn? Things are seldom what they seem. *Journal of Computer Assisted Learning*, *7*(2), 75–83. doi:10.1111/j.1365-2729.1991.tb00230.x

Justi, R. S., & Gilbert, J. K. (2002a). Science teachers' knowledge about and attitudes towards the use of models and modelling in learning science. *International Journal of Science Education*, *24*(12), 1273–1292. doi:10.1080/09500690210163198

Justi, R. S., & Gilbert, J. K. (2002b). Models and modelling in chemistry education. In J. K. Gilbert, O. De Jong, R. Justi, D. F. Treagust, & J. H. Van Driel (Eds.), *Chemical Education: Towards research-based practice* (pp. 47-68). Dordrecht, The Netherlands: Kluwer.

Kelly, R. M. (2005). *Exploring how animations of sodium chloride dissolution affect students' explanations*. Unpublished doctoral dissertation, University of Northern Colorado, Greeley, CO.

Koehler, M. J., Mishra, P., & Yahya, K. (2007). Tracing the development of teacher knowledge in a design seminar: integrating content, pedagogy and technology. *Computers & Education*, *49*(3), 740–762. doi:10.1016/j.compedu.2005.11.012

Kozma, R., & Russell, J. (2005a). Students becoming chemists: developing representational competence. In J. K. Gilbert (Ed.), *Visualization in science education* (pp. 121-145). Dordrecht, The Netherlands: Springer.

Kozma, R., & Russell, J. (2005b). Multimedia learning of chemistry. In R. Mayer (Ed.), *The Cambridge handbook of multimedia learning* (pp. 409-428). New York, NY: Cambridge University Press.

Lacina, L. J. (1984). *The determination of computer competencies needed by classroom teachers*. Dubuque, IA: Loras College. (ERIC Document Reproduction Service No. ED264 831)

Lee, H. (2007). Instructional design of web-based simulations for learners with different levels of spatial ability. *Instructional Science*, *35*, 467–479. doi:10.1007/s11251-006-9010-5

Lee, O., Eichinger, D. C., Anderson, C. W., Berkheimer, G. D., & Blakeslee, T. S. (1993). Changing middle school students' conceptions of matter and molecules. *Journal of Research in Science Teaching*, *30*(3), 249–270. doi:10.1002/tea.3660300304

Loughran, J., Milroy, P., Berry, A., & Gunstone, R. (2001). Documenting science teachers' pedagogical content knowledge through PaP-eRs. *Research in Science Education*, *31*(2), 289–307. doi:10.1023/A:1013124409567

Markic, S., & Eilks, I. (2006). Cooperative and context-based learning on electrochemical cells in lower secondary science lessons – A project of Participatory Action Research. *Science Education International*, *17*(4), 253–273.

Mayer, R. E. (2001). *Multimedia learning*. Cambridge, UK: Cambridge University Press.

Mayer, R. E. (2003). The promise of multimedia learning using the same instructional design methods across different media. *Learning and Instruction*, *13*(2), 125–140. doi:10.1016/S0959-4752(02)00016-6

Mishra, P., & Koehler, M. J. (2006). Technological Pedagogical Content Knowledge: a framework for teacher knowledge. *Teachers College Record*, *108*(6), 1017–1054. doi:10.1111/j.1467-9620.2006.00684.x

Nakhleh, M. B. (1992). Why some students don't learn chemistry. *Journal of Chemical Education*, *69*(3), 191–196.

Novick, S., & Nussbaum, J. (1978). Junior high school pupils' understanding of the particulate nature of matter: an interview study. *Science Education*, *62*(3), 273–281. doi:10.1002/sce.3730620303

Ogude, A. N., & Bradley, J. D. (1994). Ionic conduction and electrical neutrality in operating electrochemical cells. *Journal of Chemical Education*, *71*(1), 29–34.

Ogude, A. N., & Bradley, J. D. (1996). Electrode processes and aspects relating to cell EMF, current and cell components in EC cells. *Journal of Chemical Education*, *73*(12), 1145–1149.

Osborne, R. J., Bell, B. F., & Gilbert, J. K. (1983). Science teaching and children's views of the world. *European Journal of Science Education*, *5*(1), 1–14.

Pavio, A. (1986), *Mental respresentations: A dual coding approach.* Oxford, UK: Oxford University Press.

Pfundt, H. (1975). Ursprüngliche Erklärungen der Schüler für chemische Vorgänge [Initial students' explanations of chemical phenomena]. *Der Mathematische und Naturwissenschaftliche Unterricht*, *28*(3), 157–162.

Pfundt, H. (1982). Vorunterrichtliche Vorstellungen von stofflicher Veränderung [Pre-instructional imaginations of material change]. *Chimica Didactica*, *8*, 161–180.

Ploetzner, R., Bodemer, D., & Neudert, S. (2008). Successful and less successful use of dynamic visualizations. In R. Lowe & W. Schnotz (Eds.), *Learning with Animation – Research Implications for Design* (pp. 71-91). New York, NY: Cambridge University Press.

Posner, G. J., Strike, K. A., Hewson, P. W., & Gertzog, W. A. (1982). Accommodation of a scientific conception: toward a theory of conceptual change. *Science Education*, *66*(2), 211–227. doi:10.1002/sce.3730660207

Sanger, M. J., & Greenbowe, T. J. (2000). Addressing student misconceptions concerning electron flow in electrolyte solutions with instruction including computer animations and conceptual change strategies. *International Journal of Science Education*, *22*(5), 521–537. doi:10.1080/095006900289769

Schnotz, W., & Bannert, M. (2003). Construction and interference in learning from multiple respresentations. *Learning and Instruction*, *13*(2), 117–123. doi:10.1016/S0959-4752(02)00015-4

Schwartz, N., Andersen, C., Hong, N., Howard, B., & McGee, S. (2004). The influence of metacognitive skills on learners' memory of information in a hypermedia environment. *Journal of Educational Computing Research*, *31*(1), 77–93. doi:10.2190/JE7W-VL6W-RNYF-RD4M

Shapiro, A. (1999). The relationship between prior knowledge and interactive overviews during hypermedia-aided learning. *Journal of Educational Computing Research*, *20*(2), 143–167. doi:10.2190/BCKU-F3AC-CNPW-M44E

Shulman, L. S. (1987). Knowledge and teaching – foundation of the new reform. *Harvard Educational Review*, *57*(1), 1–22.

Sprotte, J. A., & Eilks, I. (2007). *Introducing the particulate nature of matter – Results from a case study on experienced German Science Teachers' PCK of models and modelling*. Paper presented at the 6th ESERA Conference, Malmoe, Sweden.

Stavy, R. (1990). Children's conception of changes in the state of matter: from liquid (or solid) to gas. *Journal of Research in Science Teaching*, *30*(3), 247–266. doi:10.1002/tea.3660270308

Taber, K. (2008). Towards a curricular model of the nature of science. *Science & Education*, *17*(2-3), 179–218. doi:10.1007/s11191-006-9056-4

Tsui, C.-Y., & Treagust, D. (2004). Motivational aspects of learning genetics with interactive multimedia. *The American Biology Teacher, 66*(4), 277–285. doi:10.1662/0002-7685(2004)066[0277:MAOLGW]2.0.CO;2

Van Driel, J. H., & Verloop, N. (1999). Teachers' knowledge of models and modelling in science. *International Journal of Science Education, 21*(11), 1141–1153. doi:10.1080/095006999290110

Van Driel, J. H., Verloop, N., & De Vos, W. (1998). Developing science teachers' Pedagogical Content Knowledge. *Journal of Research in Science Teaching, 35*(6), 673–695. doi:10.1002/(SICI)1098-2736(199808)35:6<673::AID-TEA5>3.0.CO;2-J

Wandersee, J. H., Mintzes, J. J., & Novak, J. D. (1994). Research on alternative conceptions in science. In D. L. Gabel (Ed.), *Handbook of research in science teaching and learning* (pp. 177-210). New York, NY: Macmillan.

Williamson, V. M., & Abraham, M. R. (1995). The effects of computer animation on the particulate mental models of college chemistry students. *Journal of Research in Science Teaching, 32*(5), 521–534. doi:10.1002/tea.3660320508

Yang, E.-M., Greenbowe, T. J., & Andre, T. (2004). The effective use of an interactive software programmes to reduce students' misconceptions about batteries. *Journal of Chemical Education, 81*(4), 587–595.

APPENDIX

Examples from the Internet Discussed in the Paper

(1) www.chemgapedia.de/vsengine/vlu/vsc/de/ch/13/vlu/echemie/galvanische_elemente/batterie.vlu/Page/vsc/de/ch/13/pc/echemie/galvanische_elemente/daniellapplet.vscml.html, retrieved from the World Wide Web, November 01, 2008

(2) www.chemieunterricht-interaktiv.de/en/animations/electrochemistry/daniell_cell/daniell_en.html, retrieved from the World Wide Web, November 01, 2008

(3) www.bgfl.org/bgfl/custom/resources_ftp/client_ftp/ks3/science/changing_matter/index.htm, retrieved from the World Wide Web, November 01, 2008

(4) www.visionlearning.com/library/module_viewer.php?c3=&mid=120&ut=&l=e, retrieved from the World Wide Web, November 01, 2008

Chapter 5
Three Primary Science Podcasts: Illustrating the Contextual and Staging Dimensions of Language in Podcasts

Susan Rodrigues
University of Dundee, Scotland

Graham Williamson
University of Dundee, Scotland

ABSTRACT

In this chapter the authors examine three podcasts created by the same class. This particular class generated three different types of podcasts and this allows us to explore the contextual and staging dimensions of language found within these podcasts. The podcasts were created as part of a project investigating the use of podcasting in primary science. The teachers used class time over three-four weeks to allow the pupils to generate podcasts that were then uploaded onto a site for public dissemination. The authors examine the podcasts from various perspectives in order to explore the notion of multiple literacy.

INTRODUCTION

The Partnership in Primary Science Project 3: podcasting in primary science (PIPS3), funded by the Astrazeneca Science Teaching Trust (AZSTT), began in September 2007. Podcasting is a way of broadcasting involving various media, usually audio "radio" files, via the internet. The PIPS3 project provided teachers with time and resources to explore the use of podcasting with their classes when learning science. The teachers were expected to use class time in the three-four weeks between face-to-face project team meetings to work with pupils and others to generate podcasts, but the podcasts were to be pupil- driven. Some teachers began by familiarising the pupils with the technology because though some pupils may have used MP3 players, many may not have used them to create digital records. Other teachers began with the science and the pupils developed the skills to podcast in order to showcase pupil activity.

DOI: 10.4018/978-1-60566-690-2.ch005

SO WHY PODCAST?

There are two main reasons for employing podcasts. The first reason draws on the Vygtoskian notion of higher mental actions being mediated by tools, artefacts and cultural inventions (Vygotsky, 1997), and that speech, according to Vygtosky was an important tool (Wertsch, 1990). The second reason is based on the fact that in the UK, podcasting is now a fairly common form of dissemination in the public arena. Luft (2008) reported on a study by a research group looking at the habits of web users. The Pew Research group signalled an increase in podcast use/access in the last two years. In 2006, approximately 12% of the web users had downloaded podcasts, but by 2008, this figure had reached 19% (Luft, 2008). Webster (2008) reported that similar figures were seen in a podcast consumer study conducted in 2008 by the 2008 Arbitron/Edison Media Research Internet and Multimedia study. We should point out that these figures relate specifically to the US, rather than worldwide. Interestingly a BBC report (Fildes, 2006) showed that a survey found that about 8 million British users would go in search of a podcast in the next six years. Does this imply that British people are more likely than their US counterparts to be interested in podcasts and podcasting? However, what we found interesting was how do people (surveyors/researchers) collect this type of data in the first place? Clearly there are some obvious challenges. For example, can we say that podcast download is an accurate, reliable, valid measurement when it comes to depicting the number of users? At present we really do not have a valid or reliable auditing mechanism to quantify podcast use and therefore, in reality, it is difficult to ascertain the number individuals downloading the podcast. Nevertheless, the web user podcast download data makes for interesting reading.

When we submitted the first draft of this chapter, reviewer feedback stipulated the need for the chapter to include some recent and relevant research. But trying to trace recent and relevant research that was more than anecdotal or generated for commercial purposes involving podcasts in education was a challenge. Very few studies are reported in peer reviewed academic journals. Brittain, Glowacki,Van Ittersum, and Johnson (2006) reported on three pilot case studies on the use of podcasts with a dental group. Chan and Lee (2005) reported on podcasting and suggested that podcasts provided a 'low-cost, low-barrier tool for disseminating content across the Internet (p65) while ensuring that 'podcasting allows fresh content to be delivered to students' desktops and handheld devices, as it becomes available. This makes it possible to tailor content "just in- time" to suit students' needs, and promotes a sense of currency and direct relevance of the content to the cohort…'(p66).

From an educational perspective, there is anecdotal evidence to suggest that podcasting provides pupils averse to writing with an opportunity to make public their understanding. There is also an argument for podcasting in terms of supporting and sharing pupils' creative skills. However, we have little evidence to show whether the rhetoric is matched with evidence. Nevertheless, given recent moves in Scotland to promote a new curriculum, advocating creativity, funds were obtained to provide teacher professional development with specific regard to podcasting in science.

THE CONTEXT

There were seven schools involved in the PIPS3 project. Most of the schools (five) involved two teachers, but two schools were very small and were only able to involve one teacher from each school. Each of the schools produced an unexpected variety of podcasts. Each class also hosted a synchronous chat to discuss their podcast with other pupils and parents. The topics that were selected depended upon the topic scheduled in that particular school's scheme of work. Hence there was an array of topics, including electricity, calories, healthy eating, gravity, plants, digestion, the body, and living things

In this chapter we examine three podcasts created by the same class. We focus on this particular class, as it involved the same pupils generating three different types of podcasts, and so allows us to explore the contextual and staging dimensions of language found within these podcasts. The podcasts also help to illustrate what Rogoff (1995) described as participatory appropriation: "individuals transform their understanding of and responsibility for activities through their own participation" (Rogoff, 1995:150). Rogoff's views encouraged us to look at primary pupils' engagement and the contextual factors that operated when they constructed podcasts, and encouraged us to use Systemic Functional Linguistic (SFL) theory (see Halliday, 1993, 1994; Halliday & Hasan, 1985) and the influence of genre (Giddens, 1984; Yates & Orlikowski, 1992) on staging of the podcast to look at concept presentation.

The podcasts were created by primary (P4-5) pupils to illustrate their science learning for a given topic: 'our body'. The class began by brainstorming, in pairs, what they knew about 'bodies'. They constructed a display to represent their ideas and they made a list of what they knew and what they wanted to find out. From this, as a class they constructed their big questions requiring answers. The pupils then produced fact files: they chose a body part and for their homework they researched that body part.

For the purpose of assessment, the teacher asked the pupils to work in groups and create a verse, for their body part, to the tune of 'Old Macdonald had a Farm'. This was recorded as a podcast.

The Digestive system was a subtopic within the overall body topic, and to assess their learning, the pupils were asked to produce a script about what they learnt about the digestive process, and for their homework they had to find sound effects to illustrate facets of the digestive process. This was recorded as a podcast.

The pupils were also asked to construct a list of questions one could ask a 'body part' if one was able to interview that body part. The pupils recorded the interview between the various body parts, mimicking the format deployed by a popular television celebrity chat show couple. This was recorded as a podcast.

These three podcasts provide data for analysis. The three podcasts are:

'Richard and Judy Show'
'The Organ Song'
'The Story of Digestion'

All three podcasts were well received and were the recipients of favourable online comments. From an interest and engagement perspective, the Richard and Judy podcast appeared to be the most successful. The online comments suggest that most listeners enjoyed listening to this podcast: they liked the style and they liked the humour. The story of digestion was of interest, with most comments referring to the sound effects that had been created to further illustrate particular digestion processes. The Organ song was well received, and had been performed as part of a school assembly. Overall, all three podcasts were considered successful by those involved in producing them as well as those who made up the audience.

ANALYSIS

In this chapter we examine these podcasts from various perspectives in order to explore the notion of multiple literacy. For example we look at the influence of register, using Systemic Functional Linguistic (SFL) theory (see Halliday, 1993, 1994; Halliday & Hasan, 1985) and we look at the influence of genre (Giddens, 1984; Yates & Orlikowski, 1992) on the staging of the podcast structure.

Through SFL we can explore various language levels because SFL theory allows us to look at what language does and how it does it. Hence, we can use SFL to look at the social context of the podcast and how language acts upon and is constrained by the social context of each podcast. We can explore various language levels, for example, context, semantic, lexicogrammar (syntax, lexicon and morphology) and phonology by looking at register (relations between tenor, mode and field).

For Hassan (2005), register is situationally relevant discourse, that involves interactions between, tenor, mode and field. Tenor is the social relationship between the language users, mode refers to the channel of communication and field is topics/actions. As such, register is based on what Hassan (2005;56) describes as 'recognition of regularities in types of language use.' We can also use semantics to analyse the podcasts. Semantics includes the propositional content, interpersonal semantics and textual semantics. While lexicogrammar usually considers the analysis in terms of roles, we can use it more broadly to explore communicative competence through the engagement of actors, agents, mediators, etc.

As well as SFL, the influence of genre on the staging of the podcast structure is of interest, for examining genre provides an opportunity to explore how podcasts help pupils to stage their learning and their understanding. The concept of genre can be found in literary analysis (Bakhtin, 1986) and others have used the concept of genre to explore social action (Brown, 1994; Berkenkotter and Huckin, 1995).

Yates and Orlikowski (1992, p299) defined genres as "socially recognized types of communicative action habitually enacted by members of a community to realize particular communicative and collaborative purposes". In Giddens' structuration theory, (1984) genre is shaped, and becomes shaped by, the flow of social practices with collective rules and a shared general form. The purpose of a genre is to help those within a community recognise a given situation or summon a sense of that situation.

As Yoshioka and Herman (1999) suggest, genre may be characterised by using '5W1H' questions. These can be interpreted as:

- Why? This relates to the purpose of the communication? There are in essence one of ten purposes: inform, commit, guide, request, express, decide, propose, respond, record and 'other').
- What? This relates to the content of a genre or genre system.
- Who/Whom? This relates to the participants in a genre or genre system.
- When? A particular genre is summoned so timing of genre use could therefore relate to a particular opportunity.
- Where? This relates to the place/milieu for the communication. In essence it reflects the nature of the culture shared by members of a given community.
- How? This in essence suggests that a genre is typically characterized by its form, conventions and expectations.

We use these notions of genre and SFL to explore the three podcasts created by pupils when sharing their learning of primary science.

FINDINGS

We begin by considering the genre used in each podcast. In all three podcasts, regardless of the podcast genre, the following were common:

- **Who/Whom:** The participants, in this case the pupils are clearly aware of their need to engage and entertain their beyond the classroom audience. Their product is not simply intended for their perusal, but for a wider audience, and the pupils' performance indicates an awareness of this wider audience.

- **When:** From the teacher's account, the timing was clearly related to opportunity. It was also seen to be a way of assessing and concluding a topic. Hence for example, the production of the chat show and organ song podcast came after the pupils had researched the various organs and discussed the organ functions. In most lessons a written account (possibly co-constructed by the teacher and the pupils) would be the summative product. The podcast enabled the teacher to provide the pupils with a framework (the tune of the song or the chat show format) but the construction of the verse and the interviewee response was the responsibility of the pupil.
- **Where:** The place for communication also reflects a culture that fostered sharing between the participants in a community. Hence, while the pupils performed their podcasts for their school assembly, it provided an opportunity to assess while also showcasing for a wider audience when it was podcast. In addition, though this podcast provided an opportunity to record and assess the pupils' views, and therefore was 'classroom' based, it also showcased their product for a wider audience, and therefore it was not simply limited to the classroom.

The remaining 2Ws and the 1H will be discussed for each of the podcasts.

Podcast 1: 'Richard and Judy Show'

The Richard and Judy Show podcast

Well we have some very exciting guests on today's show. Yes that's right Richard – we have lots of different organs in our body and here is our first guest.....the heart.

Hello, hello I am delighted to be here – I'm a very busy organ you know. Yes I've heard you do lots of good work in the body. Can you tell us all about it? Well I pump blood to all the different body parts and I work very fast and my friends are the lungs. That sounds like a very important job, the body is lucky to have you. Thank you for coming to talk to us. Bye

And our next guest is a very high flyer, we have been lucky to get her. She is of course the brain. Well hello, hello – I am delighted to be on the show. I am a very big fan you know. It was good of you to take time from your busy schedule. Your right – I am not sure if the body will cope to be quite honest. They can't do anything without me. Tell us a little more about what you do. I boss people around and if I was not there then nothing would work and my best friends are the eyes. Must dash, if there is nobody to tell those body parts what to do then there is no body. Bye

And here is our next guest or guests should I say – the marvellous lung bros. Inhale, exhale, inhale, exhale. It is lovely to see you. Thank you for having us on your show. So you work together. Tell us a little bit about your role in the body. Well we work all day long. We inhale air and give it to the blood what takes it all around the body and then the blood gives us carbon dioxide and we exhale so it leaves the body.

Fascinating and onto our final guests of the afternoon – the intestine twins. Yes these two have a long history together. Hello, hello – we are so happy to be invited on. Yes we thought you were only going to interview the organs from the top of the body. No, no we are not biased. Tell us a little bit about your work in the body. Well we suck out all the good stuff from the food with our villi *and then we squash it down and then it goes out and our best friend is the bladder.*

Well this is all we have time for this afternoon. Over to the ITN news at two – goodbye.

The format of the Richard and Judy Show provides a pattern of greeting/invitation and response. The invitation takes the form "Tell us what you do..." This pattern provides a 'script' or 'schema' for this text, which may not have been a written one, though certainly at one point the 'Richard' presenter introduces '...the Lung Bros.' showing that either he is not aware of the pronunciation of this contraction which he is reading out or that he is familiar with slang use to represent 'brothers'. The tenor of this podcast, in terms of the social relationship between the language users, suggests some scripting. Hence, though the children responsible for the given 'Organs' may have researched their function, the output may not necessarily reflect the scope of their research as the podcast is scripted and supports turn-taking. In terms of turn-taking, the only real hesitation is shown by the Lung Bros, who may not be struggling with what they have to say so much as having to pause to work out whose turn it is to speak.

Grammatically, the invitation, 'Tell us what...' is the only complex structure. All other sentence structures uttered by either the presenters 'Richard' or 'Judy'and the 'Organs' are simple or compound. The linking of statements is nearly all by 'and'. However, from a semantic perspective, the propositional content, interpersonal semantics (attitudes, exchange structure, speech function) and textual semantics (the structure of the message, be it theme structure or rhetorical), we suggest that the podcast demonstrates a fair degree of complexity for P4 –5 pupils.

To be specific, the programme format involves a set of formulae or frames used by the show presenters to introduce their guests, who then take their turn explaining the work that they do. These are changes of tenor and mode, from the language of the classroom to the language of TV programmes, and the interest lies in how this transformation affects the field, namely, knowledge of the science of the body.

The attractiveness and suitability of the formulae to transform tenor and mode in order to present learning of science is clear: it allows 'telling of what they know'.

The mode of communication, 'the interview by a celebrity couple' reflected the pupils' perception of a common television early afternoon chat show style. It also enables the pupils to present a particular stance between that of pupil as a product of social position within their classroom and pupil as a cultural agent. Their selection of the title for the podcast clues the listener in to the mode to expect. In effect the clue is to expect a chat show dialogue, albeit a partially scripted conversation.

The field is in essence, what is going on. So the topics and actions in the podcast form the field. From a science perspective, there is limited detail in what the organs say about themselves: they typically make one simple statement of their function, but do not pursue further grammatical development through, for example, a 'which' clause. The vocabulary used is non-technical with one or two exceptions - the names of the parts themselves and the use of the term 'vili' by the Intestines.

At the time of the project, science education in Scotland was influenced and guided by '5-14 Science Framework', LT-C2.1 (name the life processes common to humans and other animals, LT- C2.2 (identify the main organs of the human body) and LT-C2.3 (describe the broad functions of the organs of the human body) and the evolving Curriculum for Excellence (CfE) (LTScotland, 2008) in the form of 'Life and Cells - Keeping my body healthy'. The statements below were the draft outcome statements available at the time the teacher and pupils created their podcasts. In the intervening period, the CfE outcome statements have been through a consultation process and have been reviewed and some of them revised. We have not included the revised statements in this chapter, as the teachers used the draft statements. Below are the CfE statements that were available to the teacher at the time of preparing the podcasts.

Table 1. Levels (Source: LTScotland 2008)

<table>
<tr><th>Level</th><th>Stage:</th></tr>
<tr><td>Early</td><td>Pre-school and Primary 1</td></tr>
<tr><td>First</td><td>To the end of Primary 4, but earlier for some children</td></tr>
<tr><td>Second</td><td>To the end of Primary 7, but earlier for some children</td></tr>
<tr><td>Third</td><td rowspan="2">Secondary school year1-Secondary school year 3, but earlier for some. The fourth level broadly equates to Scottish Qualification Framework level 4. In most curriculum areas, the fourth level experiences and outcomes are intended to provide possibilities for choice, and young people's programmes will not include all of the fourth level outcomes.</td></tr>
<tr><td>Fourth</td></tr>
<tr><td>Senior</td><td>Secondary school years 4-6</td></tr>
</table>

Level First (SCN 111M) ' I have contributed to the development of a learning resource which shows the position and function of the major organs of the body and what I need to do to keep them healthy. I can evaluate our resource and say how it helps my learning.'

Level Third (SCN 314M) - 'Having carried out activities to support my learning, I can use my knowledge of the body systems to demonstrate how they are integrated and are essential for the functioning of the body.'

Level Fourth - SCN 426M Bio – 'By designing experiments and from research using a variety of resources I can present information on the breathing, digestive and circulatory systems and how to keep them healthy.'

Interestingly, given the levels provided by CfE, the podcasts where these primary pupils used a variety of resources to research and present information on the digestive system, could in theory be seen to meet Level 'Fourth'. This is a statement aimed at pupils in secondary schools. Of course, while we could argue that the difference may lie in the nature and depth of detail between what was achieved in the primary class and what could be achieved in the secondary class, the CfE statement gives no indication of this depth and detail.

Furthermore when one examines the podcast from a genre perspective, the complexity of the pupils' podcasts can be observed in more detail. For example, when we use the 5W1H frame of reference the following can be deduced:

- Why: For the pupils, the purpose of communication was to inform the (likely mixed- parents, other pupils, teachers) audience about some of the pertinent aspects of the function of particular organs. It also provides an opportunity for the pupils to express and record their understandings without experiencing the potential barrier of the written word.
- What: In terms of the content of this genre, the adoption of the chat show frame gives the pupils a chance to include audio clues that would otherwise not be possible with a written medium. Hence for example, though the pair of pupils have limited statements to make, their synchronisation in 'breathing' and their intonation signal to the listener how the lungs operate. Likewise, the 'haughtiness' in the tone of voice of the brain as well as the statement regarding the body's inability to function without it, clues the listener into the pupil's perceived importance of the brain. These tones and other oral clues are not possible in the written medium, and provide more evidence for the nature of understanding held by the pupils with regard to this topic. They also signal that the pupils are very aware of the medium they are using, are in essence culturally aware of the power of the medium to convey subtle messages. In many ways, what these pupils are attempt-

ing to do is invoke, in much the same way as authors do, a sense of empathy.
- How: The form, conventions and expectations of this podcast provide clues for both the participants and the listener. By introducing the podcast as 'The Richard and Judy show' the pupils provided the listener with a signal with regard to the format to expect. The interview segment between the hosts and the organs attempted to mimic the behaviour pupils associated with that of a chat show host and their guests. These forms and conventions signal expectations to both the performers and the listener.

This raises a question: is this telling a sufficient token of learning?

'The Organ Song'

The Organ song pattern is provided by the song 'Old Macdonald', which has line endings and common stem elements. As a consequence of this the only variation is in the insertion of the noun word of the organ and the verb word for its function.

The Organ Song Podcast:

In my body I have organs – organs, organs, organs
And one of those organs is the brain – think, think, think, think, think
With a message here and a message there
And lots of messages everywhere
In my body I have organs – organs, organs, organs
And one of those organs is the lungs – lungs, lungs, lungs
With an inhale here and an exhale there
Inhale, exhale breathing lungs on the (unable to ascertain)
In my body I have organs – organs, organs, organs
And one of those organs is the heart – heart, heart, heart
With a pumping here and a pumping there
Here a pump, there a pump everywhere a pumping
In my body I have organs – organs, organs, organs
And one of those organs is the liver – liver, liver, liver
With a clean, clean here and a clean, clean there
Here a clean, there a clean everywhere a clean clean
In my body I have organs – organs, organs, organs
And one of those organs is the stomach – stomach, stomach, stomach
With an eating here and an eating there
Mushing up the food everywhere
In my body I have organs – organs, organs, organs
And one of those organs is the kidneys – kidneys, kidneys, kidneys
With a wash wash here and a wash wash there
Here a wash, there a wash everywhere a wash wash
In my body I have organs – organs, organs, organs
And one of those organs is the bladder – bladder, bladder, bladder
With a store here and a spill there
Here a store, there a spill everywhere a store spill
In my body I have organs – organs, organs, organs
And one of those organs is the skin – skin, skin, skin
With a feel touch here and a feel touch there
And a feel touch everywhere
In my body I have organs – organs, organs, organs

Grammatically the song is simple in structure. The only linking word is 'and' at the beginning of a sentence. The song format involves a set of verse frames that enable the pupils to identify key elements. In addition, by using the song as a framework to represent science reporting, the pupils are adopting a change of tenor and mode, from the science reporting commonly encountered in classrooms to a form of reporting that would more likely be present in an expressive arts class. Our interest in this lies in how this transformation affects the field, namely knowledge of the science of the body.

While the change in tenor and mode to present learning of science allows 'telling of what they know', the selection of this particular song and its verse format restricts the pupils to identifying one key associative element. Hence while working in their groups to construct the verse the pupils may have had to discuss and debate which particular verb they would use to signal an association for a particular organ. As a consequence while the song provides a frame it also, in this particular case, restricts their capacity to record their understanding. In essence this is both a strength and a limitation of this song. It is a strength in that it would have required pupils to make a case for the inclusion of a particular verb, as such it is probably more akin to the nature of science in the way it is practiced. Scientists have to be able to argue their case, to support the inclusion of a particular view or to make a decision with regard to what is of importance. The verse structure forces the pupils to do this. However, if the purpose of the song was to illustrate their knowledge and understanding, the verse framework limits that function. So it may not be a comprehensive indicator of their understanding, it is more likely to be an indicator of the point they consider most important.

The mode of communication, the tune, reflected the pupils' perception of a common nursery rhyme. And as with the 'Richard and Judy show' podcast, the organ song podcast enables the pupils to present a particular stance between that of pupil as a product of social position within their classroom and pupil as a cultural agent. Their selection of the title for the podcast once again clues the listener in to the mode to expect.

The field is in essence, what is going on. So the topics and actions in the podcast form the field. From a science perspective, there is limited detail in what the organs say about themselves in this song, because the noun /verb association is in essence the only aspect that is changed in every verse.

The genre framework also allows us to explore the depth of the Old Macdonald podcast.

- **Why:** In essence the purpose of communication in this podcast was to inform, express and record the pupils' understanding of the process and behaviour for various organs.
- **What:** The content is limited to the organ and its associated verb function.
- **How:** The form, conventions and expectations of the nursery rhyme are adopted wholesale in this podcast. Hence the listener knows from the song title to expect a repetitive verse.

'The Story of Digestion'

The form found in the 'Story of Digestion' podcast is that of a choral reading. This is an interesting technique to use for presentation. It has one disadvantage in that pupils are likely to feel they have to put more effort into keeping time than into expression of meaning.

The Story of Digestion Podcast

The teeth bite the food. The (indecipherable) goes through the gut and the saliva helps mush it up into small bits and then you swallow it up and down to the esophagus.

The esophagus is a tube which comes from your mouth to your stomach. The muscles tighten just above the food so it pushes it down so it slowly travels down to the stomach. When the food gets pushed down from the esophagus, now the stomach does its job. The stomach uses its acid to melt down the food. It turns it into small bits, mushing and turning it into liquid. Also the food gets mushed up into tiny pieces. It's a wonder the stomach doesn't dissolve itself with the very strong acid. There is a special lining that stops that from happening.

The intestines divide the good food from the bad food even more. The villi *are sort of like straws and it sucks up all the good food and gets rid of all the bad food.*

Grammatically, there is a fair range of forms and the reading does give a good overall sense of the sequence of events in the digestive process. At one point the speakers lose the thread of meaning because of the sentence complexity: 'It's a wonder the ....very strong acid, very strong lining that stops...'

The use of connectives is striking here, e.g. 'Now the stomach...'; 'Also the food...'; 'Then...'; 'When the food...'. Key technical terms used, such as oesophagus etc, are all nouns. The verbs used are all common words, such as 'pressed', 'mushing', 'elt'.

This podcast is slightly different to the other two. The format is less enterprising, but the content is more detailed. Though the pupils use their research to identify key sound effects to better illustrate their understandings and associations, the science reporting is an oral form of common tenor and mode in primary science lessons. The slight variation would be introduced through the use of sound effects which would not easily translate to a written medium. The tenor and mode to present learning of science does provide pupils with an opportunity to report what they know, as they are not constrained by interviewer questions nor constrained by verse dictates. The mode of communication, the use of sound effects, reflected the pupils' associations for given elements of the digestion process.

From a science perspective, the field in terms of topics and actions in the podcast suggest there is more detail in what the various 'digestion processes' say about themselves in this oral reporting, because the sound effect as well as the language is not constrained.

- **Why:** For the pupils, the purpose of communication was to inform an external audience about the digestive process. It also provided an opportunity to creatively express and record their understandings without being constrained by the written word.
- **What:** In terms of the content of this genre, pupils could include sound effects to further emphasise their key points and to illustrate the nature of their understanding or the associations they have with particular facets of the digestion process. While this may be possible with a written medium, its currency would rely on the pupils being able to provide skills that enabled their language proficiency to enhance their descriptions of particular facets associated with digestion.

In contrast the sound effects selected by the pupils indicate their understanding of a particular digestive process while including various degrees of subtle (and not so subtle) humour. In a written medium the pupils may have had to rely on analogies or metaphors, in the podcast the pupils relied on sound effects that they associated with a particular aspect of the digestion process. For example, they included the sound of a toilet flushing for the final digestive process, the expelling of waste products. Their selection of various sound effects provide more evidence for the nature of understanding held by the pupils with regard to this topic.

DISCUSSION

The evaluations of each of the activities given above using the genre questions, '5W1H', suggest that several positive conclusions may be drawn from the experience. The general one is to do with the level of interest and engagement shown by pupils in the activities. Along with this, more particularly, is the point that the products of learning had an audience.

The analysis shows: an awareness of different relationships with the audience; an ability to vary tone; an acute sense of medium, using auditory cues and effects, as the visual were unavailable; a grasp of how to make explanation understandable, using everyday verb words in the description of digestion. The nature of the learning outcomes, in the form of podcasts, shows as a positive in that the learning experience has been given a form to which pupils can relate, in a medium in which they operate with confidence.

From a wider perspective of studies of language in learning, the role of pupil 'transformation' of concepts, relationships and processes into some kind of a genre product is significant in, for example, the work of Wray and Lewis (1997) on extending literacy, in their book of the same name and elsewhere. To put it simply, the basic argument in such a position is that learners show their learning by reshaping the 'content' learned, frequently by changing it into another genre. This jibes usefully with the outline proposed by Bereiter and Scardamalia (1987) of a model of children's writing, labeled 'knowledge - telling,' which they characterise as being simple and linear in nature, as opposed to a model of more expert writing, described as 'knowledge transformational,' because it is more sophisticated in its involvement of complex problem-solving processes.

The above are useful perspectives at a macro level on the activities considered in this study. What they do not provide however is a more detailed description of how these activities may best be developed for the benefit of pupils' learning about science, even at upper primary school level. However a number of implications of the study may afford some useful trails to follow for future study.

One implication is that pupil presentations employing elaborate techniques such as choral reading should be used with care: as the example illustrates, the 'script' may become a barrier to understanding or at least produce an example of mass recitation with little evidence of understanding.

A second is to do with the use of known formats such as songs and chat-shows. These clearly have appeal but the 'variety factor' that they work with may not best afford the room they need to allow explanations and encourage detail.

Going along closely with this last implication is the need for the formulae or frames used for such pupil learning activities to be chosen with a view to their accommodating and enabling the complexity of relationships necessary to and inherent in the topic: syntax reflects semantics. This in turn would obviate to an extent the tendency to use 'knowledge telling' as opposed to 'knowledge transformation' that may impede learning for many at this stage.

By way of practical suggestions: pupils could be asked to use a simple formula for their media based presentations that would encourage and allow the use of more depth and detail. It could be simply set as 'What's the background story?' or, with choral presentations, 'Your stand-out solo'. As for the dimension of semantic complexity, existing work on writing frames for pupils to use in structuring writing (Wray and Lewis 1995) could well be adapted and applied to the development of media texts, not least in respect of the use the linking words used to signal the direction of the text to the audience.

CONCLUSION

Though this chapter did not focus ostensibly on the interaction between the participants, through the analysis of the three podcasts, in terms of linguistics and genre, we are able to provide some insight into the nature of the interaction between these participants. We knew what the teaching and learning objectives were, we knew how the teacher and pupils related their activities to the available and other media, and we had their product.

From an anecdotal perspective, we have no doubt that the podcasts generated affective and cognitive benefits associated with predominantly audio mechanisms. The use of relatively cheap mp3 players also mean it was relatively cost-effective, and the manner in which the teacher and the pupils used the podcasts made it an ideal medium for generating examples of pupils work to address the science requirements within a school based context. But these podcasts did more than that. They also enabled the pupils to make public their understanding of science while showcasing other skills and talents and in our paper based assessment driven education system, these skills, talents and understandings are often buried under the weight of routine accountability.

From a multiple literacy perspective, the three podcasts presented in this chapter help to illustrate the role of audio communication in supporting pupil 'transformation' of concepts, relationships and processes into a genre product by reducing or removing some of the unintentional hurdles they often face in formal science lessons. The pupils were adept at facing the challenge in balancing a variety of demands while not loosing sight of the overall purpose of the task (to share their understanding of the science concepts). Their language skills in terms of awareness of genre, their science skills in terms of awareness of concepts and their technology skills in terms of awareness of function enabled them to create three distinct types of podcasts that were well received both within their local community and beyond their school walls.

Nearly twenty years ago, in the USA, information literacy was defined by The Presidential Committee on Information Literacy (1989) as the skills an individual needs to: 'recognize when information is needed and have the ability to locate, evaluate, and use effectively the needed information.' The three podcasts reported in this chapter help support the argument that these children were information literate. They were able to recognise when information was needed, and they were able to locate, evaluate and use the information to illustrate their understanding of particular science concepts. In 2004, The National Higher Education Information and Communication Technology (ICT) Initiative combined cognitive and technical skills with an ethical/legal understanding of information when they constructed their definition of information literacy (see ETS, 2005). We use our chapter to suggest that the term multiple literacy signals more clearly the role of a variety of abilities and skills, including for example the following:

- use digital technologies, (tools, software and/or networks) to support an information need, in our case science education
- access, handle, assimilate, evaluate information, in our case, science data.
- make judgments to create, share, amalgamate and/or present understanding of new information in a suitable form, in our case, podcasts.
- be able to communicate relevant personal understanding to a wider community context.

In the current environment there is a certain expectation that learners are going to have to continually increase their multiple literacy competence as technology communication, and the use of media becomes embedded in their lives. As a consequence we need to understand that they will bring to their learning environment, skills, competences, understandings, requirements and wish lists derived from, and by, their engagement with technology in their everyday lives, and that these technologies may well outstrip anything that is available within the formal learning environment. It will not be a case of a digital divide in terms of those who have and those who have not, but more a case of a literacy divide between the multiple literacies used in their formal and informal environments. This chapter illustrated how podcasting may be used to embrace and support learners and teachers as they transit between the two environments. It isn't simply a question of making the technology more available, and/

or basing our teaching on the experiences of the middle class motivated child. It is about providing experiences and opportunities that enable children to develop the range of literacies they need in order to make the most informed use of the technologies they have. We know, in reality that education will eventually involve pupils pursuing personal curricula, at probably their own pace, perhaps guided by teachers and other experts, but unless these pupils, teachers and other experts understand the challenge posed by the multiple literacy demands, then we are no closer to realising a vision that involves self motivating and ultimately self directing learners.

ACKNOWLEDGMENT

Thank you to Emma McGuigan and her class for the podcasts.

REFERENCES

American Library Association. (1989). *Presidential Committee on Information Literacy. Final Report.* Chicago: American Library Association. Retrieved October 16, 2008, from http://www.ala.org/ala/mgrps/divs/acrl/publications/whitepapers/presidential.cfm

Bakhtin, M. M. (1986). *Speech genres and other late essays* (V. McGee, Trans.). Austin: University of Texas Press. Retrieved March 30, 2009, from http://pubpages.unh.edu/~jds/BAKHTINSG.htm

Bereiter, C., & Scardamalia, M. (1987). *The psychology of written composition.* Hillsdale, NJ: Lawrence Erlbaum Associates.

Berkenkotter, C., & Huckin, T. N. (1995). *Genre Knowledge in Disciplinary Communication:Cognition / Culture / Power.* Mahwah, NJ: Lawrence Erlbaum Associates.

Brittain, S., Glowacki, P., Van Ittersum, J., & Johnson, L. (2006). Podcasting Lectures: Formative evaluation strategies helped identify a solution to a learning dilemma. *Educase Quarterly, 29*, 3. Retrieved February 24, 2009, http://connect.educause.edu/Library/EDUCAUSE+Quarterly/PodcastingLectures/39987

Brown, J. S. (1994). Borderline Issues: Social and Material Aspects of Design. *Human-Computer Interaction, 9*, 3–36. doi:10.1207/s15327051hci0901_2

Chan, A., & Lee, M. J. W. (2005). An MP3 a day keeps the worries away: Exploring the use of podcasting to address preconceptions and alleviate pre-class anxiety amongst undergraduate information technology students. In D.H.R. Spennemann & L. Burr (Eds.), *Good Practice in Practice: Proceedings of the Student Experience Conference* (pp. 58-70). Wagga Wagga: Charles Sturt University. Retrieved February 24, 2009, from http://www.csu.edu.au/division/studserv/sec/papers/chan.pdf

Education Testing Service (ETS). (2005). *Beyond Technical Competence: Literacy in Information and Communication Technology* (An Issue Paper from ETS Princeton). NJ: ETS. Retrieved June 15, 2009, from http://www.ets.org/Media/Tests/ICT_Literacy/pdf/ICT_Beyond_Technical_Competence.pdf

Fildes, J. (2006). Podcast numbers cut through hype. *BBC New Channel Website*. Retrieved April 2, 2009, from http://news.bbc.co.uk/1/hi/technology/4885010.stm

Giddens, A. (1984). *The Constitution of Society: Outline of the Theory of Structure*. Berkeley CA. University of California Press.

Halliday, M. A. K. (1993). Towards a language based theory of learning. *Linguistics and Education, 5*, 93–116. doi:10.1016/0898-5898(93)90026-7

Halliday, M. A. K. (1994). *An introduction to functional grammar* (2nd ed.). London: Edward Arnold.

Halliday, M. A. K., & Hasan, R. (1985). *Language, Context and Text: aspects of language in a social semiotic perspective*. Geelong, Australia: Deakin University Press.

Hasan, R. (2005). Language and society in a system functional perspective. In R. Hasan, C. Matthiessen, & J.J. Webster (Eds.), *Continuing discourse on Language: A functional perspective. Volume 1*. London: Equinox.

International ICT Literacy Panel. (2002). *Digital transformation: A framework for ICT literacy (A report of the International ICT Literacy Panel)*. Princeton, NJ: Educational Testing Service. Retrieved February 24, 2009, http://www.ets.org/Media/Research/pdf/ictreport.pdf

Laaser, W. (1986). Some didactic aspects of audio-cassettes in distance education. *Distance Education, 7*(1), 143–152. doi:10.1080/0158791860070110

LTScotland. (2008). *Curriculum for Excellence*. Retrieved January 1, 2009, from http://www.ltscotland.org.uk/curriculumforexcellence/outcomes/

Luft, O. (2008). *Podcasts popularity surges in the US*. Retrieved February 20, 2009, from http://www.guardian.co.uk/media/2008/aug/29/podcasting.digitalmedia?gusrc=rss

Rogoff, B. (1995). Observing sociocultural activity on three planes: Participatory appropriation, guided participation, and apprenticeship. In J. V. Wertsch, P. Del Rio, & A. Alvarez (Eds.), *Sociocultural studies of mind* (pp. 139-165). Cambridge, UK: Cambridge University Press.

Vygotsky, L. (1997). The history of the development of higher mental functions. In R. W. Rieber (Ed.), *The collected works of L. S. Vygotsky. Vol. 4*. (M. J. Hall & R. W. Rieber, Trans.). New York: Plenum Press.

Webster. (2008). *The Podcast Consumer Revealed 2008*. Retrieved February 20, 2009, http://www.edisonresearch.com/home/archives/2008/04/the_podcast_con_1.php

Wertsch, J. V. (1990). The voice of rationality in a sociocultural approach to mind. In L. C. Moll (Ed.), *Vygotsky and education: Instructional implications of sociocultural psychology* (pp. 111-126). New York: Cambridge University Press.

Wray, D., & Lewis, M. (1995). *Writing frames: scaffolding children's non-fiction writing in a range of genres*. University of Reading: Reading & Language Information Centre.

Wray, D., & Lewis, M. (1997). *Extending Literacy*. London: Routledge.

Yates, J., & Orlikowski, W. J. (1992). Genres of Organizational Communication: A Structurational Approach to Studying Communication and Media. *Academy of Management Review, 17*, 299–326. doi:10.2307/258774

Yoshioka, T., & Herman, G. (1999). *Genre Taxonomy: A Knowledge Repository of Communicative Actions*. Retrieved January 1, 2009, from http://ccs.mit.edu/papers/pdf/wp209.pdf

Chapter 6
Diffusion of ICT Innovation in Science Education

Bulent Cavas
Dokuz Eylul University, Turkey

Pınar Cavas
Ege University, Turkey

Bahar Karaoglan
Ege University, Turkey

Tarık Kısla
Ege University, Turkey

ABSTRACT

In this chapter, the authors first discuss how Roger's theory of innovation diffusion can be incorporated into ICTs in formal and informal learning and teaching environments. The authors begin by presenting the use of ICT in education in general terms, then they introduce Rogers' diffusion of innovation (DoI) theory and the related literature. This is followed by a description of a project which explored the relationship between some characteristics of primary science teachers and their attitudes toward the use of ICT in education. A national project was funded by the Scientific and Technological Research Council of Turkey (TÜBİTAK), and Ege University, Science and Technology Application and Research Center. The last section involves a discussion of the diffusion of technological innovations into science education in the light of Rogers' DoI theory.

INTRODUCTION

The awareness and attitudes of individuals towards innovation are vital factors in the constitution of information societies. The adoption of ICT in education is not as fast as the pace at which technology is developing. Technology has developed further in terms of fostering learning and teaching but that is not how it is being used in education. We need to find ways of diffusing ICT use within learning environments at the ICT development pace. Rogers' Diffusion of Innovation (DoI) theory may help with this issue (Rogers, 2003).

DOI: 10.4018/978-1-60566-690-2.ch006

PRESENT USE OF ICT IN EDUCATION

In this section we give an overview of how ICT is used in educational settings world-wide. We present a similar understanding for using ICT in education as stated by Shapiro, Roskos and Cartwright (1995): "a technology enhanced learning environment attempts to stimulate classroom activity by demonstrating and using software or tools specific to a particular discipline, by promoting high levels of interaction among students and faculty; and by involving students in simulated activities or data gathering via the Internet and remote databases. These aims require a re-conceptualization of traditional classroom design, not the mere addition of a piece of technology" (p.67).

From this point of view integrating ICT presents a significant change in educational systems however achieving this change, as anticipated, is very challenging. Furthermore, additional complexity is added to the use of ICT in education by other external factors which may be technological, social, political, economical or psychological.

The technologies involved and the approaches adopted in the use of ICT in distance education, in classroom teaching and in management of learning may vary. Technologies related to distance education can support real time interaction and include audio, audio graphics and video teleconferencing. Recently, systems that facilitate collaborative learning and support interaction with the group are gaining importance (Stacey, 1999; Sheremetov & Arenas, 2002). In both distance and in-class education, the Internet is used as a medium for meeting, discussion, exploration and a repository for course material. There is growing evidence that videoconferencing centers are being established for live lecturing and tutoring. Distance education is thought to provide a chance of equity for those who are hard-to-reach, extending learning beyond the classrooms while supporting personalising learning. Innovative practices are also emerging with mobile personal technologies which are promising to add value to distance learning. These are mostly in the area of revision for examinations, and games-based testing and practice (McFarlane *et al.*, 2008).

Smith and his colleagues (2008) report that around 80 per cent of teachers agree that technology has an impact on students' engagement in their learning, and around 60 per cent reported that it enabled tutors to better support learners' diverse needs. Even though the great majority of the teachers identify the use of ICT in education most of them fail to implement ICT in their classrooms (Cuban, 2001; Moersch, 2002; Sandholtz *et al.*, 1997). They still prefer to use conventional instruments like the board, printed materials, projectors, multimedia computers, or video players as instructional aids and prefer to use digital resources in their teaching for information retrieval, communication, word-processing and preparing slide presentations.

The use of technology in the management of learning is developing. The development is mostly in recording and reporting of learner performance; interactive and online assessments; submitting work online; using e-portfolios; and self-and peer-evaluation (Smith *et al.*, 2008; Hollingworth *et al.*, 2008). The success in the use of ICT in education also depends on the consideration of social factors. The students' learning habits, abilities in adapting to change as well as teachers, shape the use of ICT in education. In developing countries where education tends to focus on memorization, ICT is used as a tool to transfer instructional material and it is used for rehearsal. Whereas, in developed countries, finding ways of integrating the use of ICT as a tool and medium for learning, creating and sharing knowledge into education are a higher priority (Albirini, 2006; Sadık, 2005; Tella *et al*, 2007). The adoption and dissemination of ICT in education is heavily influenced by the national priorities and initiatives of the policy makers in the individual countries. Policies to introduce ICT into education are supported by the need to build a knowledge society and to foster creative and productive manpower to generate quality and equity in education. Recently online safety is being considered among local authorities. However, the focus is more on infrastructure issues than in improving learning and teaching (Becta, 2008).

Another factor that plays a role in the extent of adoption of ICT into education is the measure of its cost-efficiency. ICT should be easily accessible to the target group and shown to fulfill all the functions that are expected of it. Qualitative and quantitative assessments are needed to persuade the stakeholders to use ICT in their lessons. However, developing countries often invest in the latest technologies without considering whether the target audience is reached effectively or whether the target audience is interested in the technology (Albirini, 2006; Usun, 2004).

The psychological consideration with regard to using ICT in education suggests that creating a change in human beliefs and behaviours cannot be achieved easily. Since teachers are the end users of this technology and the prime actors in implementing it, they should be at the centre of attention. Teachers often exhibit resistance to implementing technological change that forces a change in the role of the teacher from being a repository of all learning to a manager of the teaching-learning process (Albirini, 2006; Usun, 2004). Therefore, teachers should be involved at all stages of the implementation to: develop "ownership" of the innovation and be assured that this approach is advantageous to the previous one; is compatible with their teaching practices and there will be technical help and training and be supported with instructional materials. The change is expected to occur in the style of teaching and learning as noted by Harris *et al* (2002) '.. it is not necessarily the technology that has to be innovative, but the approach to teaching and learning.' (p.35)

DIFFUSION OF INNOVATION

Nowadays researchers, policy-makers and practitioners from different fields have paid much attention to the DoI and used this model as a framework in several disciplines including political science, anthropology, education, management, information and communication technology, geography and sociology (Rogers, 2003; Damanpour, 1992; O'Neill, Pouder, & Buchholtz, 1998; Premkumar & Ramamurthy, 1994; Wright, Palmar, & Kavanaugh, 1995; Van de Ven & Poole, 1989, cited in Lundblad, 2003). Although interest in DoI has increased day by day, there is no doubt that innovations are often not diffused within, and across organizations, to achieve improvement.

The ideas about diffusion theory started with European beginnings in the social sciences and the roots of the diffusion concept are based on theories of imitation developed by Gabriel Tarde (1903). Tarde was a French lawyer and judge and also an important social scientist in his society. He discovered an s-shaped curve that governs the rate of invention and imitation diffusion within a given social context. According to Tarde, adaptation and rejection of an innovation are the key elements for the diffusion of innovation in any field. He also noted that the manner in which you could describe a change in a person's behaviour was through diffusion of innovations (Fisher, 2005). After Tarde's thoughts about diffusion, European anthropologists began to explore similar thinking to explain social change resulting from the introduction of innovations from other societies (Rogers, 1995). Ryan and Gross adopted Tarde's model as diffusion of innovations in the early 1940s in American agricultural research (Rogers, 1995). In terms of educational adaptations, the first study of diffusion was conducted by Paul Mort in 1941. Mort & Cornell, (1941, p. 3, cited in Warford, 2005) concluded that 'the succeeding waves of 'reform' which have come and passed in this century have left discouragingly little mark.'

Although diffusion of innovation research began with Tarde's observations, the most influential studies in this field have been conducted by Everett Rogers. For more than 50 years, Rogers has been described as 'the elder statesmen of change research' (Ellsworth, 2000), and 'the researcher who has done the most synthesizing of all the most significant findings and compelling theories related to diffusion' (Sury, 1997). In 1983, Rogers published a comprehensive book entitled *Diffusion of Innovation* which is regarded as a classic text on the theory of innovation adoption. Rogers identified his study by explaining why diffusion research has become so popular:

"One reason why there is so much interest in the diffusion of innovations is because getting a new idea adopted, even when it has obvious advantages, sf often very difficult. There is a wide gap in many fields, between what is known and what is actually put into use. Many innovations require a lengthy period, often of some years, from time they become available to the time when they are widely adopted. Therefore, a common problem for many individuals and organizations is how to speed up the rate of diffusion of an innovation" (Rogers, 1983, p.1, cited in Ross, 2006).

According to Parisot (1995) and Medlin (2001), Rogers' Diffusion of Innovation theory is the most suitable and effective theory to analyze and explain adaptation of technology in educational environments. Rogers (1995) defines diffusion by regarding its four primary elements that are identifiable in his massive compilation of 2925 empirical and 975 non-empirical studies. Many of Rogers's early studies mainly focused on the individual who adopted an innovation and on what factors contributed to the adoption of that innovation. Rogers' theory has produced a great deal of research focused on some important features of the diffusion process. Educational studies related to diffusion theory generally use three features of diffusion theory including innovation attributes, individual innovation, and the innovation decision process

Holloway (1997) stated that studies of diffusion and adoption help us to understand the possible reasons for acceptance or rejection of technology in education. The process of diffusion and adoption of innovations in educational context covers four specific elements: [1] *innovation* [2] *communication channels* [3] *time* [4] *social system*" (Rogers, 1995). These elements affect each other in complex patterns, and some features a given element can be arbitrarily assigned to another element. These elements are described below.

Innovations

Rogers defines the innovation term as "an idea, practice, or object that is perceived as new by an individual or other unit of adoption" (1995, p.10). The innovation may not be an 'objectively' new idea, object, or practice but rather perceived as new by the individual. Rogers generally use the concepts 'innovation' and 'technology' interchangeably since many diffusion studies include technological innovations. Rogers (2003) defines technology as 'a design for instrumental action that reduces the uncertainty in the cause-effect relationship involved in achieving a desired outcome' (p.13). According to Rogers (2003), technology mainly consists of two parts: hardware (technological tools) and software (knowledge of using a tool).

In terms of educational settings, any technological device or system can be regarded as new by the individuals. In the process of innovation adoption, technology users' (teachers, instructors etc.) rate of adoption is the most important variable to understand different rate of adoption. Rogers (1983) found that innovations have five attributes and these attributes were significantly related to the rate of individuals' adoption a new idea. The first attribute is relative advantage or perceived superiority of innovations over other alternatives or existing practices. Rogers (1983) indicated that this attribute is a good predictor of the rate of innovation adoption and include components of costs, investment returns, efficiency, and yield. The greater the perceived advantage of an innovation by an individual, the faster the innovation will be adopted. The second attribute is compatibility and this attribute indicates the degree to which an innovation is perceived as being consistent with the existing values, past experiences and the needs of potential adopters. Complexibility of an innovation is the third attribute and related to perceived difficulty of understanding or using an innovation. According to Rogers this attribute is negatively correlated with adoption processes because if individuals feel that using and understanding an innovation is too difficult, they tend not to adopt a new innovation. The fourth attribute is trialability. Trialability refers to testing innovation adoption and it allows the individual time to experiment with and understand the innovation through a phased-in approach. The last attribute is observability and it indicates the degree to which the results of innovations are visible to others.

Briefly, if any individual (teachers, teacher educators, etc.) perceive that the new innovation (technology) has greater advantage, compatibility, trialability, observability, and less complexity than others, s/he adopt this innovation quickly.

Communication Channels

The communication channel is very important in the process of diffusion and adoption of an innovation. In this process, diffusion is a specific type of communication where a message, that concerns a new idea, is exchanged. The nature of information exchange between an individual who communicates an innovation to one or others is essential to the diffusion process. The diffusion process includes four elements: (1) an innovation, (2) an individual or unit of adoption which has knowledge or experience of a new innovation (idea), (3) another individual or unit that does not have knowledge or experience with the innovation, and (4) a communication channel connecting the two units. According to Rogers (2003), a communication channel is 'the means by which messages get from one individual to another'. Rogers categorized channels into two categories: mass media and interpersonal channels. Mass media channels include radio, television, and newspapers and enable individuals to reach an audience of many. Interpersonal channels are more effective in forming and changing attitudes toward a new idea and related to acceptance or rejection of new a idea by individuals. This kind of channel involves face-to-face exchange between two or more individuals. Recently, interactive communication via the internet has become more important for diffusion of certain innovations. Most individuals evaluate an innovation through the subjective evaluations by their peers who have already adopted innovation rather than on the basis of scientific studies about consequences innovation.

Time

Rogers (2003) stated that the third element of the diffusion process, time, is simply ignored or does not matter in most behavioural studies. Including a time dimension as a variable in diffusion research is one of its strengths. Three time factors are included in the diffusion process: (1) innovation-decision process including time that an individual takes the progress from first learning about an innovation to having decided to accept or reject it; (2) relative time with which an innovation is adopted by an individual or group and (3) innovation's rate of adoption. These factors will be discussed in later sections.

Social System

The last element of the diffusion process is the social system. According to Rogers (2003), the social system is 'a set of interrelated units that are engaged in joint problem-solving to accomplish a common goal' (p.23). Individuals, informal groups, organizations and/or subsystems may be the members of a social system. In order to reach a mutual goal, all members of the social system work cooperatively by seeking to solve a common problem. The social structure of the system influences the diffusion of innovation because diffusion occurs within a social system. Rogers mainly deals with the relations between the systems' social structure and diffusion, system norms and diffusion, the roles of opinion leaders and change agents, types of innovation-decisions and the consequences of innovation.

Social systems can be categorized into two groups of norms: traditional and modern norms. Traditional norms are characterized by five attributes which are a less developed or complex technology, low levels of literacy and education, little communication between the social system and outsiders, lack of economic rationality and one-dimensional in adapting and viewing others. On the other hand, modern norms are characterized by (a) a developed technology with complex jobs, (b) strong importance placed on education, (c) acceptance of free thought and new ideas (d) strong preparation and high importance on economic considerations and (e) ability to see and understand other peoples situations. There is no doubt that the diffusion of innovations (technologies) in any system (for example the educational system) is defined by the characterization of the social systems. In the social systems, individuals do not adopt an innovation simultaneously; instead they adopt an innovation at different times. Rogers (2003) stated that innovations are diffused in a pattern that is similar to an S-shaped curve. Rogers classified individuals on the basis of when they first begin using a new idea. He used innovativeness as a criterion to categorize adopters and he proposed five adopter categories.

ADOPTER CATEGORIES

Rogers (2003) classified the adopter classification includes innovators, early adopters, early majority, late majority, and laggards. This kind of classification helps to understand human behaviour but it loses some information about individuals because of grouping. These categories were derived from 'ideal types,' that defined by Rogers (p. 282) as 'conceptualizations based on observations of reality and designed to make comparisons possible.' Each category includes specific characteristics that help to set each one apart.

The first adopter category, innovators, is defined as 'venturesome' and make up about 2.5% of the population. Innovators try to discover information about new ideas and they are very enthusiastic and keen to experience them. The remarkable characteristic of the innovators is *venturesomeness*, because of a desire for the rash, daring and the risky (Rogers, 2003). They tend to be better educated, have higher incomes, travel more, and read more than the rest of the population. Although innovators may not be respected by other members of the system they are very important in the diffusion process because they start the new idea in the system by importing the innovation from beyond the system's boundaries.

Early adopters are 'respectable' and do not have all of the resources that innovators have. In most systems, they get the highest degree of influence (opinion leadership) and have an ability to affect other people's decisions about adoption or rejection of the new idea. Early adopters also have greater rationality and intelligence, greater ability to cope with uncertainty and risk, higher aspirations, and more contact with a larger number of people (Cottrell, 1997). Rogers (2003) indicated that early adopters are considered by other members of the social system to be 'the individual to check with' before adopting a new idea. If early adopters are successful with the innovation and can express that to potential adopters, the innovation has a greater chance of full adoption among the social system.

The next category is early majority, or 'deliberate', and they 'adopt new ideas just before the average member of a social system' (Rogers, 2003, p. 283). Although they usually tend to interact with the other members of the system just as early adopters do, they are not willing to be a leader of the group. Their position is very important in the innovation-diffusion process and they adopt an innovation just before the other half of their peers adopts it. Their innovation-decision process usually takes more time than it takes the innovators or early adopters since they may sometimes deliberate before adopting a new idea. They show tendency to observe the previous members' choices and decisions and create their own when the time is right.

The late majority of adopters constitute one-third of the innovation acceptance population like early majority. Rogers (2003) indicated that late majorities are skeptical, cautious, and uncomfortable with uncertainty. Both economic necessity and peer pressure may lead them to adopt an innovation. They don't adopt any innovation until the majority of the others in their social system adopt an innovation. Almost all uncertainty must be removed from the innovation before the cautious late majority will feel safe enough to adopt.

The laggards are the last category of adopters and Rogers (2003) referred to this group as 'traditional'. They are almost the last to adopt any innovation and they are more skeptical about innovations and change agents than the late majority. They have no awareness of new ideas and they are very critical towards new ideas. Their interactions mainly occur with the other members of the social system sharing the same traditional values. Laggards' innovation-decision processes last a very long time compared to other adopters and their processes are based on limited social resources.

It is important to categorize adopters because these categorizations help us to understand why some communities accept new innovations more quickly than others. It also shows that all innovations go through a natural, predictable, and sometimes lengthy process before becoming widely adopted within a population.

INNOVATION-DECISION PROCESS

Roger's innovation-decision model has been used in many educational studies, especially on the use and non-use of instructional technology by faculty (Medlin, 2001; Zakaria, 2001; Sahin, 2006; Callava, 2007) and adaptation of technology by teachers (Fisher, 2005; Samak, 2006; Takkunen, 2008). In the analyses of an individual (or other decision-making unit) decision to adopt or reject an innovation, Rogers (2003) used an innovation-decision process which is defined by him as 'the process through which an individual passes from first knowledge of an innovation, to forming an attitude toward the innovation, to a decision to adopt or reject the innovation, to implementation of this new idea, to confirmation of this decision' (p. 168). This process consists of five stages and each stage is an information-seeking and information-processing activity. At the different stages of the process, an individual meets a series of choices that help to evaluate a new idea and decide adoption or rejection of this idea. Rogers (2003) stated that these stages might be very functional: to simplify a complex reality, to provide a fundamental understanding of human behaviour change and to the introduction of an innovation. These stages are very important for teachers because if they are aware of their needs, they should be able to make decisions about accepting and adopting the innovation. The stages of the diffusion-innovation process are knowledge, persuasion, decision, implementation, and confirmation (Rogers, 2003). This process needs time to be completed and the individual may return to a prior stage if any uncertainty forms after a decision is made.

In the first stage, the individual first learns about knowledge of the innovation's existence and some understanding of how it functions. The individual tries to answer questions: 'What is the innovation', 'How does it work? and 'Why does it work?' (Rogers, 2003). The first question represents 'awareness-knowledge' which is one of the three types of knowledge about innovation. In an educational context, this question occurs in a teacher becomes aware of an innovation and then tries to understand it through his or her in-service training process. Teachers expect to be informed about the implementation of the steps related to impressive practice of the innovation (Fisher, 2005). There are two arguments about how knowledge of an innovation presents itself. One argument suggests that an adopter looks for new innovations that agree with his or her interests, needs, or attitudes. However, the other argument claims that the adopter comes across a new innovation accidentally, 'as one cannot actively seek an innovation until one knows that it exists' (Rogers, 2003, p. 164). Rogers stated that there is no definite answer to this question regarding which comes first. The stage sequence may be related to the adopters and the type of innovation.

In the persuasion stage, the individual forms a positive or negative attitude toward the innovation (Rogers, 2003). The individual attempts to find out more information about the innovation from other members of the social system and becomes more psychologically engaged with the innovation. In an educational setting the teacher may make a decision about whether s/he likes or does not likes an innovation. In this stage, there are three important behaviours which are: 'where he or she seeks information, what messages he or she receives, and how he or she interprets the information that is received' (p. 170). It is important to understand where the teacher gets their information, what messages are being stored, and how the information gathered is interpreted. If the teachers are introduced to innovations that change their instructional practices, this kind of persuasion may lead to a change in their behaviour.

The decision stage is the third stage of the innovation-decision process. At this stage, the individual decides to adopt or reject an innovation. While adoption of an innovation refers to 'full use of an innovation as the best course of action available', rejection refers to 'not to adopt an innovation' (Rogers, 2003, p.177). Teachers' beliefs or existing practices are influenced by an innovation and they begin to experiment with the innovation to decide whether the innovation fits their needs. Individuals generally try out the innovation before adopting. According to Rogers, 'one way to cope with the inherent uncertainty about an innovation's consequences is to try out the new idea on a partial basis' (p. 177). Teachers should be encouraged to implement innovations partially since this implementation will lead to a better understanding of the use of the innovation and they will become more familiar with the strengths of the innovation. In the process of innovation adoption, extra pieces of implementation should be pursued until the innovation is thoroughly implemented. In this stage, it is important for teachers to be supported by the change agent with the technical understanding.

At the fourth stage, implementation, an innovation is put into practice. This is an indicator of change in the individual's behaviour since a decision to adopt has been made. Although the individual adopts the innovation there is a still uncertainty about the outcomes of the innovation. During the implementation stage, reinvention usually occurs. Rogers (2003) defined reinvention as "the degree to which an innovation is changed or modified by a user in the process of its adoption and implementation" (p. 180). Rogers also stated that the more invention takes place, the more rapidly an innovation is adopted and becomes institutionalized. Computer technologies are important innovations for teachers because they are suitable to reinvention. The implementation stage may last until the innovation is institutionalized.

The last stage of the innovation-decision process is the confirmation stage. At this stage, whilst the decision about an innovation has been already made, an individual seeks support for his or her decision. Rogers (2003) indicated that individuals may reverse their decisions if they are exposed to conflicting messages about the innovation. Change agents and attitudes are crucial at this stage and the individual tends to keep away from these messages.

LITERATURE REVIEW ON THE DOI REGARDING EDUCATIONAL STUDIES

Since the 1960's interest in Diffusion of Innovation has grown. The pioneer in this area, Everett Rogers and his book has been cited approximately 1,000 times in literature ranging from agriculture to educational technology. There is still constructional contribution on the diffusion of innovation theory by many researchers in different areas (Rogers, 1995). Most of the educational studies related to DoI theory highlighted important factors that effect teachers' use of computers for instructional purposes. For example, it is indicated by Rogers (1995) that new innovations particularly on the educational issues requires approximately 25 years before it's full implementation to the educational system. For example, integration of kindergartens to formal education system has just taken about 50 years.

In the field of educational technology, many researchers have utilized these theories and concepts to study the adoption and diffusion innovations. Diffusion theory has been mostly applied to the study of either artifacts, such as computers, or knowledge, such as innovative teaching techniques (Holloway, 1997). Burkman (1987) studied the relationship between diffusion theory and educational technology. In his study, perceived attributes were used to develop a method for improving instructional products that would be more appealing to potential adopters. Yates (2001) stated that utilizing the diffusions of innovations theory may be helpful for educational technologists and school leaders in explaining and even predicting patterns of adoption. 'Instructional technologists are effectively using Everett Rogers's theory of innovation diffusion in hopes of increasing the implementation and utilization of innovative and instructional products and services' (Yates, p. 3). Instructional technologies and software are, by their very nature, 'innovative' and may require an 'effective model of diffusion' to support a greater probability of adoption by teachers (Yates, p. 7).

Dooley and Murphrey (2000) explored the administrators', faculty, and support units' perspectives on the strengths, weaknesses, opportunities and threats associated with using distance education (DE) technologies. They suggested the major areas; administrative support (student/technical support and providing a seamless infrastructure and virtual presence for the distant learner), training (instructional design and pedagogy), and incentives (release time, mini-grants, continuing education stipends, and recognition in the promotion and tenure process) should be considered to increase the rate of adoption. Yates (2001) also reported that 'Instructional technologists are effectively using Everett Rogers's theory of innovation diffusion in hopes of increasing the implementation and utilization of innovative instructional products and practices' (p. 3).

Yates (2001) underlines that the process of adoption is affected by the complexity of media resources. In addition to Yates (2001) others such as Havelock and Huberman (1977), Thayer and Wolf (1984), and Rogers (1995) highlighted that schools have unique complexities in terms of diffusing an innovation.

In the literature review, much of the research indicated the importance of teacher training regarding adoption of new technologies into the teaching and learning environment. For example, Huberman and Miles (1984) reported that balancing pressure and support is a key element for successful change. Fullan (1992) highlighted that 'Pressure without support leads to resistance and alienation; support without pressure leads to drift or waste of resources' (p.91). Fullan (1992) also suggested that pressure and support as reflected in the system infrastructure served as primary diffusion tools. So, a good balance of pressure and support is vital for ICT adoption in teaching and learning environments. A national research project carried out by the CEO Forum on School Technology and Readiness indicates that only 10% of teachers were integrating technology into their teaching and learning environment supporting critical thinking and student inquiry and less than 14% of teachers were using technology to support higher-level thinking and inquiry (Moersch, 2002).

A study carried out by Berryhill (2007) investigated the relationship between adoption status and demographics, personal social network, and training source. The results showed that there was no significant relationship between the adoption/non-adoption status and the variables of age, race/ethnicity, gender, years teaching, and social communication networks. Berryhill (2007) also found that there were moderate relationships between instructional technology training and adoption/non-adoption status. Surprisingly, in the study, 90 percent of participants characterized themselves as adopters of the instructional technology provided in the Technology Classrooms. Faseyitan and Hirshbuhl (1992) studied the affects of personal attributes, organizational factors, and attitudinal factors in terms of adoption of computers for instruction by university faculties. They concluded that technological orientation of the faculty's discipline, computer self-efficacy, computer utility beliefs, and attitudes toward computer were important influences regarding adoption.

Jacobsen (1997) and Hamilton and Thompson (1992) examined the characteristics of early adopters. Jacobsen (1997) found that early adopters had good to excellent computer skills, were more self-sufficient and more willing to experiment with technology than mainstream faculty colleagues. Hamilton and Thompson (1992) concluded that early adopters in a college setting, through email and discussion boards, shared commonalities such as education, social status, social participation, cosmopolitan outlook, interpersonal communication networks, a high degree of innovation information seeking, and positive attitudes about risk and change.

OVERVIEW OF THE PROJECT

One of the claims for the diffusion of innovation to be realized is the positive attitude of the stakeholders towards innovation. Zhao and Tella (2002, p1) describe teachers as the 'gate keepers to technology, who not only determine whether it enters the classroom, but also affect how it is used in the classroom.' Anderson and Dexter (2005, p8) noted that 'annual technology expenditures have risen steadily, and many more classrooms have been wired for Internet access, but the role of ICT in education has not qualitatively changed.' It is also observed that most teachers using ICT in their teaching rarely change their classroom practices. Tyack and Cuban (1995, p126) state that: '... and whether teachers will embrace this new technology depends in good part on the ability of technologically minded reformers to understand the realities of the classroom and to enlist teachers as collaborators rather than regarding them as obstacles to progress.'

Taking teachers as the central point, we tried to find out the present attitudes of science teachers towards the use of new methods and tools in their practice and their thoughts on what can be done to promote positive attitudes and effective use of ICT in the classrooms.

The project encapsulates both quantitative and qualitative methods. The quantitative part aims at finding out the demographic characteristics of Turkish Primary School Science Teachers and then finding out the affects of selected variables (age, gender, computer ownership etc.) on teachers' attitudes toward the use of ICT in education. Further information on the project can be found at the research report and papers of the project (Karaoglan *et al*, 2007; Cavas *et al*, 2008; Kısla *et al*, 2008; Cavas *et al*, 2009).

In order to collect data an instrument was developed by the researchers after an extensive review of the literature and scales used in different educational backgrounds guided the theoretical base of the study. The instrument was administered to 1071 science teachers almost distributed uniformly throughout 7 geographical regions of Turkey. In data analysis, descriptive statistics were used to describe and summarize the properties of the mass of data collected from the respondents. Parametric statistics like ANOVA and t-test pairwise comparison were conducted to analyze any differences between teachers' attitudes and other dependent variables.

The characteristics of the participating teachers are presented Figure 1. As seen in the Figure, 41.6% of the teachers' ages range between 26 and 35. 56.9% of participants are male. The figures show that 71.7% of these teachers possessed a bachelor's degree while only 6.1 held a Master's degree. A high rate (85%) of the science teachers work in state schools and only 7.5% of the teachers had administrative duties (school director, assistant school director etc.) in their schools.

In Figure 2 statistics related to ICT experience of the participating teacher are shown. The figures show that 50% of them have used computer for at least 3 years and almost half of the teachers used computers in their courses. About 36% of the teachers have attended courses during their university education regarding ICT. Most teachers (72%) have a personal computer at home. A number of teachers had access to the Internet from their homes (52%) and from their schools (53%). 61.8% of the teachers had e-mail addresses.

The other objective of the study was to find out the effects of selected variables (age, gender, computer ownership etc.) on the attitudes of teachers' toward the usage of ICT in education. Results of this quantitative study are presented in Table 1.

As can be seen in the table, almost all of the selected variables have positive effects on the attitudes of the teachers towards ICT. Furthermore, young teachers who are aged between 20 and 35 have more positive attitudes than the teachers who are aged above 35. The results also showed that teachers' computer experience affected positively their ICT attitudes. On the other hand, no significant difference was found in ICT attitudes of Turkish science teachers in terms of gender.

The second part of the study which is qualitative involved two main objectives. The first objective was to identify the perceptions and opinions of Turkish Primary Science teachers on the impacts of using ICT in their lessons upon learning. The second objective was to determine the problems and the needs of teachers in integrating ICT into learning and teaching. A 'Semi-Structured Interview Technique' was used as the qualitative research technique. Interviews were carried out with focus groups of 5-7 science teachers in cities (Antalya, Düzce, Bozüyük, Eskişehir, Kars, İzmir, Gaziantep) located in seven different geographical regions of Turkey (Kışla *et al*, 2008; Cavas *et al*, 2008).

As a result of data analysis of qualitative research based on our "Semi-Structured Interview Technique", we categorized the results under 4 main heading: 1) To increase the interest and desire to learn, 2) to secure effective learning, 3) changing teacher roles 4) concerns about ICT.

In this study, the use of ICT in lessons was found to be effective. According to participants, the use of ICT is an important part of teaching and learning. Teachers pointed out that the use of ICT is essential to create the mental and affective conditions such as interest, desire, motivation and curiosity for the desired learning to take place. Moreover, they believe that their students' attention and enthusiasm for learning increases with

Figure 1. Characteristics of the participants

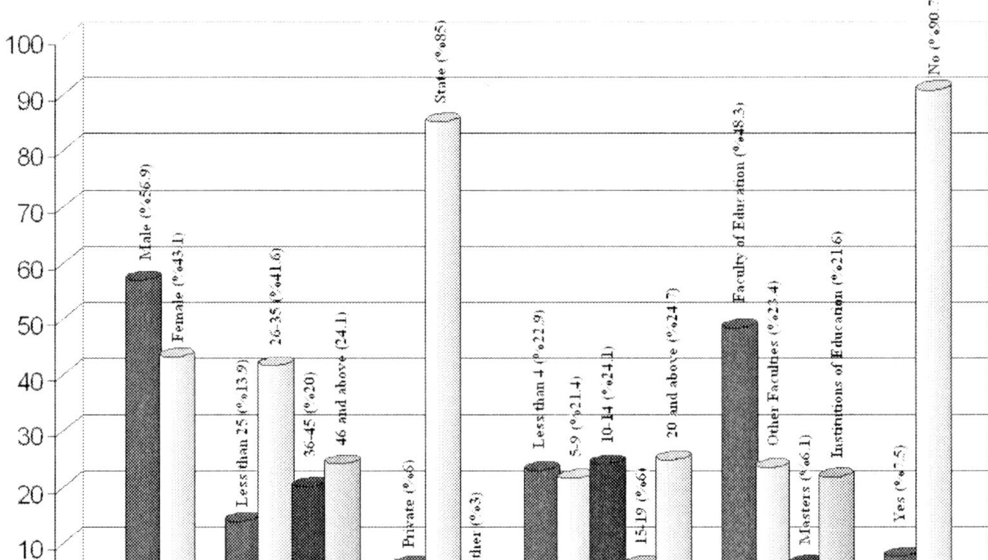

Figure 2. ICT experience of the participants

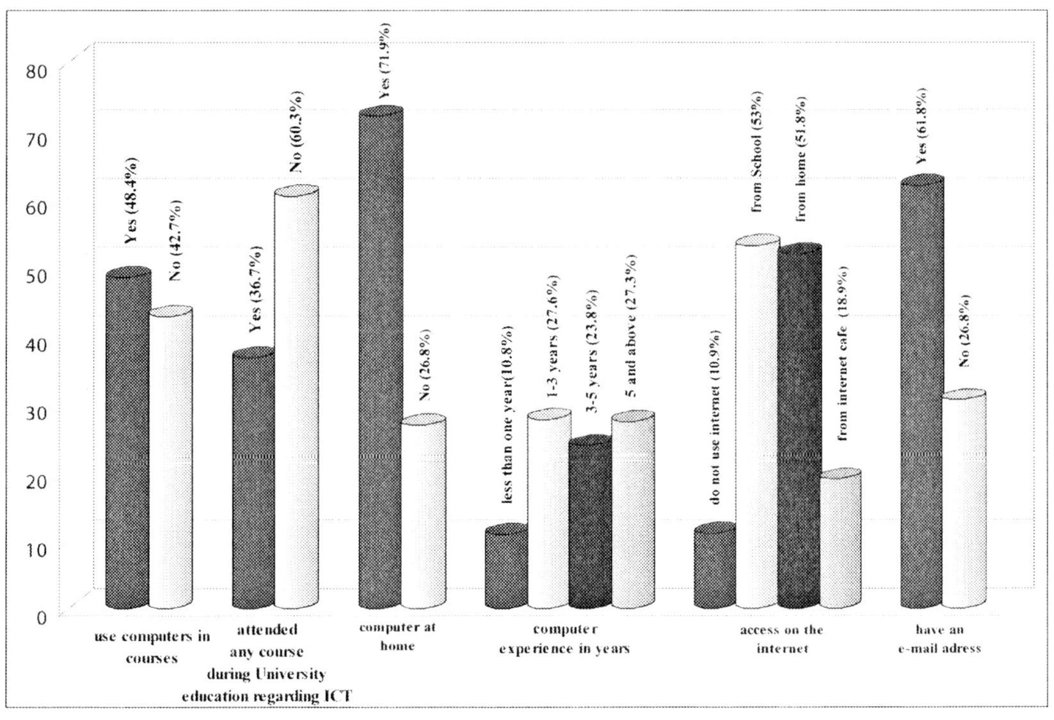

Table 1. The effects of selected variables on the attitudes of teachers' toward the usage of ICT in education

Selected variables	Significant difference	Findings
Gender	Non significant (p=.948) (p>.05)	-
Age	Significant (p=.013) (p<.05)	Teachers in the group 1 (age group: 20-35) had more positive attitudes than the teachers in other groups (36-49 / 50- +).
computer ownership at home	Significant (p=.000) (p<.01)	Teachers who own computers had more positive attitudes than the others.
experience in computer usage	Significant (p=.000) (p<.01)	Teachers in Group 4 (5 years and up) had more positive attitudes than the other teachers (0-1 years, 1-3 years and 3-5 years).
using computers in their courses	Significant (p=.000) (p<.01)	Teachers who use the computers in their courses had more positive attitudes than the others.
taking courses on ICT during university education	Significant (p=.003) (p<.01)	Teachers who took courses on ICT during university education had more positive attitudes than the others.
Using Internet	Significant (p=.000) (p<.01)	Teachers who use the Internet had more positive attitudes than the others.
Having an e-mail address	Significant (p=.000) (p<.01)	Teachers who have an e-mail had more positive attitudes than the others.

the use of ICT in lessons. The increase in students' attention and enthusiasm enhanced the students' success in science lessons.

The teachers stated that the use of ICT supports multi-dimensional, persistent and effective learning. They also mentioned that the use of ICT helps the students to concretize abstract concepts and to simplify complex subjects in their courses. For this to happen however, teachers noted that they needed well-prepared course materials which considered Piaget's developmental stages (Lin, 2002).

Teachers stated that the use of ICT changed the students' way of learning as well as the teachers' responsibilities. At the time of writing, teachers currently have two options - whether they use prepared educational materials or develop their materials by themselves with the help of ICT. The teachers specifically noted the need for mental models (how and where) to use ICT effectively in their instruction. However, even those teachers who believe that the use of ICT brings many benefits to teaching and learning have some doubts on the practical part. The major concern is students losing their creativity due to accessing ready-made materials on their computers. Teachers indicated that there are some problems with the use of ICT in the current social economic climate. Some of these problems were identified as lack of required infrastructure and technology for effective and efficient use of ICT in schools, lack of computers in students' homes and the need for appropriate materials.

CONCLUSIONS AND IMPLICATIONS

The main issues identified in this study are the identification of the processes involved in the use of ICT in education; the discussion of how Rogers' theory of DoI is linked to the adaptation of ICT innovation in learning and teaching environments, and the results of a national project that investigates how science teachers use ICT in their learning and teaching environments. In this section we try to clarify how DoI can be used in the adaptation of

ICT in science learning and teaching in light of the processes mentioned above.

Using ICT in educational environments is considered as 'new' or to be more elaborate as 'innovative'. A change in any environment cannot be expected to be instant due to established habits, beliefs and experiences. The need for change and its benefits, have to be well understood by users so there can be ownership and so that they can work towards change, for it to be realized. As noted in many studies (for example: Albirini, 2006; Uşun, 2004) pre-thinking and planning need to be done before any adoption of ICT innovation into educational environments. In the developing countries it has been noted that investments on the technological infrastructure have not materialised because the adaptation process started without thoroughly understanding the knowledge, skills and attitudes of the science teachers with respect to the use of ICT in their practice. The results of the national project discussed in this chapter have also shown that although a lot of money was invested on the adaptation of innovation process in schools in Turkey, the outcomes have not been found satisfactory due to the fact that the teachers, who are the implementers, have not been given sufficient consideration. In this respect, for the innovations in education to be welcomed, the basics of DoI theory have to be considered thoroughly in the implementation process. Science teacher involvement, pre- and in-service training, teacher support in learning-teaching material and teaching models, reorganization of the curriculum, technological infrastructure and building communication channels are vital to create understanding, ownership, confidence and applicability.

Computer access in classrooms is important for the successful adoption of computers in instructional environments (Sahin & Thompson, 2006). According to Medlin (2001) the accessibility and availability of computers was an important factor affecting the use of computers for instructional purposes. Rogers (2003) indicated that trialability and observability are two attributes of an innovation which might increase the rate of adoption of the innovation. If science teachers are aware of computer technologies and have an opportunity to access computers, their level of using technology in their courses might also rise.

The results obtained within the scope of the project mentioned in this chapter have shown some affirmative aspects regarding the application of ICT innovation in science learning and teaching environments. For example, the finding that science teachers aged between 20 and 35 have more positive attitudes than their colleagues in the older age range may be interpreted as a signal that in future the adaptation of ICT innovation will be smoother. If we consider the results of the project at length in general, the need for in-service training to enhance the knowledge, skills and experience of the science teachers is apparent even though the science teachers have positive attitudes toward ICT. This situation is explained with DoI theory as before the adaptation process of the innovation the understanding and internalization is required by the individuals. Without any doubt, ICT innovation in science learning and teaching is implemented by the science teachers as the end users. Within this process, some teachers who succeed in internalising ICT innovation find it important and meaningful and consequently use it effectively and efficiently in their lessons.

Some traditional science teachers or those in the 'Late Majority' or 'Laggard' categories of DoI theory may not accept integration of ICT innovation in to their courses. In the event of adaptation of ICT innovation to the science teaching and learning environment, it is very important to determine science teachers' decision processes on the acceptance or rejection using adaptor categories of the DoI. It will help in-service science training educators to prepare and organize training courses according to science teachers' characteristics according to the adopter categories. These types of training courses will shine light on the understanding of science teachers' differences in terms of knowledge, skills, and attitudes.

The analyses of these outcomes will also provide new perspectives and approaches to plan effective budget possibilities.

One of the main elements of DoI theory is the social systems. In diffusion studies, social systems can be categorized into two subgroups: traditional and modern systems. Social systems can be categorized as: traditional and modern norms. Individuals taking a part in traditional social systems have limited knowledge about complex technologies, low levels of ICT literacy and low communication level with the other individuals in the group. However, individuals in the modern social systems are able to accomplish complex jobs, value educational and instructional activities, are open to new ideas and thoughts, and understand other individuals' thoughts and situations. Rogers (2003) found that individuals who are members of a modern social system adopt new innovations into their daily lives more easily. In the light of those thoughts, some important measures should be taken by teachers to involve them in modern social systems and communicate with others effectively. It is important to create digital and real environments so that science teachers, especially in developing countries, can exchange ideas with other colleagues at national and international level.

Huberman and Miles (1984) indicated that the balancing of pressure and support is a vital element to achieve change successfully. They also revealed that 'Pressure without support leads to resistance and alienation; support without pressure leads to drift or waste of resources' (Fullan, 1992, p.25). It is thought that if a teacher does not internalize or accept the ICT innovation for his or her teaching, s/he probably will carry on with traditional methodologies.

Communication channels like portals, web pages, journals, newsletters and newspapers may play an important role to pursue teachers to adapt ICT innovation into education. This is also mentioned in other studies (Berryhill, 2007; Yates 20011). For this reason, new and effective communication channels should be set up for teachers.

The adaptation process of ICT to science teaching and learning environments is surely laborious, exhaustive and requires long term budget sources. It is also requires close collaboration between researchers, teachers, stakeholders and necessitates much training, and a creative mind-set to reach the final destination of benefit to schools, society and people. Even though technology infrastructures are transferred to the school systems, the teachers still require help from researchers and decision makers to incorporate the ICT innovation into the school education system. As a final remark, DoI theory might be used and applied to achieve educational goals related to ICT innovation in science education.

ACKNOWLEDGMENT

This study was funded by The Scientific and Technical Research Council of Turkey (SOBAG-104K034) and Ege University, Science and Technology Application and Research Center (2005/BIL/027).

REFERENCES

Albirini, A. (2006). Teachers' attitudes toward information and communication technologies: the case of Syrian EFL teachers. *Computers & Education*, *47*, 373–398. doi:10.1016/j.compedu.2004.10.013

Anderson, R. E., & Dexter, S. (2005). School technology leadership: An empirical investigation of prevalence and effect. *Educational Administration Quarterly*, *41*(1), 49–82. doi:10.1177/0013161X04269517

BECTA. (2008). *Harnessing Technology Review 2008: The role of technology and its impact on education.* Retrieved January, 1, 2009, from http://publications.becta.org.uk/display.cfm?resID=37346

Berryhill, A. H. (2007). *A correlational study of adoption of instructional technology by higher education faculty and their social communications network.* Unpublished doctoral dissertation, Mississippi State University.

Burkman, E., & Gagne, R. M. (Eds.). (1987). *Factors Affecting Utilization. Instructional technology: Foundations.* Hillsdale, NJ: Lawrence Erlbaum.

Callava, T. (2007). *Mainstream social science faculty uses and attitudes toward information technologies.* Unpublished doctoral dissertation, Capella University.

Cavas, B., Cavas, P., Karaoglan, B., & Kışla, T. (2009). A Study on Science Teachers' Attitudes toward Information and Communication Technologies in Education, *The Turkish Online Journal of Educational Technology, 8*(2). Retrieved January 1, 2009, from http://www.tojet.net

Cavas, P., Cavas, B., Kışla, T., & Karaoglan, B. (2008). The use of ICT in science education: a case study. In *Proceedings of the XIII.IOSTE Symposium*, Aydin, Turkey.

Cottrell, J. (1997). The diffusion of innovations: Applying change theory to academic computing. In *Proceedings of the ACM SIGUCCS*, Monterey, CA.

Cuban, L. (2001). *Oversold and Underused: Reforming Schools Through Technology, 1980-2000.* Cambridge, MA: Harvard University Press.

Damanpour, F. (1992). Organizational size and innovation. *Organization Studies, 13*(3), 375–402. doi:10.1177/017084069201300304

Dooley, K. E., & Murphrey, T. P. (2000). *How the perspectives of administrators, faculty and support units impact the rate of distance education adoption.* Retrieved January 1, 2009, from http://www.westga.edu/~distance/ojdla/winter34/dooley34.html

Ellsworth, J. B. (2000). Surviving change: A survey of educational change models. *Syracuse, NY: ERIC Clearinghouse on Information and Technology.* (ERIC Document Reprodcution Service No: ED 443 417)

Faseyitan, S. O., & Hirschbuhl, J. (1992). Computers in university instruction: What are the significant variables that influence adoption? *Interactive Learning International, 8*(3), 185–194.

Fisher, V. F. (2005). *Rogers' diffusion theory in education: the implementation and sustained use of innovations introduced staff development.* Unpublished PhD Thesis, University of Nebraska, Lincoln, NE.

Fullan, M. G. (1992). *Successful school improvement.* Buckingham, UK: Open University Press.

Hamilton, J., & Thompson A. (1992). *The adoption and diffusion of an electronic network for education.* (ERIC Document Reproduction Service No. ED 347 991)

Harris, S., Kington, A., & Lee, B. (2001). ICT and innovative pedagogy: examples from case studies in two schools collected as part of the Second Information Technology in Education Study (SITES) in England. In *Proceedings of the British Educational Research Association Annual Conference*, University of Leeds. Retrieved from http://www.leeds.ac.uk/educol/documents/00001906.htm

Havelock, R. G., & Huberman, A. M. (1977). *Solving educational problems.* Paris: UNESCO.

Hollingworth, S., Allen, K., Hutchings, M., Abol Kuyok, K., & Williams, K. (IPSE). (2008). *Technology and school improvement: reducing social inequity with technology?* Coventry, UK: Becta. Retrieved January 1, 2009, from http://partners.becta.org.uk/index.php?section=rh&catcode=_re_rp_02&rid=14541

Holloway, R. E. (1997). Diffusion and adoption of educational technology: A critique of research design. In D. H. Jonassen (Ed.), *Handbook of research for educational communications and technology* (pp. 1107-1133). New York: Simon & Schuster Macmillan.

Huberman, M., & Miles, M. (1984). *Innovation up close*. New York: Plenum.

Jacobsen, M. (1997). *Bridging the gap between early adopters' and mainstream faculty's use of instructional technology*. Calgary, Alberta, Canada: University of Calgary. (ERIC Document Reproduction Service No. ED423785)

Karaoglan, B., Cavas, B., Cavas, P., & Kı la, T. (2007). *Fen Bilgisi Öğretmenlerinin Bilgi ve İletişim Teknolojilerini Kullanma Bilgi ve Becerilerinin Araştırılmasına ve Geliştirilmesine Yönelik Bir Araştırma*, Araştırma Raporu, SOBAG- SBB-104K034, Tübitak.

Kışla, T., Çavaş, P., Çavaş, B., & Karaoglan, B. (2008). Turkish Science Teachers' Attitudes toward ICT in Education. In *Proceedings of the International Computer and Instructional Technologies Symposium,* Aydin, Turkey.

Lin, S. (2002). Piaget's developmental stages. In B. Hoffman (Ed.), *Encyclopedia of Educational Technology*. Retrieved January 1, 2009, from http://coe.sdsu.edu/eet/Articles/piaget/start.htm

Lundblad, J. (2003). A review and critique of Rogers' diffusion of innovation theory as it applies to organizations. *Organization Development Journal*, *21*(4), 50–64.

McFarlane, A., Triggs, P., & Wan, Y. (2008). *Researching mobile learning: interim report*, Coventry, UK: Becta. Retrieved January 1, 2009, from http://partners.becta.org.uk/index.php?section=rh&catcode=_re_rp_02&rid=14204

Medlin, B. D. (2001). *The factors that may influence a faculty member's decision to adopt electronic technologies in instruction*. Unpublished doctoral dissertation, Faculty of the Virginia Polytechnic Institute and State University.

Moersch, C. (2002). *Beyond hardware: Using existing technology to promote higher level thinking*. Eugene, OR: ISTE.

Mort, P. R., & Cornell, F. G. (1941). *American schools in transition: How our schools adapt their practices to changing needs, a study of Pennsylvania*. New York: Teachers College.

O'Neill, H. M., Pouder, R. W., & Buchholtz, A. K. (1998). Patterns in the diffusion of strategies across organizations: Insights from the innovation diffusion literature. *Academy of Management Review*, *23*(1), 98–114. doi:10.2307/259101

Parisot, A. H. (1995). *Technology and teaching: The adaptation and diffusion of technological innovations by a community college faculty*. Unpublished doctoral dissertation, Montana State University.

Premkumar, G., & Ramamurthy, K. (1994). Implementation of electronic data interchange: an innovation diffusion perspective. *Journal of Management Information Systems*, *11*(2), 157–186.

Rogers, E. M. (1983). *Diffusion of innovations* (3rd ed.). New York: Free Press, Collier Macmillan.

Rogers, E. M. (1995). *Diffusion of innovations* (4th ed.). New York: The Free Press.

Rogers, E. M. (2003). *Diffusion of Innovations* (5th ed.). New York: Free Press.

Ross, M. K. (2006). *Bridging the gap: a multi-case study of the adoption and implementation of instructional technology in higher education.* Unpublished doctoral dissertation, Vanderbilt University.

Ryan, B., & Gross, N. (1943). The Diffusion of Hybrid Seed Corn in Two Iowa Communities. *Rural Sociology*, *8*(1), 15–24.

Sadik, A. (2005). Factors influencing teachers' attitudes towards personal use and schools use of computers: New evidence from a developing nation. *Evaluation Review*, *2*(1), 1–29.

Sahin, I. (2006). *Instructional computer use by COE faculty in Turkey: Application of Diffusion of Innovation.* Unpublished doctoral dissertation, Iowa State University, Ames, Iowa.

Sahin, I., & Thompson, A. (2006). Using Rogers' Theory to Interpret Instructional Computer Use by COE Faculty. *Journal of Research on Technology in Education*, *39*(1), 81–104.

Samak, Z. A. (2006). *An exploration of jordanian english language teachers' attitudes, skills, and access as indicator of information and communication technology integration in Jordan.* Unpublished doctoral dissertation, Florida State University.

Sandholtz, J. H., Ringstaff, C., & Dwyer, D. (1997). *Teaching with technology: Creating student-centered classrooms*. New York: Teachers College Press. (ERIC Document Reproduction Service No. ED 402 923)

Shapiro, W. L., Roskos, K., Cartwright, G. P., Hirschbuhl, J. J., & Bishop, D. (Eds.). (1995). Technology-enhanced learning environments. *Computers in Education*. Guilford, CT: Dushkin Publishing Group/Brown and Benchmark Publishers.

Sheremetov, L., & Arenas, A. G. (2002). EVA: an interactive Web-based collaborative learning environment . *Computers & Education*, *39*(2), 161–182. doi:10.1016/S0360-1315(02)00030-1

Smith, P., Rudd, P., & Coghlan, M. (NFER). (2008). *Harnessing Technology: Schools Survey 2008 Report 1: Analysis*, Coventry, UK: Becta. Retrieved from http://partners.becta.org.uk/index.php?section=rh&catcode=_re_rp_02&rid=15952

Stacey, E. (1999). Collaborative Learning in an Online Environment. *Journal of Distance Education, 14*(2), 14–33.

Surry, D. W., & Farquahr, J. D. (1997). Diffusion theory and instructional technology. *Journal of Instructional Science and Technology*, *2*(1), 24–36.

Takkunen, C. L. (2008). *Learning to teach with technology: New teachers' perspectives on using educational technology: After participating in a PT3 grant initiative*. Unpublished doctoral dissertation, Capella University.

Tarde, G. (1903). *The Laws of Imitation* (E. C. Parson, Trans.). New York: Holt.

Tella, A., Tella, A., Toyobo, O. M., Adika, L. O., & Adeyinka, A. A. (2007). An Assessment of Secondary School Teachers Uses of ICT's: Implications for further Development of ICT's Use in Nigerian Secondary Schools. *The Turkish Online Journal of Educational Technology, 6*(3). Retrieved January 1, 2009, from http://www.tojet.net

Thayer, W. R., & Wolf, W. C. Jr. (1984). The generalizability of selected knowledge diffusion/utilization know-how: A case of educational practice. *Knowledge*, *5*(4), 447–467.

Tyack, D., & Cuban, L. (1995). *Tinkering toward utopia*. Cambridge, MA: Harvard University Press.

Usun, S. (2004). Factors Affecting the Application of Information and Communication Technologies (ICT) in Distance Education. *Turkish Online Journal of Distance Education, 5*(1), Retrieved January 1, 2009, from http://tojde.anadolu.edu.tr/tojde13/articles/usun.html

Van de Ven, A., Poole, M. S., & Angle, H. (Eds.). (1989). Methods for Studying Innovation Processes. In *Research on the Management of Innovation: The Minnesota Studies*. New York: Oxford University Press.

Warford, M. K. (2005). Testing a Diffusion of Innovations in Education Model (DIEM). The *Innovation Journal. The Public Sector Innovation Journal, 10*(3), 32.

Wright, R. E., Palmar, J. C., & Kavanaugh, D. C. (1995). The importance of promoting stakeholder acceptance of educational innovations. *Education, 115*, 628–633.

Yates, B. L. (2001). *Applying diffusion theory: Adoption of media literacy programmes in schools*. Paper presented at Instructional and Developmental Communication Division, International Communication Association Conference, Washington, DC, Retrieved February 8, 2008, from http://www.westga.edu/~byates/applying.htm

Zakaria, Z. (2001). *Factors related to information technology implementation in the Malaysian Ministry of Education Polytechnics.* Unpublished doctoral dissertation, Virginia, Polytechnic Institute and State University, Blacksburg.

Zhao, Y., & Tella, S. (2002). From the special issue editors. *Language Learning & Technology, 6*(3), 2–5.

Chapter 7
The Role of ICTs in Primary Science Education in Developing a Community of Learners to Enhance Scientific Literacy

Beverley Jane
Monash University, Australia

Marilyn Fleer
Monash University, Australia

John Gipps
Monash University, Australia

ABSTRACT

The purpose of this chapter is to show how information communication technologies (ICT) facilitated communication between primary pre-service teachers that enabled a 'community of learners' to develop children's scientific literacy. Cultural-historical theory was used to frame a study that sought to explicitly go beyond thinking as being individualistic, and to show how thinking can also be considered as a collective endeavour. In particular the study identifies how thinking forms part of a 'community of learners' both virtually and in reality within classrooms. The study was able to make visible child and pre-service teacher interactional sequences that brought together everyday concepts and scientific concepts to support concept formation in science. The study revealed the dialectical relations between everyday concepts and scientific concepts for moving from an interpsychological level to an intrapsychological level. The collective, rather than the individual orientation, made such a perspective possible. Importantly, the use of ICTs facilitated communication between members of the collective.

DOI: 10.4018/978-1-60566-690-2.ch007

INTRODUCTION

Learning in most Western communities has traditionally been conceptualized within the framework of the individual and the collective (see Rogoff, 2003), and this kind of dualistic thinking can also be found within the literature and everyday discourses surrounding virtual technologies. Schultze and Rennecker (2007) argue that 'virtuality-reality' dualisms cast a divide between the synthetic world and the world of reality, and this type of framing is unhelpful because it places them in opposition to each other. Their work suggests that the interface between reality and the virtual world should be the focus of attention in research. In building upon this research trajectory, the literature also points to the need for a re-conceptualization of the pedagogy that is afforded by the interface between the worlds of virtuality and reality (see Wei, 2007). These writings are important for understanding the broader sociological and philosophical arguments about the market/service economy, the place of digital technologies, and the dialectical interactions they afford (particularly Baudrillard's 1988 earlier critiques of a Marxist capital economy). Whilst we have been inspired by these philosophical and sociological critiques, we believe they do not go far enough in teasing out the relations between psychological functioning and the potential pedagogical interface afforded through the dialectical relations between virtuality and reality.

In the study reported in this chapter the dynamic interface between reality and the synthetic world is framed from a cultural-historical perspective. The study aimed to examine the dynamic interface between the reality of classroom teaching in science with young children, and the pre-service teachers' accounts and analyses of this as a learning community in the virtual world of WebCT. In particular, we seek to make visible the psychological functioning of the children in the classrooms alongside that of the pre-service teachers as operationalised through the pedagogy they enact in classrooms. Learning in science is explored, not from an individualistic perspective, but from a 'community of learners' orientation (Lave & Wenger, 1991). In particular, the case study described here shows how ICTs, when integral to assessment, can assist pre-service teachers to enhance the scientific literacy of children. Cultural-historical theory was used to frame the study that sought to explicitly go beyond thinking as being located within the individual child, by identifying how thinking forms part of a 'community of learners'. In the first section of the chapter the theoretical perspective is presented, and a description of the study design follows.

INTRODUCTION

The influential writing of Vygotsky (1987) showed that to understand a particular individual requires an understanding of the cultural-historical context pertaining to that individual. Cultural-historical theory highlights those contexts that shape social relations, community values, and past practices that in turn influence what participants pay attention to in their communities. When considering issues in science education some relevant Vygotskian concepts are mediated action, psychological functioning, and everyday and scientific concepts.

MEDIATED ACTION

Law and Bijker (1992) argue that people and structures are *both* products, "they are created and sustained together" (p. 293). That is, they are dialectically related to each other. Traditionally, they argue that either people or structures have dominated the explanation for why things are the way they are in society. For instance, Wajcman (1991) contends that many believe that the technology determines and shapes people - a technologically deterministic perspective. However, others have

Figure 1. Vygotsky's concept of mediation

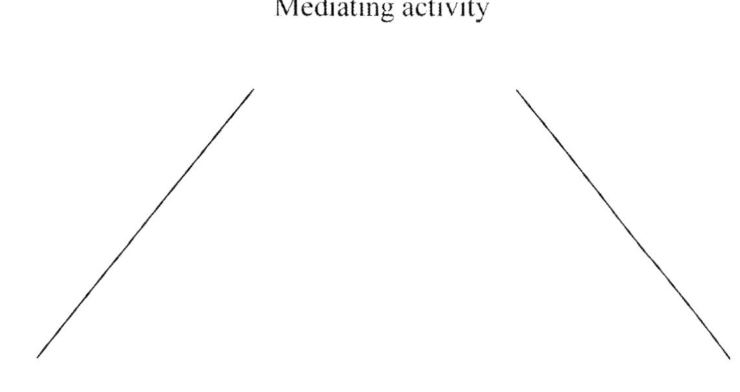

argued for the opposite, that people through their interactions with each other and within society as a whole, generate and shape the technologies that are created and used (Siraj-Blatchford, 1997; MacKenzie & Wajcman 1991). These readings place structure (technology) and agency (people) in opposition to each other, creating a dualism. These sociological studies and theoretical positions are important for understanding the dynamics within society in relation to the social shaping of technology. Advanced readings of the concepts discussed here all point toward a dialectical approach to understanding this problem (se Bijker & Law, 1992). However, within cultural-historical theory, the dialectical reading of people and objects is framed within the concept of mediation (as tools). When objects afford a particular function, or generate a particular activity, they are no longer objects, but tools imbued with social meaning. Vygtosky (1997) showed this concept as a figure, where mediation is symbolised as the relation between objects and people, in the form of tools (see Figure 1). This was the outward projection on reality or the material world. He also showed the relations internally, as the use of signs. This reading is helpful because it puts centre stage the interface between the world of virtuality and reality discussed, and provides a platform from which to examine concept formation for children and pre-service teachers. As Vygotsky (1997) notes:

'The tool serves for conveying man's [sic] activity to the object of his activity, it is directed outward, it must result in one change or another in nature. The sign changes nothing in the object of the psychological operation, it is a means of psychological action on behavior, one's own or another's, a means of internal activity directed toward mastering man himself; the sign is directed inward.' (Vygotsky, 1997, p62).

Vygotsky (1997) theorised that higher mental actions are mediated within activities by tools,  artefacts and cultural inventions (such as writing, schemes, diagrams, maps, drawings and all sorts of conventional signs). Other tools recognised in socio-cultural-historical discourse include ICTs, paintbrushes, calendars and symbol systems (John-Steiner & Mahn, 1996). For Vygotsky, speech was the most important tool (Wertsch, 1990). These tools and signs are social in origin and used initially to communicate with others, and later to mediate interaction with self. Most of these tools bring a cultural history with them. By learning to manipulate these artefacts and the practices in which they are used, people integrate the experiences of their society. Säljö (1998, p55) articulates "...the world is pre-interpreted for us by previous generations, and we draw on the experiences that others have made for us". Vygotsky's important theory related to tool-based mediation is that it is "first inter-mental and then becomes intra-mental as children learn to regulate the mediational cultural tools with their own social and mental activity" (Lantolf, 2003, p350).

INTERPSYCHOLOGICAL AND INTRAPSYCHOLOGICAL FUNCTIONING

According to Vygotsky thinking progresses as a dialectical relation between interpsychological functioning and intrapsychological functioning. This means that thinking first occurs on the social plane (*between* people engaged in joint socio-cultural activity) in the form of a collective conscious, where people participate in activities and contribute to collective thinking at that moment in time. Importantly, individuals may also use tools or engage in individual activities where the contexts support them to remember or to act as a collective, even if at that moment the person is alone. Intrapsychological functioning occurs on an individual plane (*within* the child). When a child or a person can consciously consider the concepts or activities with understanding, then intraspychological functioning has said to occur. That is, when the 'social' becomes 'the individual', psychological development has taken place. Hedegaard (2001) describes this internalisation in the following way.

'In Vygotsky's theory, learning is a social process that takes place between people. He conceptualized learning as internalisation of social interactions in which communication is central. Learning takes place in social interaction in a specific context which comes internalised by a person. By internalisation, Vygotsky did not mean copying but transforming the external interaction to a new form of interaction that guides the child's actions. Internalisation does not directly mirror the external social relations; it is a transformed reflection.' (Hedegaard, 2001, p16-17)

Hedegaard's (2001) reading of the motion from interpsychological to intrapsychological functioning foregrounds the importance of the movement or transformation of the social as a result of the individual acting upon, and also shaping the social world. The concept of internalisation has been used in several theoretical approaches to describe how shared thinking (or interpsychological functioning) results in changes in the thinking of the individual (intrapsychological functioning).

Rogoff (1998) is critical of the concept of internalisation because it implies a 'boundary' between the individual mind and the external social world. She argues that such a boundary differentiates the 'social influences' in cultural-historical theory. For instance, people construct the internalisation process as simply supporting a mainstream developmental view, with the social or cultural dimensions of development being simply 'added on' to their analysis. That is, they still view learning and development as a transmission approach occurring through a framework of what is appropriate for particular ages and stages of development, as conceptualized from a European heritage perspective only. Differences in research designs and views of development are simply bolted on as the cultural criterion in the study design, without considering that development across and within cultural communities is diverse as well as culturally specific. From a cultural-historical perspective, learning and development are creative processes that occur through a changing participation in activities, not via internalisation across a boundary. Rogoff's *transformation of participation* recognizes that individuals continually develop and change using the understanding gained through participation in shared endeavours in socio-cultural activities. "In the process of participation, individuals *change*, and their later involvement in similar events may reflect these changes" (Rogoff, 1998, p689). This reading of internalisation is important when discussing ICT because in this context the specific technological frame of the social group, such as pre-service teachers using WebCT, must be examined in order to "explain the interactions within and between social groups that shape the artifacts; these technological frames shape and are shaped by these interactions" (Bijker, 1992, p76). This dialectical reading is important for understanding how everyday concepts are transformed into scientific literacy, and how scientific literacy concepts shape or transform everyday practice.

EVERYDAY AND SCIENTIFIC CONCEPTS

Vygotsky emphasised that children acquire concepts via a process that is mediated by speech. Through a series of experiments he demonstrated that children gain important concepts inside school as well as outside and that they are related (Panofsky, John-Steiner & Blackwell, 1990). Vygotsky used the term *scientific* (*academic* or *scholarly*) *concepts* when he referred to ideas explicitly introduced by adults in school, and sometimes by parents (van der Veer & Valsiner, 1993). Scientific concepts form a logical system in a particular discipline, are generalisable, sometimes removed from material experience, and exist within a hierarchical network of related concepts.

In contrast, *spontaneous* or *everyday concepts* develop within children's daily lives due to their interaction with adults, peers and the non-social environment. These two distinct types of concepts are also interdependent. Ordinarily, the development of scientific concepts in school depends on a previously developed set of word meanings that stem from the child's everyday experiences, and this spontaneously acquired knowledge mediates the learning of the new scientific concepts (Panofsky, John-Steiner & Blackwell, 1990). For example, children will not readily understand scientific concepts associated with *electrical circuits* unless they have the everyday concepts of things such as *batteries*, *bulbs* and *switches*. In turn, when children learn scientific concepts, these change their understanding of *batteries*, *bulbs* and *switches*.

Vygotsky (1987) explains that there are strengths and weaknesses associated with scientific and everyday concepts. Scientific concepts are explicitly taught, so children can consider their relationship with other concepts. A potential difficulty associated with scientific concepts is that they may be beyond the children's personal experience, lack meaning and relevance, and therefore are easily forgotten. Whereas everyday concepts have come from personal experience and not memorisation, making them less likely to be forgotten, but they are only applicable within specific contexts or activities.

Vygotsky's theoretical writing about concept formation is helpful for the present study because this work makes explicit the importance of simultaneously capturing the everyday practices of pre-service teachers and children in classrooms, whilst examining the scientific knowledge that the pedagogical experiences afford. It is the dialectical relations between the children's everyday concepts and pre-service teachers' introduced scientific concepts that can be made visible through the use of ICT. It is also the system of relational concepts, illuminated through WebCT in the pre-service teachers' postings that generate and sustain the concept of a 'community of learners'. This latter idea is discussed further in the next section.

CULTURAL-HISTORICAL RESEARCH AND COMMUNITIES OF PRACTICE

Researchers who use a cultural-historical framework focus not only on cognition, but also seek to examine the social contexts and engagements in which conceptual processes take place. They explore the kinds of social engagements that provide the context for learning to occur. Taking a cultural-historical perspective, learning is viewed as a process that is situated in the participation framework, not in an individual mind. Variations in participation structures have been noted most strongly across cultural communities (see Rogoff, 2003). Classrooms represent a particular kind of participation structure. Although most Western communities organise schooling by placing emphasis on the child, learning is not individualistic and requires co-participants.

Cultural-historical researchers are aware of the contradictions between Western concepts and cultural beliefs/practices in particular communities, as well as the disjunction between Western concepts which are dis-embedded and concepts which arise naturally within a community of practice. Schools in particular, exemplify this disjunction, "...the organization of schooling as an educational form is predicated on claims that knowledge can be decontextualized" (Lave & Wenger, 1991, p40). These researchers describe schools as social institutions and places of learning that constitute very specific contexts. Therefore, analysis of school learning as situated requires a multi-layered view of how 'knowing and learning' is part of social practice.

Lave and Wenger (1991) put forward the concept of legitimate peripheral participation as an analytical tool for understanding how learning occurs. They suggest that initially learners reside on the periphery and then gradually become fully participating members of the community of practice. "Legitimate peripheral participation refers both to the development of knowledgeably skilled identities in practice and to the reproduction and transformation of communities of practice" (Lave & Wenger, 1991, p55). Consequently it is important to document the changing participation of members of each community of practice, and to analyze how "communities of practice are engaged in the generative process of producing their own future" (Lave & Wenger, 1991, p57-58). The nature and organization of learning communities for children is also important. Young children embedded in the 'real world' do not 'learn from talk', but rather 'learn to talk'. Their participation in the community of practice provides momentum for their learning. Over time as they participate, they move from the peripheral to become full members of the community of practice, and their participation transforms their role and helps shape the group.

Wenger (1998) recognised that we belong to several communities of practice at any one time and "develop our own practices, routines, rituals, artefacts, symbols, conventions, stories, and histories" while being part of a family community, a work community and school community (p. 6). However, many of the artefacts and processes that frame our communities of practice go unnoticed. Moreover concepts of practice can be both explicit and tacit. Wenger (1998, p47) contends that what is represented can include "language, tools, documents, images, symbols, well-defined roles, specified criteria codified procedures, regulations, and contracts that various practices make explicit for a variety of purposes". Assumptions are made about the "implicit relations, tacit conventions, subtle cues, untold rules of thumb, recognizable intuitions, and specific perceptions, well-tuned sensitivities, embodied understandings, and shared world views" (p. 47).

Rogoff (1998) has built upon these ideas by demonstrating that people move *through* rather than *to* understanding. In this process of participatory appropriation it is possible to document the transformation of understanding. Participatory appropriation occurs when "individuals transform their understanding of and responsibility for activities through their own participation" (Rogoff, 1995, p150). This view contrasts with the traditional perspective discussed earlier of regarding the process as one of "internalization in which something static is taken across a boundary from the external to the internal" (Rogoff, 1995, p151). Traditionally schooling has been framed from the perspective that learning is a static border crossing process (rather than a transformation) whereby important historical and contextual factors are not taken into account. The teaching and learning context in which children must take their newfound 'knowledge' is generally not considered when children's learning is examined. Similarly, the assessment process has traditionally not factored in the importance of these contextual elements, nor documented the appropriation process. Instead, analysis has been limited to the static transformation across a border to a predetermined outcome. In contrast, Rogoff (1995, p153) identifies that: 'The appropriation perspective views development as a dynamic, active, mutual process involving people's participation in cultural activities. Therefore examining a group of learners by paying attention to their social engagement and the contextual factors that are operating may lead to a fuller and richer understanding of conceptual change.

MOVING TOWARDS A CULTURAL-HISTORICAL PERSPECTIVE FOR RESEARCH IN SCIENCE EDUCATION

In the main, research in science education has described the research context and then focused specifically on conceptual understanding, thereby omitting from the research context those engagements that allow particular concepts to be articulated or identified. It has been common practice for most science education researchers to document what an 'individual' thinks about a science concept, and the groups being researched are generally viewed as homogeneous – for instance, as a group of ten year olds. The problem with this practice is its inconsistency with the ideas of Lave and Wenger (1991) who argue that all members of a learning community "contribute inseparable aspects whose combinations create a landscape – shapes, degrees, textures – of community membership" (p. 35).

Some researchers in science education have started to consider the potential of Vygotsky's work for framing their studies (see Roth, 2000). A cultural-historical perspective on research in science education moves the research focus from the individual to the group of learners, and concept formation is viewed as something held within a community of practice and enacted through social engagement. Fleer and Robbins (2004) showed that when researchers take a cultural-historical perspective, extended time in the research context enables a mapping of the movement of children's thinking, instead of documenting the end point. In the study that is the focus of this chapter, the researchers valued the appropriation of science concepts within the context of social relations and other contextual factors operating in the classroom.

CASE STUDY OF A CULTURAL-HISTORICAL APPROACH TO TEACHING AND LEARNING PRIMARY SCIENCE

Context of the Study

The study that is the focus of this chapter took place in three government primary schools in Victoria, a State in Australia, with most children coming from middle class European heritage families. Partnerships between the schools and nearby Monash University gave 110 pre-service teachers opportunities to teach science to children in primary classrooms for one-hour per week for one semester. In each school the participants included children in four composite Prep-1-2 classes (aged six and seven years) and four composite Year 5-6 classes (aged ten and eleven years). At least four pre-service teachers (second year students undertaking a Bachelor of Primary Education or Bachelor of Early Childhood Education) were assigned to each class of 20-25 children. The science programmes involved three hours a week in a local school and consisted of a 60 minute tutorial by university science staff (the authors), a 30 minute coffee break with time for lesson preparation, a 60 minute teaching session with children, followed by a 30 minute debriefing session. Tutorials took the form of inquiry-based learning with hands-on activities that the pre-service teachers could later implement with their small group. As the semester progressed tutorials also focused on a range of teaching approaches and theoretical underpinnings. Class teachers were only required to be present, but sooner or later most of them put down their paperwork and got up to see what was keeping their children so engaged.

There were identifiable differences in the children's approaches to learning and the pre-service teachers took these differences into account when planning their lessons. The children from the Prep-1-2 classes enjoyed learning by choosing to participate in activities that interested them. Accordingly the pre-service teachers designed a range of activities and organized these as stations around the room. Children were then encouraged to move freely to the activity of their choice. This arrangement required many resources but reduced the need for multiple sets of equipment and materials because the children could rotate from one activity to another, depending on their interest. A room full of hands-on science activities was very motivating for the children. After teaching their group of children, and during the follow-up sharing sessions, the pre-service teachers assigned to the Year 5-6 classes took notice of the success of the activities implemented with the Prep-1-2 classes. In particular their peers stressed the importance of providing adequate resources for each child to engage in the hands-on activities. Consequently the pre-service teachers began to plan in groups and communicated with one another (such as who would bring what materials and equipment) via the WebCT discussion forum set up for the unit.

Use of ICTs

ICTs were an integral part of the science education unit. The main assessment task required the pre-service teachers to use ICTs to create a PowerPoint presentation that depicted their group of children's thinking in science. The pre-service teachers recorded verbatim accounts of the children's prior views, documented how the children participated in the activities, wrote down observations of how the children were learning, and included reflections on what they and the children had learnt. This assessment task relied heavily on ICTs such as digital photographs and short video clips, taken with written permission from the children's parents/guardians.

WebCT was used for three main purposes, all of which were important because of the decentralised nature of the unit. The first purpose was to provide information such as staff contacts, resources, unit guides and details of schools. The second purpose was to provide a forum for the pre-service teachers at the different schools to communicate with university staff and with each other. (Use of WebCT for this purpose was uneven, with many using it frequently and effectively, but some not making much use of it at all). The third purpose was to act as a channel for the submission of the PowerPoint presentations, worth 70% of the assessment for the unit.

As the pre-service teachers were novices to science teaching they influenced each other a great deal and a 'community of learners' naturally developed. Science content, sources of content, and ideas about teaching strategies that had proved effective, were exchanged during the preparation and debriefing times and also through an online discussion forum associated with the unit. WebCT facilitated frequent communication between the pre-service teachers when they were not at the school, and also between those at different schools who may not see each other during the week. Typical comments posted to the on-line discussion forum were:

'I intend to do ... what do you think?' 'Why not try...' "I found this great website on ...'

'I tried ... it was ok but next time I would ...'

'If I bring these items, will you bring the rest?'

'Does anyone know of any good references on...?'

Data Collection

A cultural-historical approach to data collection meant that the data was generated through a 'community of learners' framework, and ongoing documentation (and analysis) of student postings was an important dimension of the study. Data were collected in several ways (Methods 1-3 shown below). Expansive data have been deliberately included in this chapter to show the depth of material gathered through a range of methods, but also as examples of the main findings, which are discussed later under the heading of 'The findings of the study'.

Method 1. During their teaching sessions the pre-service teachers took digital photographs, asked questions and made observations to obtain the information required for their PowerPoint presentations.

For example, one of *Pre-service teacher 1's* slide showed a Prep girl participating in a science activity, observations and reflections.

Rainbows and Reflections (week 3)

C was the first to figure out how to do it, and was excited to show the others what to do. Before we knew it everyone was making their own rainbows.

"You put the torch like this, shining into the water. It has to point at the mirror, and then the light goes onto the table at make a little rainbow." (C explains to Beverley, quite confidently, how to make a rainbow).

We can see that C learned through a discovery approach while the others had the procedure transmitted to them from C. It is an example of how individuals influence learning.

Allowing the others to test out C's method, as well as its limitations and discuss their findings provided deeper understandings than that of just hearing about it from C. (Pre-service teacher 1)

Another slide showed a photograph of two children as they painted, the finished paintings and the following text:

Rainbows and Reflections (week 3)

The children moved on to do some paintings. I asked them to do a painting of the outside. It could be anywhere, at home, at school, at the park, or anywhere else they like to go.

As they painted I asked them about the weather in their paintings.

R: "I'm at home and it has been raining from the clouds, but now the Sun is out and there is a rainbow. I like how the rain feels so do my flowers. They're daffodils, and the rain helps them to grow."

C: "This is my friend's house, and it is raining lots and lots from all these clouds. The Sun is still out so there is a rainbow."

D: "It's a sunny day at my house. Here is my favourite tree. It's good for climbing. It needs the Sun to grow. The Sun has made the air hot and that's why my tree goes like that [leaning to the side] – it's making the wind."

T: "My house is raining too, and there's the Sun and that's made a rainbow and it's making my flowers grow."

There were a lot of scientific concepts being explained here. C was the first to add a rainbow, and soon more appeared on other paintings. When D started talking about his tree the others began adding flowers and the conversation became about how all the parts of weather help plants grow.

These conversations led to the week 4 Topic of plants, which included some elements of weather. (Pre-service teacher 1)

The final slide showed *Pre-service teacher 1's* understanding of the value of a 'community of learners' environment for science learning.

Conclusion

It is through interactions with peers and/or adults (interpersonal), within a variety of contexts (contextual), along with the child's everyday concepts (individual) that children co-construct knowledge (Rogoff's socio-cultural theory as described by Fleer, Jane & Hardy, 2007, p183). This is supported by Olitsky (2007, p33) "Learning is also a social activity in which interaction with others is essential for students to participate in the language and practices of particular communities or disciplines."

The learning environment should be a happy and safe place, in which children interact with peers, teachers, and parents in an uninhibited manner without fear of taunts and rejection (Groundwater-Smith et al., 2006, p116). Learning is a reciprocal process in which children and adults learn from each other. A pedagogy of listening provides a way to ensure that each student is valued and is a respected member of the learning community and that their interests are considered during the planning process. When planning a teacher must carefully consider the individual, the context of the lesson, the materials used, the interactions that will occur and the discoveries expected to be made. If the learning only relates to things that occur inside the classroom, then so will the concepts (Fleer, Jane & Hardy, 2007).

Play is an effective tool to develop children's concepts of the world as it allows them to apply knowledge to varying situations that are meaningful to them whilst providing opportunities to share, construct, value, test out, manipulate materials, discover, listen, explore, negotiate, imagine, research, build concepts, consider others, debate, experience success and failure and problem solve. Many elements of play are included in the teaching approaches suggested in the text and this is part of what enables children to participate in the lessons successfully.

The role of the teacher is to facilitate learning by providing materials, encouraging creativity, asking questions and challenging ideas, whilst modelling behaviours that promote respect, acceptance and resilience. The teacher must also provide a range of activities to cater for the diverse preferred learning styles of the members of the learning community.

When we operate as a community of learners and honestly value each individual within the group the possibilities are endless. (Pre-service teacher 1)

Method 2. The authors observed lessons in the classrooms and playground to see the types of activities that were taking place and also asked the children questions about what they were doing. We took photographs in order to capture the interactions between the pre-service teachers and the children, and recorded field notes.

While we exposed the pre-service teachers to a variety of teaching approaches, including transmission, we noticed that as they got to know their group of children the great majority of them settled on approaches involving discovery, interaction and plenty of 'hands-on'. Practical experiments, making things, worksheets, book research, reading stories, discussions, games, excursions to the playground were all used. The classes tended to be rather noisy, although we observed that almost all of this noise appeared to be 'on-task'. The pre-service teachers encouraged their children to speak out and to write down their ideas, and they took account of the children's interests while planning their lessons. An observation made by some of the class teachers was that this small-group interactive work had made a difference to the literacy skills of the less advanced children.

Method 3. We recorded verbatim comments made by pre-service teachers and university science staff during debriefing sessions. These data provided us with additional material on how mediation was occurring between staff and pre-service teachers throughout the full semester.

Data Analysis

Rogoff's (2003) three foci of analysis were used to investigate how scientific literacy developed for the participants in this study. Rogoff uses three lenses to map the transformation of understanding among participants. The first lens concentrates upon the individual and what the individual may think. The second lens examines the social relations among groups of people as they appropriate concepts and give meaning to them through their interactions. The third lens draws attention to the cultural and contextual factors that may be operating within the community of practice. For example, the pedagogical approach adopted by the teacher, the resources available to the children, or the cultural belief systems or worldview of the children and the teacher. Although three lenses were used on all the data sets, the lenses foreground one aspect at a time, but other cultural-historical aspects were held in place. This ensured that we were able to not only document thinking, but could do so within the context of the artefacts available (3rd lens) and the collaborations between children and teachers (2nd lens). The three lenses enabled the analysis to examine the *personal* (students were required to submit a Power Point detailing the development of science ideas by their group of children), *interpersonal* (ideas flowed in all directions between children, pre-service teachers, teachers, university staff and sometimes even visiting parents) and *cultural* (achieving a cultural change in how participants perceived science and how it might be effectively communicated).

A focus on group thinking, rather than simply examining what an individual thinks, allowed for an analysis of how knowledge is held within a group and how knowledge becomes something that group members can influence or interrogate. Noting how ideas are introduced, move about, are discredited or built upon, provides greater insight into teaching and learning in science. In addition, alternative views can be better understood as they are contextually located in the group thinking or group investigations. The everyday or experimental contexts are made visible in the data in relation to the science concepts being explored. In line with cultural-historical theory, thinking always begins as a collective enterprise, and participation is 'external' (rather than internally understood) and supported by the collective. Documenting group thinking provides a better understanding of interpsychological functioning in relation to scientific literacy. In the study of pre-service teachers teaching science in schools participation as a 'community of learners' was documented at both the interpsychological level (collective) and in relation to how pedagogical practices enacted by the pre-service teachers supported intrapsychological functioning of the children (individual level). ICTs made visible this dialectical motion in psychological functioning, but it also facilitated collective learning of the pre-service teachers through on-line discussions about the PowerPoints of pedagogical practice and student concept formation.

Findings of the Study

In undertaking a cultural-historical analysis of the data generated through the three methods described above, it was possible to see how ICTs enabled pre-service teachers to make visible the full journey of children's learning in science as it moved and changed over time, rather than making the end of learning the focus of attention (as was shown in Student 1's summary PowerPoint presentation). Pre-service teachers could position themselves as learners in this process (rather than having to be experts in science pedagogy), and could make public their views on the importance of how practice informs theory, as is shown below in *Pre-service teacher 2's* final PowerPoint slide.

Reflection on my own Learning

Through having actual hands-on experience myself in a school, I found that science is exactly the same i.e. it needs to be hands-on. Children benefit a great deal more when the lesson and topic are interesting, motivating and where they can actually get involved themselves. This enables them to discover and construct their own meanings and in a sense become mini-scientists themselves. I would like to continue using a hands-on/discovery approach when teaching science as it gives the power to the children and sees them as the pivotal part of science teaching. (Pre-service teacher 2)

Importantly, the use of ICT gave the pre-service teachers a reason for documenting the children's understandings in science. The PowerPoint slides of children's experiences in science learning enabled the pre-service teachers to record both what the children said, and the kinds of scientific concepts they were bringing to the experiences that the pre-service teachers were providing in the classroom. ICTs enabled the pre-service teachers to not just document individual and group thinking, but to collectively analyse the data they were generating about the children's thinking, and determine what this meant in relation to theories of teaching and learning in science. This can be seen in the example below, where *Pre-service teacher 3* acknowledges the importance of the children's existing concepts and mini theories.

What I Learnt

Throughout this unit I have become increasingly aware of the vast understandings children have of their world and bring with them to the classroom. What I found particularly interesting is the sophisticated language that the students attempted to use when talking about their understandings of light and sound. For example, the use of terms such as: 'intensity of heat', 'force fields around sound waves', and 'vibrations being amplified'. I felt I perhaps underestimated the capacity students have in gaining this knowledge and speaking scientifically. This showed me the importance of assessing their prior knowledge and understandings when going into a new topic, especially in terms of determining children's existing mini-theories and matching their understandings to school science (Pre-service teacher 3).

As is well understood in the literature, many of the pre-service teachers did not feel confident or competent in teaching science at the beginning of the semester, and few had imagined that science would or could be relevant to their daily lives and an enjoyable experience. As a result of the pre-service teachers' early contact with children in classrooms, and their ongoing group planning and implementing of a range of science experiences for children, they began to transfer their learning needs onto the children's needs, as shown below.

The students had a fantastic time conducting the food experiments. They were excited because the experiments belonged to them: it was their ideas that formed this lesson. I wanted to show the students that science wasn't 'hard work done by those geeky guys in white coats and glasses'. I wanted to show them how science is a part of our daily lives and a subject that everyone can be actively involved in. I know I succeeded at this task when a student approached me and declared:

"I thought that Science was just Chemistry stuff. This is so fun! I want to be a scientist when I finish school." (Pre-service teacher 4)

Although the PowerPoints were shown as individual student data generation, analysis and critique, the experiences were undertaken as a collective community. The ongoing documentation of children's comments in relation to the learning experiences provided by the pre-service teachers occurred as a group. Their planning was collective, their teaching collaborative, and their reflections were done as a group (see further below). This collective ownership of the experiences in the science education unit can be seen within the following slide, which shows the tension between the standard individual orientation (I) and the collective we (our).

The End

This concludes our journey through science. Through the quotes in this PowerPoint presentation we are able to see how the students' thoughts before and after experiments changed because of the results. Usually one or two students expected the outcome to occur but often there was some debating prior to the experiment that actually led to the students being more interested in the results.

I believe the students learned from our experiments and they were always interested in what we would be doing each week. Their beliefs about science and science stereotypes were challenged, but in a way that they found fun and you can tell that from their responses that are occasionally joking yet contain some serious thought shifts.

Looking back on my time at the school I can see that the lessons the students enjoyed the most were the ones where there was a large amount of hands-on exploration with materials. Students did enjoy debating (with) each other and I found that they were able to solidly back up their own theories (Pre-service teacher 5).

Working as a 'community of learners' was an important orientation that needed to be established by the staff teaching in the unit. Through organizing science learning in the unit with the pre-service teachers as 'ongoing', and related to the 'children's everyday life and concepts', the responsibility moved from the individual to the group. This can be seen by Bev's comments below to the group early in the semester, after a teaching session by the pre-service teachers in the classroom, where the children had investigated the topic of light. The full transcript is shown below (as indented) with analytical comments made throughout the text (not indented):

Bev: We have seen some wonderful activities. You all are bringing in lots of equipment and resources for the children. When you come back here after teaching the children you are so engaged, many of you have been writing comments on the children's work. There is good evidence of a 'community of learners', where you are not told what to do, but are self-motivated. This tutorial group developed in a short time into a 'community of learners', whereby you are working together and responding to what the children are doing, not just going in and doing the activity and end of story.

Pre-service teacher: We get them to do the work in their special books, so we can go back to that. We fill in a table in their books so we can refer back to it. If they are doing drawing we write down what they say, so we can go back to it.

Bev: How are the children responding, are they looking for your comments?

Pre-service teacher: This one knows what reflection is. He can see himself in the mirror and draw himself in the mirror.

This part of the transcript shows how pre-service teachers were encouraged to collectively analyse the children's comments in relation to scientific concepts. This is important for supporting the pre-service teachers to examine the concepts children are exploring, so that they may continue to do this level of analysis with the data they are gathering in groups and placing on-line for further analysis and discussion. The dialectical relations between everyday concepts and scientific concepts is continually foregrounded by Bev, as she makes explicit the home context and everyday thinking that is being brought into the classroom by the children:

Bev: Research in the past has shown that learning that happens in school remains completely separate from the children's home experiences. So what children learn in school is not taken out of the classroom. But what we are seeing here is, what is happening in the home is being shown here.

Pre-service teacher: We did the puppet last session and this week a child made a puppet box with net and string. She made that at home and brought it in to show us.

Bev: The girl is building upon ideas you have introduced. She wants to bring it back and share with you, which shows connected learning. Children are making connections between their everyday world and the classroom.

Pre-service teacher: I had a boy at the end of the session say "I have got shiny Lego at home and it reflects light, can I bring it along next week"? He wanted to bring all the shiny stuff to share with us.

They all made their own little kaleidoscopes, ones that reflected really well. We have ours with an overhead transparency. It's got the shiny plastic inside. We had sequins stuck to the end. It was more of a spiral kaleidoscope.

Bev: Sharing something that went well.

Pre-service teacher: We made periscopes and this week there was more hands-on and making something, rather than playing with something.

In the next segment of the transcript we see how Bev moves the pre-service teachers towards thinking about the scientific concepts associated with light. She begins this discussion by inviting all the students to analyse the data they had gathered through their field notes and photographs.

Bev: Did you notice anything different?

Pre-service teacher: The children learnt more. They were talking about why the angles, and why it would reflect. If it is flatter it wouldn't reflect. They were using them all sorts of ways. They were using their imagination more and really experimenting.

Bev: Experimenting because of their curiosity.

Pre-service teacher: We got them to write about it afterwards and they seem to understand that the mirrors were reflecting from each other. We looked at writing with mirrors and they were expecting it to be the same. They looked at how you had to write differently.

Bev: POE strategy – Get them to predict or guess, and then observe and see what happens. So in this case what would happen with the writing and then do the activity. Did you get them to explain?

Pre-service teacher: They did their writing all ways to see how to get it go the right way.

Bev: A good way to challenge the children's naïve views. Give them an activity in which they can see the difference between the observations and the prediction. This is a powerful strategy.

Pre-service teacher: We did a similar thing with the girls. We had mirrors and asked them to write their names and it came out backwards; then to write out backwards so it would go the right way in the mirror. A lot of them would write it backwards and write their letters the same way. They had to work out how to turn the letters around. Then we gave them plastic mirrors and for fun to see if they would wipe their images. Then we gave them kaleidoscopes.

Bev: You made it a game.

Pre-service teacher: They realised things were opposite in the mirror; it reinforced it better.

Bev: Like mirrors in fun parks. So you let them have fun doing that. Play is an important part.

By bringing together both everyday practices and concepts with scientific thinking, Bev nudged the pre-service teachers forward in their own thinking by considering the pedagogy that they were adopting at the time when children's thinking was moving. In this example, she highlights the use of play in science. This is immediately considered by the pre-service teachers, as is shown below.

Pre-service teacher: When you were saying play, that's what we did. We got plastic bottles and they had to do their own experiments. They put coins in the water and could see it reflecting on the top. There was one girl who filled a bottle of water up and began shining the torch through it and then putting tinsel over it. For the whole hour she used the shiny stones, CD underneath it, and put rainbows in the bottle. When I took it away there was (sic) no more rainbows. We did the same thing with the names. So some who were getting bored - they shone torches down there and mirrors down there – reflection with about six things at once.

Bev: Children were playing with everyday materials. What are you going to do next?

Pre-service teacher: They are really interested in rainbows so we will do something that relates to that. Last few times the activities were really structured, and this session worked so much better because the children designed their own experiments.

Bev: An interactive approach engaged them more. Did they come up with questions?

Pre-service teacher: They did at the start and they were giving more reasons why, (rather) than asking questions. One child wasn't engaged but then he used his car in the experiment.

Bev: So you engaged him. Allowing him the freedom was important.

Pre-service teacher: We had spoons, bowls, marbles, glass and glitter to see if they could see their reflection, reflection in the water; in the spoon – one side is better for seeing themselves, one side darker and one lighter. With one bowl of water and marbles we discussed what they thought and then said 'go for it'. We had glitter mixed with water and spoons coated with glitter. We had water everywhere and they experimented for ages. They had a half circle and tried to make a full circle with the mirror.

One boy did some great experimenting and I discussed with him and scribed for him and filled an entire page with his ideas, and then we made a table of his ideas. If I hadn't scribed for him I wouldn't have known what he knew, and thought that he was just mucking around. I realise that children are thinking even if they are not writing.

Another boy did the experiments and had stuff in front of him. Then he sat down and instead of doing a table he started writing. He looked at his materials and wrote two full pages. His teacher was really surprised. We had no game plan but just let them go. We were worried but it really worked well. They were experimenting with the mirror cards and didn't know that if they sat down it went up, so they found that out.

Bev: Should you have a variety of ways of recording things?

Pre-service teacher: We didn't say - you use the table, but "what is it for? You could turn it over and use the blank paper" – so it was up to them. Our original plan was to find out what they know about what reflects. My group had eight people and they all went off in completely different directions. Some wanted to do kaleidoscopes, some walked around, others did something very different, so I needed to have three different activities. One girl had a theory about how it works, so she drew a picture of her kaleidoscope. I had a back up plan for the boys who were not as interested - to make a periscope.

Bev: It was good that you didn't make the boys do the kaleidoscope. So you have shown to be comfortable with the uncertainty. That tells you something about your own development and philosophy of teaching that it has shifted. You are now letting go, allowing the children to focus on what they are interested in.

In the final part of the transcript Bev focuses the pre-service teachers thinking on their own learning about science teaching, where she actively brings together her observations of the pre-service teachers' practices in the classroom in relation to effective pedagogy that has become more student-centred (rather than only concept oriented).

This example taken from the weekly transcripts of the staff interacting with the pre-service teachers, highlights the structure of the unit that was undertaken in building a community of learners in the debriefing sessions after the small group science teaching in the primary classrooms. The approach taken mirrors what the pre-service teachers also undertook in small groups, and as individuals, and posted on the WebCT site for further discussion and reflection – where a virtual community of learners was also building. The focus of attention in the discussion and in the postings was on the psychological functioning of the children as related to the pedagogy the pre-service teachers were enacting in the classrooms. This was important for supporting the children's learning, but also the learning of the pre-service teachers who were working hard to build confidence and competence in science teaching.

This study has shown how group thinking, fostered by ICTs, is a significant aspect of individual learning. It was found that, as suggested by Vygotsky (1997), the interpsychological context significantly influenced what children paid attention to, and that in addition to the pre-service teachers the children themselves significantly shaped the direction of that attention and therefore activity. Through examining *group thinking* it was possible to determine the extent of the influence of 'the other person'.

In the past, the influence of children upon each other has rarely featured in study designs in science education because the research has been driven by constructivism which foregrounds individuality and overlooks the dimensions of the collective.

BEYOND SITUATED COGNITION

A cultural-historical perspective on research foregrounds the context in which cognition takes place. An analysis of the data gathered in this study suggests that much of the learning was embedded within the immediate situation of the materials. The children brought their prior experiences to the context and could 'play with new ideas' within a collaborative context.

Vygotsky (1987) has argued that everyday concepts, based on everyday experiences, provide a pathway for scientific concepts. The dialectical relations between everyday concepts and scientific concepts constituted for Vygotsky (1987) the basis of concept formation. The study made visible the myriad of contexts where everyday concepts and scientific concepts were brought together for the children. By focusing on 'group thinking' pre-service teachers could create, celebrate and fore-ground, the zones of opportunity where everyday concepts and scientific concepts could be brought together. Concept formation is optimised when the everyday practices of children are understood using scientific concepts, as the children are able to transfer their newly found knowledge to other everyday contexts, and transform their everyday practice. In this research, it was important to examine closely the situated nature of cognition in order to determine the extent of the dialectical relations between everyday concepts and scientific concepts within 'group thinking'. The study has shown that 'group thinking' facilitated concept formation, as collectively the children could draw upon more everyday contexts in which scientific meaning could be embedded.

The collective orientation to the study design and the focus on 'group thinking' by the pre-service teachers (who were using cultural-historical theory rather than constructivism to frame their interactions), not only allowed for a range of pathways/discussions to emerge, but more everyday contexts could be drawn upon by the children. What emerged was the need to allow children and pre-service teachers time to explore ideas and contexts within a scientific framework at the interpsychological level. In the 21st century the ICTs provide an important contributor at the interpsychological level through making visible the dialectical relations between context and concepts, so important for building scientific literacy.

Teachers of science should expect a vast amount of interpsychological functioning from children before intrapsychological functioning is possible. Vygotsky (1987) argued that once the foundations for everyday concepts are laid, a rapid transformation occurs, as scientific concepts and everyday concepts move more quickly towards each other. Giving time to the social relations and explorations of everyday concepts and contexts within school environments positions the children psychologically for higher order mental functioning. The ICTs allowed this process to be made visible and provided a virtual space for anlaysing and determining which contexts allowed concepts to be built so that scientific literacy could be easily afforded in primary school classrooms.

CONCLUSION

Through using cultural-historical theory to frame the study reported here, we have gone beyond the view that thinking is located within the individual, and to consider the value of a 'community of learners'. The unit of analysis for this study was the transformation of understanding within the group of learners rather than within the individual. As suggested by Karpov (2003, p268) science education has tended to "overlook that a theory of science learning has to include not only individual cognitive development but also the situational and cultural factors facilitating it". Because the pre-service teachers and the researchers focused on collective rather than individual thinking, the dialectical relations between everyday concepts and scientific concepts were facilitated. This research has shown the significance of everyday concepts and contexts for higher psychological functioning. By foregrounding everyday concepts, rather than seeing them as problematic, the pre-service teachers were able to actively draw upon the children's everyday contexts and to not only explore these further (at interpsychological level) but to value them as important for moving children to intrapsychological functioning. In addition, the study revealed the dialectical relations between everyday concepts and scientific concepts, as outlined by Vygotsky (1987), for moving from an interpsychological level to an intrapsychological level to support concept formation. The collective, rather than the individual orientation, made such a perspective possible. Importantly, the use of ICTs facilitated communication between members of the collective.

The findings of this study provide a beginning point for understanding the nature of learning in science from a cultural-historical perspective. The re-orientation from an individual to a collective perspective has been helpful for moving beyond the belief that everyday concepts get in the way of scientific concepts. Rather, the complexity and importance of everyday concepts, as discussed by Vygotsky (1987) were framed from a broader interactional and pedagogical sequence organised by the pre-service teachers. This study has shown the significance of a collective focus of attention for mapping concept formation from a cultural-historical perspective, and in turn offers new insights into the nature of everyday concepts.

ACKNOWLEDGMENT

We appreciate the enthusiasm of the Monash University Primary pre-service teachers and the children who participated in the study. We thank the principals and teachers at the schools involved for their support and the opportunity to run the science programmes in schools.

REFERENCES

Baudrillard, J. (1988). Simulacra and simulations. In M. Poster (Ed.). *Jean Baudrillard. Selected writings* (pp. 166-184). Oxford, UK: Stanford University Press/Polity Press.

Bijker, W. E. (1992). The social construction of fluorescent lighting or how an artifact was invented in its diffusion stage. In W.E. Bijker & J. Law (Eds.), *Shaping technology/building society. Studies in sociotechnical change* (pp. 75-104). Cambridge, MA: MIT Press.

Bijker, W. E., & Law, J. (Eds.). (1992). *Shaping technology/building society. Studies in sociotechnical change*. Cambridge, MA: MIT Press.

Fleer, M., Jane, B., & Hardy, T. (2007). *Science for children: Developing a personal approach to teaching* (3rd ed.). Australia, NSW: Pearson Education.

Fleer, M., & Robbins, J. (2004). "Yeah that's what they teach you at uni, its just rubbish": The participatory appropriation of new cultural tools as early childhood student teachers move from a developmental to a sociocultural framework for observing and planning. *Journal of Australian Research in Early Childhood Education, 11*(1), 47–62.

Groundwater-Smith, S., Ewing, R., & Le Cornu, R. (2006). *Teaching challenges and dilemmas* (3rd ed.). Australia: Thomson Learning.

Hedegaard, M. (2001). Learning through acting within societal traditions: Learning in classrooms. In M. Hedegaard (Ed.), *Learning in classrooms: A cultural-historical approach* (pp. 15-35). Aarhus, Denmark: Aarhus Unwood Press.

John-Steiner, V., & Mahn, H. (1996). Sociocultural approaches to learning and development: A Vygotskian framework. *Educational Psychologist, 31*(3/4), 191–206. doi:10.1207/s15326985ep3103&4_4

Karpov, Y. V. (2003). Vygotsky's doctrine of scientific concepts. Its role for contempary education. In A. Kozulin, B. Gindis, V.S. Ageyev, & S.M. Miller (Eds.), *Vygotksy's educational theory in cultural context* (pp. 65-82). New York: Cambridge University Press.

Lantolf, J. P. (2003). Intrapersonal communication and internalisation in the second language classroom. In A. Kozulin, B. Gindis, V. S. Ageyev, & S. M. Miller (Eds.), *Vygotsky's educational theory in cultural context* (pp. 349-370). Cambridge, UK: Cambridge University Press.

Lave, J., & Wenger, I. (1991). *Situated learning: Legitimate peripheral participation*. New York: Cambridge University Press.

Law, J., & Bijiker, W. E. (1992). Postscript: Technology, stability, and social theory. In W.E. Bijker & J. Law (Eds.), *Shaping technology/building society. Studies in sociotechnical change* (pp. 290-308). Cambridge, MA: MIT Press.

MacKenzie, D., & Wajcman, J. (Eds.). (1985). *The social shaping of technology*. Milton Keynes, UK: Open University Press.

Olitsky, S. (2007). Promoting student engagement in science: Interaction rituals and the pursuit of a community of practice. *Journal of Research in Science Teaching, 44*(1), 33–56. doi:10.1002/tea.20128

Panofsky, C. P., John-Steiner, V., & Blackwell, P. J. (1990). The development of scientific concepts and discourse. In L. C. Moll (Ed.), *Vygotsky and Education: Instructional implications of sociohistorical psychology* (pp. 251-267). New York: Cambridge University Press.

Rogoff, B. (1995). Observing sociocultural activity on three planes: Participatory appropriation, guided participation, and apprenticeship. In J. V. Wertsch, P. Del Rio, & A. Alvarez (Eds.), *Sociocultural studies of mind* (pp. 139-165). Cambridge, UK: Cambridge University Press.

Rogoff, B. (1998). Cognition as a collaborative process. In D. Kuhn & R. S. Siegler (Eds.), *Handbook of Child Psychology* (5th ed.) (Vol. 2, pp. 679-744). New York: John Wiley.

Rogoff, B. (2003). *The cultural nature of human development*. Oxford, UK: Oxford University Press.

Roth, W.-M. (2000). Autobiography and science education: An introduction. *Research in Science Education, 30*(1), 1–12. doi:10.1007/BF02461649

Saljo, R. (1998). Thinking with and through: The role of psychological tools and physical artifacts in human learning and cognition. In D. Faulkner, K. Littleton, & M. Woodhead (Eds.), *Learning relationships in the classroom* (pp. 54-66). London: Routledge.

Schultze, U., & Rennecker, J. (2007). Reframing online games. Synthetic worlds as media for organizational communication. In K. Crowston, S Sier, & E. Wynn (Eds.), *IFIP International Federation of Information Processing, Virtuality and Virtualization* (Vol. 236, pp. 335-351). Boston: Springer.

Siraj-Blatchford, J. (1997). *Learning technology, science and social justice: An integrated approach for 3-3 year olds*. Nottingham, UK: Education Now Publishing Co-operative.

Van der Veer, R., & Valsiner, J. (1993). *Understanding Vygotsky: A quest for synthesis*. Oxford, UK: Blackwell.

Vygotsky, L. (1997). The history of the development of higher mental functions. In R. W. Rieber (Ed.), *The collected works of L. S. Vygotsky. Vol. 4* (M. J. Hall & R. W. Rieber, Trans.).New York: Plenum Press.

Vygotsky, L. S. (1987). Thinking and speech. In R. W. Rieber & A. S. Carton (Eds.), *The collected works of L. S. Vygotsky, Vol. 1, Problems of general psychology* (N. Minick, Trans.) (pp. 39-285). New York: Plenum Press.

Wajcman, J. (1991). *Feminism confronts technology*. London: Allen and Unwin.

Wei, K. (2007). Sharing knowledge in global virtual teams. How do Chinese team members perceive the impact of national cultural differences on knowledge sharing? In K. Crowston, S. Sier, & E. Wynn (Eds.), *IFIP International Federation of Information Processing, Virtuality and Virtualization* (Vol. 236, pp. 251-265). Boston: Springer.

Wenger, E. (1998). *Communities of practice: Learning, meaning and identity*. Cambridge, UK: Cambridge University Press.

Wertsch, J. V. (1990). The voice of rationality in a sociocultural approach to mind. In L. C. Moll (Ed.), *Vygotsky and education: Instructional implications of sociocultural psychology* (pp. 111-126). New York: Cambridge University Press.

Chapter 8
Multiple Literacies and Environmental Science Education:
Information Communication Technologies in Formal and Informal Learning Environments

Ruth Hickey
James Cook University, Australia

Hilary Whitehouse
James Cook University, Australia

ABSTRACT

A project by James Cook University's School of Education created an online learning environment targeted at rural and regional schools in Far North Queensland. Pre-service teachers worked with practising teachers and children to develop learning activities which were shared through the BirdNet website. The site hosts a wide range of learning activities for bird identification, building school gardens, as well as professional learning tools such as lesson plans and integrated units of work. Project successes indicate that innovation, creativity and place-based learning can support high levels of both ICT and scientific literacy in all participants. The challenges faced included those resulting from technical issues, effects of distance, child-safety provisions for an on-line environment, and entry level of skills for participants. The value of informal learning by pre-service teachers, freed from the formal learning assessment regime, is endorsed as a valid sustainable, strategy which can be adopted by teacher educators.

INTRODUCTION

In Australia, as in many other nations, there is real concern about science education (Ainley, Kos & Nicholas, 2008; Tytler, 2008) especially how to position science as meaningful enterprise for the generation of children presently in primary (elementary) school. Research continues to show children are making decisions about career possibilities by the age of 9 years (Ford, 2007). The Australian

DOI: 10.4018/978-1-60566-690-2.ch008

Chief Scientist, Dr Jim Peacock, has written that, in contemporary times, "science education should not be prescriptive - it is about the spark of excitement that stems from discovery ... Teacher confidence and professional development is just as important as student learning materials" (Tytler, 2008, p. 5).

There are similar concerns about information communication technologies (ICTs) education. One report (Anderson, Lankshear, Courtney & Timms, 2008) of geographical disadvantage indicates that not only do studies show school students in rural and regional areas tend to achieve at lower levels than those in large metropolitan areas in the disciplines of science, ICT and mathematics, but that their teachers in rural and regional Australia experience geographical and professional isolation which affects their capacities to address such disparity. Primary teachers in rural and remote areas have much higher levels of unmet need with regards to professional development, resources and collaboration with colleagues than do teachers in regional urban and metropolitan centres particularly in relation to ICTs (Lyons, Cooksey, Panizzon & Pegg, 2005; Tytler, 2008).

In this chapter we discuss a unique project that employed ICTs and digital pedagogies for place-based, environmental science education within a teacher education programme in tropical far northern Queensland, Australia. This online, community project, known as BirdNet (Whitehouse & Hickey, 2007) was managed through the School of Education, James Cook University (JCU) Cairns, and was initially funded by an Australian Schools Innovation in Science, Technology and Mathematics (ASISTM) project grant in 2006/07 (Australian Government Department of Education Employment and Workplace Relations, 2006). The purpose of all ASISTM project grants was to fund innovations for enhancing science, technology and mathematical learning. Innovations were expected to have momentum beyond the point when federal funding finished. The total number of participants over three years was 65 pre-service teachers enrolled in Education degree programmes, two teacher educators, a project manager, two information technology support technicians, 130 primary students and 17 teachers and principals from nine regional and rural primary schools and one rural environmental education centre.

The original aim of the project was to establish partnerships between a teacher education institution and a suite of geographically diverse primary schools. The project concept was to address the relative isolation of children in small, rural and regional schools, who typically have limited numbers of children of their own age or ability level to interact with during learning activities. Similarly, their teachers may work with one part-time colleague, or two full-time staff, and be unable to attend centrally-located professional development due to the required two or three days absence (some communities are not serviced by air, and staff must drive long distances on unsealed roads). Pre-service teachers were positioned as a motivational force for learning, by bringing their skills, in ICTs and developing knowledge of environmental issues, into classrooms. They were supported by community environmental educators (such as ornithologists and botanists), volunteer organisations, and staff at commercial tourism sites. In the first two years of the project, 14 pre-service teachers accepted the opportunity to develop learning activities about birds and conservation using digital pedagogies. They worked directly in small rural schools with teachers and children and with different university and community groups as opportunities arose. They produced a range of place-relevant, curriculum materials accessible online.

Online communities can be built from formal learning (when pre-service teachers are assessed on their project completion) and informal learning (when community environmental educators work in schools to provide professional development for teachers, children and pre-service teachers). Informal learning is characterised as having significant proportions of pre-service teacher autonomy in their choice of project scope, focus, context, strategies and sequence. This type of learning is under-utilised in teacher education. This is the case at the project university, where policy mandates assessment activities within formally described limits. However, the informal learning opportunities afforded to pre-service teachers were one of the most exciting dimensions of this project. As a reality check to counter unrealistic expectations of the scope of ICT and environmental science-based projects, in this chapter, we raise concerns that emerged from our analysis of the practicalities of the project. These concerns are presented as challenges, which will have wider implications for effective use of ICTs in environmental science education.

We found it important to proceed with an understanding that providence for serendipitous events and individual suggestions played a part. In doing this work, we discovered there were many (predictable and unpredictable) obstacles in attempting to integrate ICT learning and teaching with environmental science education, pre-service teacher education and rural school community engagement. For example, BirdNet was first conceived as a community engagement project with small and isolated primary schools in rural areas. When two of the larger, regional schools and an environmental education centre asked to join the project we readily agreed because cyclones, flooding, and staff turn-over were taking their toll on progress in the small schools. The project developed as we learned more, becoming more nodal and networked in character. Different schools became involved with pre-service teachers as opportunities arose and possibilities were conceived.

The pre-service teachers who participated in BirdNet were enrolled in a teacher education programme that results in a Bachelor of Education or a Graduate Diploma of Education. Environmental education (Gooch, Rigano, Hickey & Fien, 2008), scientific conceptual development (Hickey, 2007) and assessment of children's conceptual growth (Hickey & Anderson, 2003) form part of their experiences. Most graduates are destined for careers in non-metropolitan areas. It is our responsibility, as teacher educators for a regional, rural and remote workforce, to prepare pre-service teachers to be resilient and self-reliant when it comes to their learning for teaching science and ICT education. BirdNet aimed to increase future primary teachers' confidence to learn and teach environmental science through engagement with ICTs.

As university staff based on the Cairns campus in the School of Education, the authors were the project initiators. A part time, project manager, a web-manager and a technical advisor supported us. The project manager was a retired teacher of considerable experience. She mentored and guided pre-service teachers to identify and refine projects; advised on the content of pre-service teachers' projects; facilitated the creation and maintenance of links to schools; visited some project schools to present prizes and help to plant bird gardens; and oversaw the budget expenditure. The web-manager was employed to build the website and to upload contributions. The School of Education contributed a technical advisor who worked with the web-manager to provide point-of-need advice on ICTs, responding to queries from pre-service teachers and teachers and troubleshoot problems with network speed, compatibility and memory requirements for BirdNet projects. He advised on software, web design, functionality, equipment specifications, connections, upgrades, and equipment costs to support pre-service teachers' work with students and teachers in project schools.

The ASISTM project collected qualitative and quantitative data from students and teachers in all participating schools to assess improvement in digital pedagogical knowledge. While qualitative data show BirdNet had an impact on digital pedagogical knowledge, the quantitative evidence is less conclusive given the low numbers of respondents. When the federal funding ended, our original school and university partnership proved

Figure 1. Groups participating in the BirdNet online community. © 2009. Dr. Ruth Hickey. Used with permission.

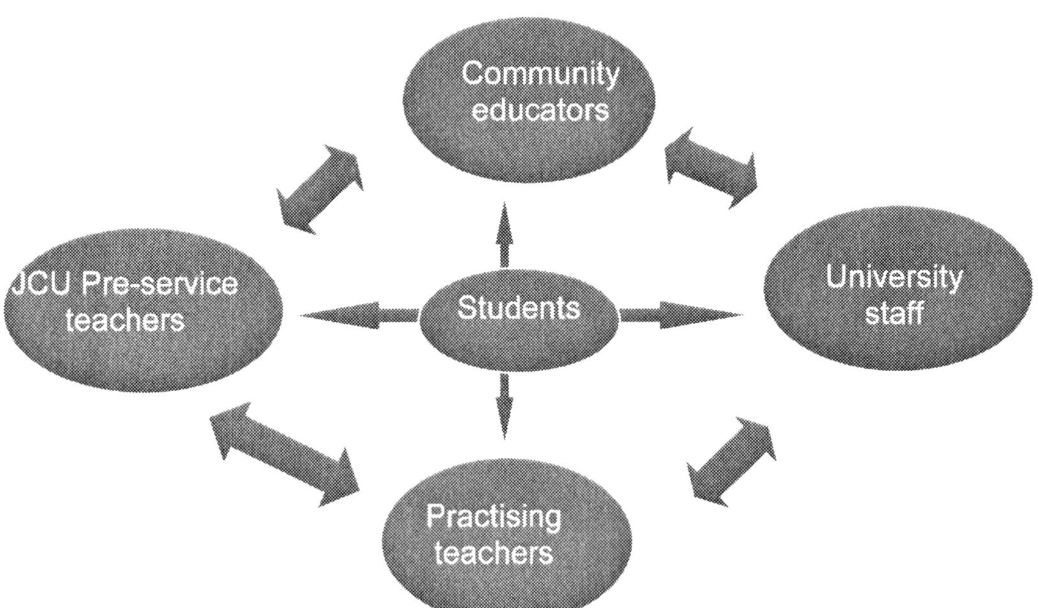

just too difficult to maintain. There were problems of high levels of school staff turnover (both of teachers and principals) in the rural schools. The larger regional schools face so many competing priorities and demands it became difficult to maintain the BirdNet partnership without funding for teacher release time, academic release time, travel and consumables. Fully conceived, cross-institutional partnerships need to be supported by funding. Consequently, in the absence of a budget, we made a decision to re-focus the project on pre-service teacher professional learning in a range of formal and non-formal education settings. This meant the BirdNet website (Hickey & Whitehouse, 2008) in its current iteration, is larger and more focused on pre-service teachers as developers and users than originally intended.

From the inception of BirdNet in 2006, we collected data from a variety of sources, including participant interviews and project documents, to research the project, which integrates the historically differentiated fields of environmental science education and ICT education within teacher education. We brought together pre-service teachers, community environmental educators, in-service classroom teachers, university lecturers, and primary school children from rural and regional schools.

BACKGROUND

Australia is a big country. In the state of Queensland, there are huge distances between towns and cities. The real strength of employing ICTs is that digital environments give educators and learners in geographically distanced regional and rural regions access to education innovation. Some of the most pressing innovations are needed in relation to climate change education and biodiversity conservation (Lowe, 2006). So it makes sense to work out how ICTs can be meaningfully employed within a teacher education program to enable innovation in locally focused, environmental science education. BirdNet was developed within the concept of place-based learning (Grunewald

& Smith, 2008) and we implicitly and explicitly applied place-based pedagogical models (Knapp, 2008) to working with large primary schools in a regional centre and very small rural schools scattered across a large tract of country. We saw it as important that BirdNet addressed the well-documented 'underachievement of tertiary education in helping people understand and appreciate the environment and their role' (Palmer & Birch, 2005). We were inspired by research documenting the interactive capacities of web-based education that offered enormous potential for teacher training and professional development in the environmental sciences (Briano, Midoro & Trentin 1997; Simons, 2002). Online learning flourishes when digital pedagogies support the formation of learning relationships and when exploration, original research and risk-taking are encouraged (Martin & Madigan, 2006). Our own experience is that pre-service teachers respond well to learning environmental science topics online (Whitehouse, 2008).

In reviewing 'the way forward' for science education in Australia, Tytler (2008, p. 64) recommended that 'school science should be linked more often and more closely with local and wider communities and science should be studied in community settings that represent contemporary science practices and concerns'. It is also Tytler's (2008, p. 64) view that 'content should not be restrictive but needs to allow room for initiatives built around local conditions'. We chose to focus on birds as northern Australia has a great diversity of bird life that includes endemic, pest and migratory species. Birds can spark people's excitement and "twitchers" (bird watchers) from all around the world visit northern Australia. Birds are accessible to primary school study, in that they show up in school grounds and neighbourhoods. Many Australian birds are brightly coloured and highly observable and lead interesting, and often courageous, lives (Barber, 1993). A number of species are directly threatened and/or vulnerable to climate change, meaning important abstract socio-ecological concepts, such as environmental change and biodiversity conservation, can be introduced to primary children through the study of birds.

The Australian Government Department of the Environment and Heritage's (2005) commitment to education for sustainability is demonstrated through its document, *Educating for a Sustainable Future – A National Environmental Education Statement for Australian Schools,* colloquially known as 'the NEES', which provides the first-ever nationally agreed description of the nature and purpose of environmental education for sustainability (EEfS) through all 13 years of Australian schooling (Preparatory to Year 12). It highlights a range of approaches to quality teaching and learning (and administration practices) to enhance the ecological sustainability of schools and their communities. The statement is a coherent and well-researched resource for educators and was our reference point in developing BirdNet as inclusive of EEfS principles. The statement articulates a vision for a whole of school approach to sustainability principles where teachers are 'enthusiastic about teaching and about developing effective relationships with their students, committed to the goals of education for sustainability, life-long learners, adaptable [and] open to new ideas and teaching strategies' (Australian Government, Department of the Environment and Heritage, 2005, p.8). In terms of content knowledge, the document argues, that, similar to science, EEfS learning

'often involves abstract concepts ... [and] teachers need to be selective in matching content to developmental needs and preferred learning styles of students. In many cases it will be most appropriate to teach these concepts though concrete studies of local, national or global examples in order to make abstract concepts meaningful to students' everyday lives and practical experiences' (Australian Government, Department of the Environment and Heritage, 2005, p.17).

Bird studies are a perfect vehicle for teaching and learning about ecological principles, human and avian interdependence and biodiversity conservation.

The SiMERR *National Survey on Science, ICT and Mathematics Education in Rural and Regional Australia* (Australian Government, Department of Education, Science and Training, 2006) recommended that institutional partnerships in regional areas are formed to support ICT and science education. Integrated projects such as BirdNet are difficult to first conceive and then implement as they challenge disciplined fields of knowledge and people's expectations of what can be known and innovated within and beyond formal teacher education programmes. In the beginning, we were never certain whether we were using the context of bird conservation to develop ICT literacy, or using context of ICT learning to develop scientific knowledge and environmental action literacy.

We came to understand that integrated projects across formal and informal learning contexts are inductive and distributive in practice. What it is that you are really doing emerges throughout, as all the people involved contribute in different ways and exit the project with a range of developed understandings and a variety of literacies. Success requires that many opportunities for inquiry learning be made available and be explicitly recognised as such. We hung onto this principle as the project mutated in response to a variety of external pressures, we have optimistically named as 'challenges' and which we discuss in the next section of the chapter. Data shows the project did achieve EEfS goals, as this pre-service teacher wrote in 2008:

Before I became involved in BirdNet, I had little idea of the value of education for sustainability in schools. My knowledge has increased since I began researching the key concept [of] loss of biodiversity. Since participating in the project team's field trip to the Mareeba Wetlands and conducting online research, I have a far greater awareness of resources, both geographical and web-based available to educate students about the environment (Pre-service teacher A).

Place-based education is described by it proponents as 'both an old and new phenomenon' (Gruenewald & Smith, 2008, p. 1). All education prior to the advent of formal schooling was local and place-based. The shape and content of curriculum in this globalised era of mass schooling has, to a great extent, generally ignored the importance of locale in shaping people's lives, knowledges and understandings. However, it is becoming increasingly clear in respect to global problems, such as climate change and biodiversity conservation and the many other challenges for science education that solutions are local and are to be found in our locales. The challenge is to harness the globally applied and available software and hardware of ICTs to create learning spaces that are local and meaningful in content and processes. Place learning is a useful framework for constructing multidisciplinary, experiential and participatory learning. It can be seen as 'horizontal' in character and analogous in digital learning pedagogy. 'Horizontalisation' is a term coined by Wenger (2005, in Mayes & Fowler, 2006) to describe the establishment of knowledge sharing and knowledge creating networks. Place-based education, because it is located within local communities, has also been characterized as networked and connected, and as 'using multiple instructional contexts and using interdependent learning groups to solve problems' (Knapp, 2008, p. 23).

There are many challenges to educators interested in moving towards a place-based education. Gruenewald and Smith (2008, p. 347) set out one dilemma as follows:

'For a variety of reasons, state-controlled ... schools continue to focus the attention of teachers and learners on standardized and decontextualized learning objectives that often have little direct

bearing on the lived experience of diverse people and places. Place-based education endeavors to reverse this trend by asking educators to include more local experience, inquiry, action and reflection in the practice of teaching and learning. This does not mean that place-based educators necessarily seek to replace standards-based schooling. Rather, advocates for place-based learning aspire to make this powerful approach to learning available at least part of the time to learners in all communities.'

This includes challenging 'the common prejudice against the local' (Gruenewald & Smith, 2008, p. 348) and, in Australia, a lingering educational prejudice against the regional, rural and remote. Pre-service teachers, who have developed a connection to a regional place, are thought to be more likely to invest their intelligence and efforts in that region (Lyons et al, 2005). The tenets of place-based education align with the State agenda of keeping skilled educators in the regions, as one means of reducing the chronic problem of staff mobility. Also, place-based education meets the State and National agenda for responding and adapting to the impacts of climate change. All the evidence indicates the Australian continent is going to be badly affected, which means, perhaps, that local scale learning about bird biodiversity may be of long-term importance (Olsen, 2007).

However,One of the most exciting features of the concept of place-based education is the value it accords to creativity, which cannot be underestimated in supporting science learning in teacher education. Science educators recognise the necessity to communicate the 'joy' and 'excitement' of science, and to stress the importance of scientific endeavour as creative (Tytler, 2008). Project-based learning is not only creative, it is motivational. Projects can also 'develop core skills which are difficult to achieve in a [standard] classroom setting, yet are considered essential in the modern commercial and business world' (Lethbridge & Davies, 1995, p. 455).

Analysis and Summary of the Project Outcomes

The informal learning opportunities proved to be most effective for achieving professional growth outcomes for pre-service teachers. They rose to the challenge of creating environmental science curriculum materials grounded *in* place. The curriculum materials now online include: how to create bird gardens in school grounds; planning for bird-attracting planting across a range of habitats and different regional environments; a photographic survey of seasonal and resident birds in the Mareeba Wetlands; a Bird Wonder Web Quest; a series of bird identification challenges; a photographic survey of Hastie's Swamp; the problems associated with the Indian Myna *(Acridotheres tristis)* an introduced avian pest species and ways to eradicate it; how to create a natural learning space; and making name plaques for school vegetation to identify local and introduced species.

The Project's web sites support children to identify birds in their backyards, and one provides step-by-step procedures to take their own photographs and to contribute their work to the website. Another asks students to complete a web quest to support knowledgeable local action, which helps students learn about the cassowary, an endangered animal of Far North Queensland, and why its ecosystem is important to its survival. Another activity teaches students and practising teachers ICT skills and how to publish bird research projects online for access by peers, parents and community (for example a Year 5's study of the crimson finch). Some pre-service teachers submitted entire planned units of work, including step-by-step directions and resources for a ten-week study of the evolution of birds.

Both context and creativity is critical to the success of integrated environmental science projects. Creative freedom and free choice can (re)generate interest. One pre-service teacher was struggling to finish her Bachelor of Education degree when she made inquiries how to participate in BirdNet. Her project of creating a visual tour of the Mareeba Wetlands reserve (Mareeba Wetland Foundation, 2009) had a profound effect on her professional confidence. She said had she not had this opportunity to work as an autonomous, and self-directed, participant in this open-ended project, she may never have discovered how truly exciting contextual learning can be. She found herself absorbed working within her local place. She borrowed a camera and joyously photographed her visual tour. She showcased her work to a group of appreciative overseas teachers. She engaged with university lecturers. She took BirdNet into her practicum classroom and discovered primary age children responded strongly to her ideas. We witnessed her growth as an educator and realized a real strength of BirdNet was the impact informal learning opportunities had on pre-service teacher confidence and capability for employing digital pedagogies. When we, as teacher educators, remained open and responsive to the different ways and means through which pre-service teachers engaged with BirdNet, opportunities for informal learning multiplied. Pre-service teachers drew on their own capacity for supporting creativity, particularly in relation to their level of ICT skills competence. Digital technologies were employed as 'tools for creation', as explained by one pre-service teacher in describing his role in a primary school (Whitehouse, 2007):

As a pre-service teacher for BirdNet, one of my tasks was to provide multimedia training to the staff and students on possible ways of building their resources and findings into a digital presentation. This included photography, video, web pages, advanced Powerpoint presentations, and [learning to use] Macromedia Flash [and] Joomla. We needed to use multimedia as a tool of creation and not just as a tool of use (Shane).

Pre-service teachers learned the skills they needed *in situ*. We didn't send pre-service teachers into schools with an ICT skill set or training beyond that received as part of normal teacher education. This wasn't a deliberate strategy. Of course, we would have liked to run additional training but teacher education is not currently well resourced within the Australian tertiary sector and teacher educators find themselves doing what they can. The following (Whitehouse, 2007) is an extract from a pre-service teacher's report who worked with a small rural school. Her aims were to increase primary students' knowledge of and capability of acting for local ecology 'hand-in-hand' with increasing student knowledge of and capabilities for using electronic technology skills:

We began the project by looking in depth at the local birds common and endemic to the schools immediate area. We looked at birds in the school grounds, we set out feeding stations at the school, we took digital photographs of the Welcome Swallows as they returned to nest, we observed migratory birds, we took a field trip to the local swamp, students kept records of the birds around their homes and farms, we listened to birdcalls and we looked at the types of trees that provide food for local birds.

The students then turned their attention to communicating their learning, and this is the part of the BirdNet project that I found most valuable and unique to my experiences in schools throughout my [Bachelor of Education] course — the project did not culminate in some arbitrary piece of assessment that had no value outside of an academic exercise, but rather the assessment had currency beyond the classroom and the students were motivated to communicate their learning in meaningful ways with a larger audience. Technology was a very appropriate way for them to achieve this communication.

One class (a Year 3/4/5 composite) put together a Powerpoint presentation about the birds and the local trees that these birds use as a food source. The other class (a Year 6/7 composite class) put together a website featuring the 50 most common local birds. Their webpages were not the most amazing ones in the world, but they were webpages. These students then presented their finding to the local community. Students loved doing this project, were really proud of all that they had learned and were very happy to share their learning. Web design was an area of need for this school. Until this project there was no-one at the school able to teach web design, making this project very valuable from the school's perspective (Kelly).

This pre-service teacher spent an average of five hours a week for 18 weeks in the school working on the BirdNet project. When formally interviewed for research data collection, she said, 'it did take a lot of work, but it was the most (sic) easy hours to spend' as she was 'really happy to do it'. What she found most valuable in terms of her own pre-professional learning was, 'I had a huge amount of autonomy. I could do [the project] how I wanted to do it [as] I am a very self-directed person'. Also valuable was finding, 'the kids really enjoyed it and as [the project] started to flow, I really looked forward to getting down there every week, I looked forward to seeing the kids and finding out what they had learnt'. When she first arrived at the school, she found there was no staff member with ICT teaching skills and an opportunity existed for her to learn and teach web design.

It was a very small school. There [were] only three teachers [who] expressed to me their need for someone to teach web sites to students. They did have a teacher aide that does ICT, but she also wasn't familiar with web design and they really, really wanted someone to teach kids web design so that those students would have that experience. They were very happy to be involved in the Project, as long as it meant I would teach web design. So I took a deep breath and went, 'Oh my goodness, I don't know anything about web design'. But I sat down and from day one ... we headed in that direction. We decided to use Front Page, even though [the staff] would have preferred to use Dreamweaver, but that would have been an additional expense. [The school] had Front Page, it was bought and paid for, so we used it [as the program met] the needs of the school ... As I said, I wasn't huge in the whole world of ICT, I'd never done it. I didn't know what to expect. I thought it would be more like a chore, like, OK, we are going to use birds as a way to leverage this [web site design] activity. But it wasn't a chore... the kids were very eager to learn (Kelly)

BirdNet provided opportunities for informal learning that enhanced pre-professional competence and confidence towards handling the demands of multiple literacies of employing ICTs for creative purposes in primary science education. The BirdNet experiences are different to those pre-service teachers experience within their formal school practicum, where they are placed only for a few weeks "in somebody else's class" and on which their performance is formally assessed against a uniform set of criteria. Integrated projects that enable pre-service teachers to spend additional time in schools and the opportunity to act autonomously to make unique contributions do build their confidence. This young woman said BirdNet, "gave me that sense of the steps involved to get to the destination. In Practicum, you know, you are there for a few weeks but then you don't to get to follow things through. You don't always get to [gain] a sense of completion."

It was the role of the pre-service teachers to act as a mentor and encourage student-directed projects. Children and classroom teachers went 'birding' in school grounds and in local swamplands. They learned how to use ICTs (emailing, accessing internet sites, loading material onto sites, and webpage software) and digital pedagogies (GPS, photographs, recording bird sounds, converting word to pdf posters, and interactive games) and learned how to create bird-friendly gardens in their respective schools. It proved a successful strategy to address the element of teachers' competency with ICTs and science to position pre-service teachers as skilled contributors to a school's learning activities.

As pre-service teachers learned about bird conservation science, and how to use innovative digital pedagogies (often in response to school need) they created low-key and non-threatening situations when they shared their understanding and expertise with school-based teachers and children. This was critical for isolated and small schools (one with 6 children) where teachers reported experiencing difficulties in accessing professional development, which was usually offered at regional centres. This low key strategy aligns with research on ways to counter student disengagement from science (Bore, 2006).

DISCUSSION OF THE CHALLENGES FACED

Challenge 1. Low Level of ICT Competency

A core innovation was having pre-service teachers bring their knowledge of ICTs and science to the schools, instead of the more traditional strategy of asking practising teachers to travel to a central location. However, the first challenge we encountered was the complexity of working in a web-based project when pre-service teachers' and practising teachers' unexpectedly low levels of skill in meeting technical requirements (e.g., file size, converting between formats) had to be targeted for project-based learning. The logistical elements of managing diverse levels of ICT expertise had major impacts on the building of our online community. Variable competencies both with using ICTs and employing digital pedagogies meant we had to rely on participants' willingness to learn how to use the technologies as they progressed with their individual projects in schools and at university.

This issue was paired with a re-conceptualisation of the value of pre-service teachers. Instead of positioning them as unskilled and as 'only a student teacher' pre-service teachers in the project were repositioned as contributors (e.g., working with teachers to enact projects such as building bird-attracting gardens), and at times, experts in using digital pedagogies (e.g., taking digital photographs and processing them to suit a website), web design (e.g., adding children's work to the BirdNet site) and storing of units of work and learning activities (e.g., using the site as a digital repository on the school's intranet).

We originally intended to match pre-service teachers' interests and skills teachers with a school's context and requirements. However, getting the right school placement proved unexpectedly problematic, due to the extensive range of variables involved in the process. So much depended on which teachers offered to take pre-service teachers, at which year levels, placements being too short in time to adequately support project requirements, sick leave, unexpected transfers of school staff, impact of pre-service teachers' travel considerations, and difficulty finding and covering the cost of accommodation for rural placements. The successes depended on the flexibility of both teachers and pre-service teachers to mutually adjust to each other's expectations.

Challenge 2. Low Level of Digital Pedagogical Knowledge

Because pre-service teachers did not receive focused training at high-skill levels in ICTs, the web-manager had to take on the role of quality control for the wide range of material forwarded to him for inclusion on the site. As interest in the project expanded, he concluded:

JCU needs to send out IT trained people that in turn will help train these schools that are sadly lacking this type of real world knowledge. Schools are looking for ICT expertise and this is a great opportunity to make teachers more employable. (Web-manager)

The web-manager identified the level of training and range of skills pre-service teachers needed to knowledgeably contribute material to a website:

The projects were brilliantly compiled and resourced and will provide schools with some amazing learning resources. But the main issue that we need to address [is] how these projects are structured and designed for a web based project or a CD presentation. Most of these projects did not meet any real criteria for them to function effectively on the web environment. In turn our [pre-service teachers] will probably pass this unsatisfactory knowledge onto their students in the classroom. We need to show our [pre-service teachers] how to prepare their work for the e-learning environment and then they will be able to pass these skills onto their students and fellow teachers. (Web-manager)

The web-manager's suggestions for specific content of workshops for pre-service teachers to understand how to prepare their resources for presentation on the web or CD included: how to reduce file size so work is suitable for web pages; how to edit images by techniques such as cropping so they are effective on a web page; awareness of download times and the need to reduce file size; understanding of frames and layout so they can give instructions to the web-manager on how to set up the page; a clear understanding of the principles of site navigation, and how to refer to site navigation plan in their submitted work; how to convert documents to .pdf and familiarity with a range of formats such as .rtf; how to upload files to a CD; how to manage backups; how to name files in a coherent manner; and overall, 'the best technique to present their resource in the best format' (Web-manager) built on an understanding of how webpages are constructed.

Included in the project budget was provision for the web-manager and project manager to visit each school. However, this only happened once, with one of the closest schools, when direct hands-on assistance to classroom teachers who wanted to work more effectively with their project pre-service teacher was provided. Travel and overnight stays were not feasible for project staff, but a recommendation for similar projects is to select staff able to undertake such demands.

Challenge 3. External Attacks on BirdNet Site

Our choice of server proved to have an unanticipated, and severe, impact on the project. Due to university server limitations, and on advice by JCU technical staff, we contracted an outside server who provided a domain name for the project, with a password protected by the web-manager. The site proved prone to repeated 'attacks' which resulted in users being directed to sites advertising pills, photographs and other services of an 'adult' nature: 'Looks like the re-direction has been hacked again. I will attempt to change the DNS so it can't be hacked and [will] get a new password' (JCU Technician). Each time it was solved, within a few days the site was hacked again, apparently because of the term 'bird' in the site name of BirdNet. We innocently thought the name indicated an interest in avian fauna. We were mistaken. As a stop-gap measure, the material was used on a CDROM, but it took months to move the site to the relative safety of the university server, and resume widespread use. This caused cessation of any on-line community for a considerable time.

Challenge 4. Online Safety Issues for Children

A further challenge was management of online safety for children. We intended the BirdNet project to build an online community for school students, to share work with others in similar schools. Sharing would be through the Managed Internet Service (The State of Queensland, 2003) that gives each child her/his own email address and storage. Work shared included publishing their photographs on the site; emailing each other about their learning; and sharing bird observations, digital photographs, drawings, reports and field notes. Despite our awareness of the need for control of publication of images of young children, for screening emails of child-to-expert or child-to-child, formulating policy and procedures to achieve this, it was a significant hurdle, and proved a major impediment to allowing children to post their work on the BirdNet public-access site. This forced the adoption of complex and awkward practices of media-release forms, teachers intercepting student-to-student emails, children using the principal's email due to limits on student email file sizes, and a cumbersome process of vetting submitted work. Our difficulty was increased as it was only after the commencement of the project that authoritative resources about internet safety became readily available (Australian Government, 2007; Emslie, (n.d.)).

We developed a mechanism for managing and securing permission for release for images published on the site. Under Queensland education legal requirements (The State of Queensland, 2009a) publication of any child's work (e.g., a drawing, story, report or photograph) requires release forms to be signed by parents and/or school principal, and retained by the university (as site manager) for as long as the image remains on the web site. This is linked to copyright issues, where ownership of the work remains with the artist. Based on the limitations of security on the website, we could not unconditionally guarantee that images would not be copied and used by others. Schools are appropriately concerned about children's images (still and video) being published in a public access website (and parents may not wish for a child's face to be on the internet). Understandably, parents were concerned when they understood the university could not guarantee these images were 'safe' online.

Similarly, teachers proved reluctant to allow children to communicate with each other via email (e.g., to share photographs of bird visitors) because of concerns about safety. They were even more worried about children freely engaging in 'email-an-expert' which was a corner-stone of getting children to interact with community environmental educators. The result of these requirements was a reduction in the amount of children's work that was published on the site: few images of children engaged in learning activities and no examples of child-to-child emails to bridge the gap between isolated students.

We did consider providing a site password to legitimate users. However, management of passwords across multiple sites with many users quickly means it becomes an annoyance rather than an effective barrier as teachers may post the password publicly to help children remember it and for parents at home to view the site to share in their child's interest in being published on the web. We decided a better strategy to respect teachers' and parents' concerns was to post only images of children with their faces obscured (e.g., photograph backs of heads, hands only, or wearing fun masks); and to make routine the filling in and storage of permission forms and make this non-confrontational. Most importantly, to instruct pre-service teachers in child-internet-safety procedures, such as 'stop, block and tell' (Aftab, 2006) so they can help children develop their own protective strategies. The project manager had to secure informed consent and release forms for every image of an individual child and act as gatekeeper and receiver of student contributions. Predictably this constrained and slowed communication and made this aspect of the project unsustainable due to budgetary limits. What we learned from this experience is that child safety issues with respect to ICTs must be regarded as necessities and that the management of child safety must be budgeted properly.

Challenge 5. Low Level of Science Education in Primary Schools

Reluctance to teach primary science is a national concern. In Rennie, Goodrum and Hackling's (2001) national study *Science Teaching and Learning in Australian Schools*, teachers identified a lack of science background knowledge as one of the key obstacles to effective delivery. It was found that science was taught on average 59 minutes per week in Australian primary schools. The review indicated that primary teachers emerge from pre-service teacher programmes lacking confidence in the requisite knowledge and skills to be effective science educators. To address this issue, the BirdNet project planned to reconstruct pre-service teachers' preconceptions of 'science as laboratory' by utilising and encouraging outdoors activities that suited the local context (e.g., visiting swamps, building gardens), learning with children (adopting the 'let's see if we can find out what sort of bird that was' approach). To raise the status of science, BirdNet activities appeared in the local press (e.g., children receiving awards at school ceremonies). To engender a stronger interest in science and ICTs, children participated in a competition to design the website logo, and contributed their photographs. Pre-service teachers reported that children were very excited to see their own work on the internet, and for some, to know how it was loaded onto the site. BirdNet had an impact, at the time, to improve science learning as it was designed and funded to do. But we have no data to indicate whether higher levels of environmental science teaching in project schools occurred, or that an interest in birds was more than sporadically maintained.

Developing scientific literacy, as defined by the OECD (2007) for the PISA (Programme for International Student Assessment), and the *National Science Curriculum Framing Paper* (National Curriculum Board, 2009), is not about educating students to be career scientists. In light of a global context of ever-growing billions on a warming planet with resource bases under stress and economies reliant on fossil fuels, there is consensus that all of us need a degree of scientific literacy in order to critically engage in the key debates and media of the day, and to make appropriate, evidence-based decisions for our local environments, communities and own lives.

Challenge 6. Sustainability

A key challenge was to make BirdNet sustainable, given projects of this type are dependent on external funding. A condition of the ASISTM project was the identification of deliberate strategies to make the project sustainable, after funds were expended. We could not do much about the high level of staff turnover in participating schools, which was why we abandoned this partnership model in 2008 – the energy and time spent maintaining the partnership in 2006 and 2007 was funded. But partnerships need more than just goodwill. They need coherent and consistent goodwill and staff knowledge. A way had to be found to keep the website being used after the project manager completed her contract. How can university staff reduce costs in terms of work load, time commitment and management? How could the project be low-maintenance but still attract pre-service teachers to become involved? How can it be structured in a way that minimises demands on university staff? We initiated a range of strategies to keep the project going, but with manageable input from university staff. These included:

- displaying a permanent, attractive hallway advertisement in the School of Education showing photographs of birds and the linkages between participant groups

- minimising familiarisation load for lecturers for new pre-service teachers to the project by directing them to review the existing BirdNet website for ideas on how to contribute
- establishing an tradition or expectation that pre-service teachers 'always get involved' in this school-based, conservation and environmental project
- embedding participation into ongoing core curriculum or participating in elective subjects (e.g., an JCU elective 'Environmental Education in the Tropics')
- pre-service teachers working with project schools as part of an internship (e.g., a semester-long, project-oriented subject of pre-employment work experience)
- enrolling in an independent study subject (e.g., pre-service teachers elect to develop sufficient ICT skills and science understanding to upgrade the existing website)
- completing days at project schools towards community education service as required by degree regulations
- pre-service teachers alerting the local press of special events at project schools
- encouraging professional goal-setting by pre-service teachers to improve their personal level of competence in ICT skills as integral to learning as indicated by the state registration board (Queensland College of Teachers, 2006) to improve readiness for employment.

Another aspect of sustainability was due to operating in the pre-service teacher environment. An important consideration was the decision taken to not use highly trained and skilled professionals to build and manage the website, because this would not maintain its entry-level appearance. It is ineffective to ask learners at 'entry level' to manage their own contribution to a website which is at a sophisticated professional level. This makes it more feasible for unskilled pre-service teachers to engage in contribution to a website (a key goal of the project) but a downside is that the site has navigation flaws, links are not all coherent (e.g., not all go back to the home page), framing can be irregular, attached audio files become lost and colours may be unattractive. It is a delicate balancing act to keep the site as a teaching tool without making it so unworkable that users do not re-visit it.

Challenge 7. Weather

Lastly, there was the challenge with the weather. You may consider low levels of ICT competency, low levels of digital pedagogical knowledge, online child safety issues, lack of environmental science knowledge, resistance to teaching science in primary schools, very high levels of small school staff turn-over and site hacking were obstacles enough, but as that iconic advertising line of the 1980s says: "Wait, there's more!" The final obstacle we learned to plan for (and cope with, by employing principles of resilience and adaptability towards ensuring the sustainability of the project itself) was sporadic disruption caused by the weather. At the beginning of the project, in 2006, two Category 4 cyclones, Larry and Monica, swept through the region. Cyclones are remarkably democratic in that they affect everyone by disrupting basic services including electrical power. Every participating school was affected. It took some time for the project to gain its feet again as local school communities had higher priorities that year. Severe flooding events in 2007 and 2008 also impacted on BirdNet projects (one carefully constructed bird garden suffered from a metre of water when a local creek flooded.)

FURTHER RESEARCH DIRECTIONS

A key focus for future research is using the *Smart Classrooms Professional Development Framework* (The State of Queensland, 2009b, p. 1) as a guide and goal-setting tool for pre-service teachers engaging in digital pedagogies. Building on the framework's definition of 'digital pedagogies which move the focus from adding ICT tools on to teaching and learning strategies, to a way of working in a digital world' this major employer of graduates encourages teachers, of all subject areas, to strengthen their use of ICTs for teaching and learning.

The direction of current research by the authors is in engaging pre-service teachers in targeted professional development during their on-campus studies. In the first phase, a group brainstorms a list of software (e.g., Word, Powerpoint, Photoshop, Flash, Access, Excel, etc.) and ICTs (digital microscopes, digital thermometers, email, video, etc.). These are examined to identify how each can be used pedagogically, to support and enhance student learning (e.g., the simple mechanics of using photoshop software to crop students' photographs of birds) and by students (who record their sightings in a database and manipulate data for a range of audiences and purposes). For each item in the list, pre-service teachers undertake an informal, personal audit of their perceived level of current skill, using a rating scale from 0-5. The lowest score (0) indicates a response of 'never heard of it, no idea how to use it', through 'seen it, tried to use it, but I'm hopeless' (score of 2) to 'very comfortable, able to show others how to use it' (score of 4). The highest score (5) refers to technical competence. Next, each identifies a goal for personal learning, for example, it may be to select Photoshop (with a current skill level of 0), and set a personal learning goal to reach score of 4 during the semester's work in BirdNet. During the semester, pre-service teachers work in project teams to build their own, and their team's skills, as they complete their contribution to the BirdNet website. Pre-service teachers integrate this new skill into their pedagogy during professional experience.

This research will study the opportunities teacher-education institutions can grasp to instill confidence in pre-service teachers use of ICTs, personal goal-setting, practical competencies linked to an appreciation of the broad scope of digital pedagogies, strategies to incorporate ICTs into their current studies and planned learning experiences upon graduation. Implementation issues relate to lecturer competence in using a broad range of digital pedagogies (Savery, 2002), adequate resourcing for a diverse range of software and instrumentation, and building in sufficient time for pre-service teachers to develop from unskilled to competent during their time at the university, particularly with the competition from other fields of study. It will utilise notions of the 'technology-pedagogy chasm' developed by the Australian Flexible Learning Framework (Jasinski, 2007, p3) which cautions that e-learning technologies are adopted faster than e-learning pedagogies, and the crucial role of teacher educators to both model good practice and that "incremental innovations that complement existing practice are more likely to be adopted and embedded than more radical approaches that require significant change of practice".

CONCLUSION

We have provided an evidence-based analysis of using ICTs as a strategy to overcome the geographies of the Australian continent, and documented an educational application of these technologies. We have shared some challenges and concerns for managing integrated and community focused teacher education, within the goal of enhancing place-based, environmental science learning in small and relatively isolated primary schools and in a regional teacher education institution.

Teacher educators know that the pedagogy which forms part of pre-service teachers' experiences will influence their own practice as future teachers. Integrated projects are not easy to innovate, but they may be necessary now more than ever to meet the future in science education. Through the promotion of multiple literacies of ICTs as embedded pedagogy within place-based learning in environmental education, both informal and formal education can mutually engage the next generation of teachers.

The challenge is to successfully embed digital pedagogies and ICTs as part of innovative science education. Jenkins (2006, p170) argued that digital learning, "presents education institutions with both challenges and opportunities to the support of students. To meet these challenges and reap the potential benefits it is important that a holistic approach is taken ... There will be no one-size-fits all approach." Our over-riding conclusion about science education is to agree with Tytler and Symington's (2006, p.1) view of the need for "radical surgery [to be] performed on the school science curriculum" to effect innovation in the education of science teachers: "while teacher educators traditionally contribute to discussion and activity related to the curriculum, we argue that their greatest contribution to improving the situation is through innovation in teacher education".

REFERENCES

Aftab, P. (2006). *Stop, block and tell!* Retrieved February 11, 2009, from http://safety.xanga.com/2006/06/12/stop-block-and-tell/

Ainley, J., Kos, J., & Nicholas, M. (2008). *Participation in Science, Mathematics and Technology in Australian Education* (ACER Research Monograph No. 63). Melbourne, Australia: ACER.

Anderson, N., Lankshear, C., Courtney, L., & Timms, C. (2008). 'Because it's boring, irrelevant and I don't like computers': Why female secondary school students avoid professionally-oriented ICT subjects. *Computers & Education, 50*(4), 1304-1318. Retrieved February 11, 2009, from http://www.sciencedirect.com/science?_ob=ArticleURL&_udi=B6VCJ-4MY0TNR-2&_user=972264&_rdoc=1&_fmt=&_orig=search&_sort=d&view=c&_acct=C000049659&_version=1&_urlVersion=0&_userid=972264&md5=432f65a0b2421d22a5b452835904c080

Australian Government. (2007). *A teacher's guide to Internet safety: How to teach Internet safety in our schools*. Retrieved February 11, 2009, from http://www.netalert.gov.au/__data/assets/pdf_file/0019/1819/01427-A-Teachers-Guide-to-Internet-Safety.pdf

Australian Government Department of Education. Employment and Workplace Relations. (2006). *Australian school innovation in science, technology and mathematics (ASISTM): BirdNet Creating and online community for far north Queensland students*. Retrieved February 11, 2009, from http://project.asistm.edu.au/special/successful.asp?r=8&state=QLD

Australian Government, Department of Education, Science and Training. (2006). *The SiMERR National Survey on Science, ICT and Mathematics Education in Rural and Regional Australia*. Retrieved February 2, 2009, from http://www.une.edu.au/simerr/pages/projects/1nationalsurvey/

Australian Government Department of the Environment and Heritage. (2005). *Educating for a sustainable future: A national environmental education statement for Australian schools*. Retrieved February 2, 2009, from http://www.environment.gov.au/education/publications/sustainable-future.html

Barber, T. X. (1993). *The human nature of birds*. London: St Martins Press.

Bore, A. (2006). Creativity, continuity and context in teacher education: Lessons from the field. *Australian Journal of Environmental Education, 22*(1), 31–38.

Briano, R., Midoro, V., & Trentin, G. (1997). Computer mediated communication and online teachers training in environmental education. *Journal of Information Technology for Teacher Education, 6*(2), 127–146.

Emslie, P. (n.d.). *Safety resources*. Retrieved February 11, 2009, from http://education.qld.gov.au/learningplace/cx/docs/resourcelist/safety.doc

Ford, C. (2007). Boys are doctors; girls are nurses: Sustaining the stereotypes of career aspirations. *Redress,* (December), 2-7.

Gooch, M., Rigano, D., Hickey, R., & Fien, J. (2008). How do primary pre-service teachers in a regional Australian university plan for teaching, learning and acting in environmentally responsible ways? *Environmental Education Research, 14*(2), 175–186. doi:10.1080/13504620801951715

Gruenewald, D. A., & Smith, G. A. (2008). Creating a movement to ground learning in place. In D. A. Gruenewald & G.A. Smith (Eds.), *Place-based education in the global age* (pp. 345-358). New York: Lawrence Erlbaum Associates.

Hickey, R. (2007). Understanding how children learn science. In V. Dawson & G. Venville (Eds.), *The Art of teaching primary science* (pp. 43-61). Crows Nest, NSW, Australia: Allen & Unwin.

Hickey, R. L., & Anderson, N. (2003). The science education website: Making judgements about science understandings. In C.P. Constantinou, & Z.C. Zacharia (Eds.), *Computer-based Learning in Science: Vol. 1. New Technologies and Their Applications in Education* (pp. 746-753). Nicosia: University of Cyprus.

Hickey, R. L., & Whitehouse, H. (Eds.). (2008). *BirdNet2008. Cairns, Australia: James Cook University*. Retrieved February 11, 2009, from http://www.soe.jcu.edu.au/birdnet2008/

Jasinski, M. (2007). *Innovate and integrate: Embedding innovative practices*. Canberra, Australia: Australian Flexible Learning Framework and the Australian Government Department of Education, Science and Training. Retrieved February 11, 2009, from http://innovateandintegrate.flexiblelearning.net.au/docs/Innovate_and_Integrate_Final_26Jun07.pdf

Jenkins, M. (2006). Supporting students in e-learning. In A. Martin & D. Madigan (Eds.), *Digital literacies for learning* (pp. 162-171). London: Facet Publishing.

Knapp, C. E. (2008). Place-based curricular and pedagogical models. In D.A. Gruenewald & G.A. Smith (Eds.), *Place-based education in the global age* (pp. 5-28). New York: Lawrence Erlbaum Associates.

Lethbridge, D., & Davies, G. (1995). An initiative for the development of creativity in science and technology (CREST): An interim report on a partnership between schools and industry. *Technovation, 15*(7), 453–465. doi:10.1016/0166-4972(95)96594-J

Lowe, I. (2006). *Sustainable natural resource management. Inaugural Rick Farley Lecture 2006, Sydney Conservatorium of Music, Sydney, NSW, Australia*. Retrieved February 11, 2009, from http://www.acfonline.org.au/uploads/res/RFL1-Ian_Lowe.pdf

Lyons, T., Cooksey, R., Panizzon, D., & Pegg, J. (2005). *Science, ICT and Mathematics education in rural and regional Australia: The SiMERR national survey*. Canberra, Australia: University of New England and the Australian Government Department of Education, Science and Training.

Mareeba Wetland Foundation. (2009). Mareeba savanna and wetlands reserve. Retrieved February 11, 2009, from http://www.mareebawetlands.org/index.html

Martin, A., & Madigan, D. (Eds.). (2006). *Digital literacies for learning*. London: Facet Publishing.

Mayes, J. T., & Fowler, C. J. H. (2006). Learners, learning literacy and the pedagogy of e-learning. In A. Martin & D. Madigan (Eds.), *Digital Literacies for Learning* (pp. 26-33). London: Facet Publishing.

National Curriculum Board. (2009). *National science curriculum framing paper*. Retrieved January 13, 2009, from http://www.ncb.org.au/verve/_resources/National_Science_Curriculum_-_Framing_Paper.pdf

Olsen, P. (2007). The state of Australian birds. *Wingspan Supplement, 14*(4), 1-17. Retrieved February 11, 2009, from http://www.birdsaustralia.com.au/images/stories/downloads/soab/soab2007s.pdf

Organisation for Economic Co-operation and Development (OECD). (2007). *Programme for International Student Assessment (PISA) 2006 Science competencies for tomorrow's world*. Retrieved February 11, 2009, from http://www.pisa.oecd.org/dataoecd/30/17/39703267.pdf

Palmer, J. A., & Birch, J. C. (2005). Changing academic perspectives in environment education research and practice: Progress and promise. In E. A. Johnson & M. J. Mappin (Eds.), *Environmental Education and Advocacy* (pp. 114-136). Cambridge, UK: Cambridge University Press.

Queensland College of Teachers. (2006). *Professional Standards*. Retrieved February 11, 2009, from http://www.qct.edu.au/Publications/ProffesionalStandards/ProfessionalStandardsForQldTeachers2006.pdf

Rennie, L. J., Goodrum, D., & Hackling, M. (2001). Science teaching and learning in Australian schools: Results of a national study. *Research in Science Education, 31*(4), 455-498. Retrieved January 15, 2009, from http://www.ingentaconnect.com/content/klu/rise/2001/00000031/00000004/00383551

Savery, J. R. (2002). Faculty and student perceptions of technology integration in teaching. *The Journal of Interactive Online Learning, 1*(2), 1-16. Retrieved February 11, 2009, from http://www.ncolr.org/jiol/issues/PDF/1.2.5.pdf

Simons, S. (2002). Participatory online environmental education at the Open University, UK. In W. D. Filho (Ed.), *Teaching sustainability at universities: Towards curriculum greening*. Bern Switzerland: Peter Lang.

The State of Queensland (Department of Education, Training and the Arts). (2003). *Management Internet Service*. Retrieved February 11, 2009, from http://education.qld.gov.au/schools/mis/

The State of Queensland (Department of Education, Training and the Arts)/ (2009a). *Conditions for including personal information on publicly accessible school internet websites (part of ICT-PR-004)*. Retrieved February 11, 2009, from http://education.qld.gov.au/strategic/eppr/ict/ictpr004/conditions.pdf

The State of Queensland (Department of Education, Training and the Arts). (2009b). *Smart Classrooms Professional Development Framework*. Retrieved February 11, 2009, from http://education.qld.gov.au/smartclassrooms/strategy/tsdev_pd.html

Tytler, R. (2008). *Re-imagining Science Education: Engaging students in science for Australia's future* (Australian Education Review Number 51). Camberwell, Victoria, Australia: Australian Council for Educational Research.

Tytler, R., & Symington, D. (2006). Redesigning science teacher education to reflect the nature of contemporary science. In *Proceedings of the annual meeting of the National Association for Research in Science Teaching*, San Francisco. Retrieved February 11, 2009, from http://www.deakin.edu.au/alt/edsmf/steme/?q=user/36

Whitehouse, H. (2007). *Final project report for BirdNet: Creating an online community for far north Queensland students*. Unpublished report for the Australian School Innovation in Science, Technology and Mathematics (ASISTM). Cairns, Australia: James Cook University.

Whitehouse, H. (2008). "EE in cyberspace, why not?" Teaching, learning and researching tertiary pre-service and in-service teachers' environmental education online. *Australian Journal of Environmental Education*, *24*, 11–21.

Whitehouse, H., & Hickey, R. L. (Eds.). (2007). *BirdNet2007. Cairns, Australia: James Cook University*. Retrieved February 11, 2009, from http://www.soe.jcu.edu.au/birdnet2007/

Chapter 9
Enhancing ICT Application in Science and Mathematics Education:
The Malaysian Smart School Experience

Suan Yoong
Sultan Idris University of Education, UPSI, Malaysia

Lee Yuen Lew
Long Island University, C. W. Post Campus, USA

ABSTRACT

This chapter reviews the Malaysian experience in implementing the Smart School Flagship initiatives, notably in the implementation of information communication technology (ICT) application in science and mathematics education. From a macro perspective, this chapter takes stock of the achievements of the Smart School Flagship in enabling ICT infrastructure and Internet connectivity in Malaysian schools. It attempts to appraise current trends and practice, clarifies emerging issues or challenges that schools face in trying to improve the ways in which ICT is applied to enhance teaching and learning, and identifies promising good practices so that general lessons may be drawn that are of interests to Malaysia and other countries. It does not claim to comprehensively cover every aspect of the initiatives but aims to contribute to current thinking about this topic by presenting a practical and pragmatic evaluation of some of its key features.

INTRODUCTION

Throughout history, we have witnessed how new technologies transformed society in every walk of life: the work place, trade and commerce, communication, entertainment, and education. As technologies race forward, potential applications of technology constantly stimulate imaginations, offering numerous exciting possibilities in every field. Technological innovations in the new millennium are characterized by unprecedented rates of change and improvement, often in the strides of geometric progression. The rapid spread and infusion of information communication technologies (ICT), especially through the Internet, into every phase of contemporary life, and the optimistic ex-

DOI: 10.4018/978-1-60566-690-2.ch009

pectations held by most people towards the new innovations have without doubt, fueled tremendous growth and enthusiasm in the application of ICT in education worldwide.

With the promises of ICT being brightly sketched and the potential impacts of ICT in education highly heralded, the rapid upgrade in the development of hardware and software means that great deals are now possible. Moreover, because hardware and software prices have dropped significantly, institutions of learning are upgrading their ICT infrastructure and facilities. There have been significant developments in teacher expertise and confidence in ICT, and ICT resources are also increasingly becoming available to support teaching and learning. Nonetheless, there remains a considerable gap between the aspirations of experts and the realities of the classroom. The application of ICT to teaching and learning, specifically in science and mathematics education, hardly reflects the state-of-the-art of current technological advances. Indeed, in the rush to embrace new technologies, one may risk ignoring pertinent issues that have far reaching implications for practice.

Malaysia is one of the newly industrialising countries which has benefited from a very rapid rate of economic growth. Educational development has figured prominently as an integral part of the government's developmental policy as spelled out in every Malaysia Plan (Malaysia, 2006). Developments in Malaysia in the last three decades have amply demonstrated that education and training are powerful social engineering tools towards achieving its educational goals. In the early nineties, Malaysia expressed its aspiration to fulfill the objective of full industrialization by the year 2020 in a new policy document called *Vision 2020* (Mahathir, 1991). One of the strategic challenges spelled out in the document was the need for Malaysia to establish a scientific, progressive, innovative, and forward looking society by 2020 – a society that is not only a mere consumer of technology but also a contributor to the scientific and technological civilisation of the future. Attaining such an objective requires human resource planning to train ICT literate and highly skilled human power that has broad based education at the primary and secondary level.

Toward this end, a master plan known as the Multimedia Super Corridor (MSC) was formulated (MSC, 1996) whereby the upgrading of the country's ICT infrastructure, connectivity, and penetration in all fields of application, including education, became an integral part of the national development policy. Correspondingly, the Smart School Flagship initiative (Malaysia, 1997a; 1997b; 1997c) was formulated to equip all Malaysian schools with computers and other ICT facilities, especially Internet access, so as to implement ICT application in support of education in general, and science and mathematics education in particular.

THE MULTIMEDIA SUPER CORRIDOR (MSC) PROJECT

Major international forums have given considerable attention to the role that ICT can play in social and economic development (G8 Heads of State, 2000; OECD, 2001, 2006; World Bank, 2003; United Nations, 2005). This role is most pronounced in the developed countries where technology has permeated businesses, schools, and homes and changed the way people work, learn, and play. ICT is viewed as an engine of growth for the global economy and its potential to enhance the quality of life of the people is highly acknowledged. In an increasingly globalising economy, nations fear lagging behind and loosing their competitive edge if they do not respond to human resource development that is capable of meeting the demands of the information age. Thus, attention is focused on nurturing a workforce skilled in ICT through education, training, and lifelong learning in order to increase national productivity. Naturally, more and more countries are embracing ICT as a strategic driver to support and contribute directly to the growth of the economy.

Malaysia's *Vision 2020* document (Mahathir, 1991) called for sustained, productivity-driven growth which is achievable by embracing ICT as key enablers of national development to transform the nation into a knowledge-based economy and society. Consequently, the multi-billion dollar Multimedia Super Corridor (MSC) project was launched in 1996 by setting up a high growth centre and dynamic hub in the vicinity of the capital city to attract global ICT industry and help Malaysia leapfrog into the information and knowledge age. MSC now encompasses other parts of Malaysia, the core sectors of which have been equipped with high-capacity global telecommunications and logistics networks (MSC Malaysia, 2008).

Malaysia initially promoted the MSC project on a high key, since it was seen as a forerunner to large-scale, government-supported ICT hub projects in Asia. In recent times, as other successful government-driven ICT hub projects have sprung up across the region, it has toned down somewhat with some re-branding. Nonetheless, the government announced that the MSC target had been met, since by June 2008, more than 2,000 foreign-owned and home-grown Malaysian companies focusing on multimedia and ICT products, solutions, and services had evolved and R&D investment had totaled more than RM814 million (Fierce Wireless Daily, 2008). A large percentage of the population and a large area of the nation now has access to radio, TV, video players, PC, mobile phone, and broadband internet connection. Over the years, Malaysia has established itself as having one of the more advanced ICT environments in the developing world, and has the second highest mobile penetration in South East Asia (after Singapore). Coming into 2008 just over 90% of the 27 million people in Malaysia had a mobile telephone service. Growth in fixed-line services has continued to 'flat-line' with a penetration of only about 16%. Internet take-up in Malaysia has been surprisingly restrained, with broadband growth being low. Malaysia's broadband penetration rate stood at 1.5 million subscriber mark, representing around 16% household penetration rate or 5% of the population penetration rate. This figure, remains well behind the regional leaders (Singapore) where broadband household penetration is running at above 50% (Fierce Wireless Daily, 2008).

THE MALAYSIAN SMART SCHOOL PROJECT

With the country's ICT infra-structure master plan in action, Malaysia would correspondingly strengthen the use of ICT in educational institutions, notably at the primary and secondary school levels. Indeed, among the 7 planned MSC innovative flagship applications[1] was the Smart School Project master plan for ICT application in education (Malaysia, 1997a). The professed intention was to radically transform the education system through reinventing the core teaching-learning processes in Malaysian schools by making ICT an enabler and not just a driver in order to make the teaching and learning processes interesting and enriching. With ICT-integrated curricula, pedagogy, and assessment practices, the professed goal was to enhance the roles of students, teachers, administrators, parents and the community, notably by empowering the students to be responsible for their own learning and making them technology savvy in the long run (Azian, 2006; Chan, 2002).

Leveraging on the establishment of national ICT infrastructure in the education system, the essence of the Smart School flagship application involved the delivery of the following main components, as was spelled out in the *Smart School Integrated Solution* (SSIS) blueprint (Malaysia, 1997b):

- Browser-based Teaching and Learning Materials (and related print materials) for 4 subjects, namely Bahasa Melayu (National Language), English Language, Science and Mathematics;
- A computerised Smart School Management System (SMSS) that encompasses a whole range of school functions such as school governance, student affairs management, educational resources management, financial and technology management;
- Technology Infrastructure comprising ICT and non-ICT equipment, Local Area Networks, and a virtual private network that connects the pilot schools, the Ministry's Data Centre and Help Desk;
- Support Services in the form of a centralised Help Desk, and service centres throughout the country to provide maintenance and support;
- Specialised Services such as systems integration, project management, business process reengineering, and change management.

The ICT industries were invited to propose innovative solutions with regard to these main components for delivery to the Smart School pilot schools. Instead of the normal tendering process, the Ministry of Education (MOE) produced 5 'Concept Requests for Proposals' documents (Malaysia, 1997d-f) outlining the high-level functional requirements and pedagogical demands envisioned in the Smart School Blueprint to enable private ICT sectors to deliver their innovative solutions.

The pilot phase (1999-2002) of the Smart School Project involved 87 trial schools divided into 3 categories or models of technology specification (Table 1). Of these, only 4 were primary schools. Level A (Full Classroom Model) schools consisted of 6 newly built schools (2 secondary and 4 primary), each being provided with 520 computers. The computers were distributed as follow: 7 computers each in all the classrooms and science laboratories, 36 computers each in 4 computer laboratories, and the remaining computers in teachers' room, resource centre, and the administrative office. Level B (Laboratory Model) schools consisted of 41 existing fully residential secondary schools and 40 day secondary schools, with each school being provided with 37 computers. Twenty one of these computers were used to set up one computer laboratory, and the rest were placed in the resource centre and administrative office. The single Level B+ (Limited Classroom Model) school was provided with 81 computers, of which 15 classrooms and the science laboratory

Table 1. Smart school project tryout specifications

Level B (Laboratory Model)	Level B+ (Limited Classroom Model)	Level A (Full Classroom Model)	Data Centre	Help Desk
37 computers	81 computers	520 computers	10 computers	13 computers
2 notebooks	2 notebooks	5 notebooks	1 notebook	2 notebooks
3 servers (communications, databases, applications)	3 servers (communications, databases, applications)	6 servers (communications, databases, applications)	3 servers (communications, databases, applications)	5 servers (communications, databases, applications)
Fast Ethernet backbone (100 baseT) with 128/64 kbps leased line	Fast Ethernet backbone (100 baseT) with 128/64 kbps leased line	Fast Ethernet backbone (100 baseT) with 512/256 kbps leased line	COINS leased line (2 Mbps)	COINS leased line (2 Mbps)

* COINS (Corporate Information Superhighway) is a fast, open, nationwide and globally connected broadband multimedia communications network with a capacity of up to 10 Gbps that connects all the pilot schools, the Data Centre and Help Desk

were equipped with 6 computers each, and the remaining computers were placed in the resource centre and administrative office (Table 1). Many of the level B schools had already had a substantial number of computers either provided by the Ministry of Education from previous computer in education projects, or donated by the parent-teacher association.

By the end of the pilot tryout period (December 2002), the pilot schools had been connected through the Smart School portal (BESTARInet) using both broadband and ISDN, 1494 courseware titles for the 4 subjects were supplied, a computerised and integrated Smart School Management System was delivered, a Help Desk and Data Centre was set up, and administrators, teachers, and IT coordinators in all the pilot schools were given in-service training (Azian, 2006). Some pilot schools in the remote areas were given satellite and wireless connection to the Smart School Network because of problems with landlines in their areas. Each pilot school in the Smart School Network was also provided with its own Local Area Network through a structured cabling system.

Funding for the Smart School Project was provided by the government with an allocation of RM400[2] million, of which RM100 million was for training administrators and teachers and RM300 million was allocated for the implementation of SSIS in the 87 pilot schools. The contract to develop the applications under SSIS was awarded to TSS (Telekom Smart School) in 1999. TSS is a private consortium led by the country's main telecommunications company, Telekom Malaysia. The Multimedia Development Corporation (MDeC), a private company wholly owned by the Malaysian Government to manage and market the MSC, also collaborated with the MOE to ensure the implementation of the smart school flagship applications.

During the post-pilot phase (2002 to 2005), the MOE also undertook other ICT initiatives, among these were the Computerisation project, Electronic Book project, Teaching of Mathematics and Science in English, Training of teachers to teach Mathematics and Science in English, Universal Service Provision Project, and Educational TV. The school computerisation project, which begun in March 2000, planned to build one to three computer laboratories, each with 20 computers in every school, depending on the its size. To date, computer laboratories have been built for over 2000 schools. The Electronic Book Project was initiated in 2001 and 2491 e-books had been developed by the company that won the contract and supplied to 35 pilot schools. The Universal Service Provision project was initiated in 2004 with the intention to connect all schools in Malaysia via Internet access by providing basic infrastructure, including electricity and telephone lines. The first phase of the project involving 220 schools in East Malaysia (Sabah and Sarawak) had been completed in 2006 and steps had been underway to merge them into the broader SchoolNet Project (Chan, 2006).

While the implementation of the Smart School Tryout Project was in progress, the Malaysian government made an abrupt and controversial decision in 2002 to use English to teach Science and Mathematics in all school. The rationale was basically market driven, hinging on the immense economic value of English language as an effective mean to access scientific knowledge and skills. Since most scientific books, journals, research or technical papers, 'popular science' writings and the Internet are predominantly in the English Language, access to scientific information via English is fast becoming the worldwide norm today. With globalisation and economic demands, the government leaders & corporate sectors feared that Malaysia might lose its competitive edge if the nation's large workforce did not have the necessary competency in English to deal with the challenges of the ICT age, notably to effectively access scientific knowledge and skills, which are predominantly in English. Hence, improving the standard of English, Science, and Mathematics became a major aspiration of the Malaysian government, and the government's solution was to use English to teach Science and Mathematics. With the policy of teaching Science and Mathematics in English implemented in all the schools in 2003, Ministry of Education (MOE, 2003, a,b,c,d) ICT initiatives were immediately shifted to the development of science and mathematics teaching courseware in English and intensive teacher training to ensure that science and mathematics teachers were fully equipped to teach these two subjects using the English Language.

EVALUATION OF THE SMART SCHOOL INITIATIVES

Since the Malaysian government laid the foundation for ICT applications in education through the Smart School initiatives in 1996, no systematic study on the effectiveness and impact of the SSIS, notably the impact of ICT applications on student performances, had been planned or conducted prior to implementation. Given the large investment and experimentation involved, what was worrying to educators was that standard practices of systematic curriculum development process requiring proper formative and summative evaluations prior to large-scale implementation had been ignored in the development and implementation of the Smart School initiatives. The scale of the pilot was very large and costly. The MOE planned to carry out collaborative surveys to monitor the implementation of the Smart School project twice, in 2000 and in 2001 (Malaysia Ministry of Education, 2001a, 2001b). However, the pilot period was hugely subsumed by development activities for application, installation, testing and integration and the actual use of the SSIS total solution was not fully released to schools until after March 2003. As a result, the monitoring exercises shifted focus to evaluate aspects of school management and teaching and learning that should take place in any school. In the post-pilot phase (2002–2005), the MOE and MDeC commissioned three perception-based evaluation studies: a benchmarking study by an independent consultant, Frost and Sullivan (2005), a study by local academics, and a technology review by the MOE and TSS.

The results of these and other local studies revealed that the Smart School Project faced many big challenges, notably in infrastructure readiness, technology obsolescence, connectivity, course quality and delivery, teacher training, and change management (Malaysia, 2001a; MDeC, 2005). According to the study by researchers from five local universities in March-April 2003, most schools reported that the number of computer laboratories was inadequate and a number of schools did not have enough computers for students to use. Hardware problems related to the LAN, PCs and servers were also reported (MDeC, 2005). These included unexpected increase in expenditure for materials and utilities. Maintenance of hardware was problematic and breakdowns were frequent. The MOE Help Desk faced problems in addressing queries within a stipulated time-frame. Infrastructure and technology problems could not be addressed effectively as there was no dedicated manpower to assist in such tasks. By the time the pilot project ended in 2002, the specifications were no longer sufficient to support the applications. The chosen bandwidths and server capacity were also insufficient to support the Smart School applications software and communications requirements (MDeC, 2005). As Teoh (2005) puts it, 'the Smart School Pilot Project was introduced and ended with broadband still in its infancy in 2005'.

An analysis of the preventive maintenance record in one of the rural secondary schools revealed that the use of ICT equipment was low, despite a large sum of public funds being used for the purchase. It was found that of the nine LCD projectors procured by the school, only six were available at the time of inspection. The total number of operation hours recorded by the machines was 174 hours for a period of two years. On average, each projector was only used for about 29 hours in the two-year duration which is very low (MHS, 2005). The technology review by MOE/TSS (MDeC, 2005) found that, from the start, schools provided with Level B ICT infrastructure had difficulty sharing laboratory facilities between classes and there was not enough access to the browser-based courseware supplied. Thus, the teacher had to resort to using stand-alone CD-ROM courseware, which meant that student progress could not be monitored and tracked on the SSMS.

Although the instructional materials were received well by both teachers and students, the frequency of their use was limited as some of the lessons could not adequately cater for the students' needs and did not reflect a complete curriculum. On average, the teachers used the computer laboratory for teaching and learning only four times a month. There were limitations in using lesson planning software. According to the MOE/TSS review, the Smart School learning courseware was under-utilised. This was due mainly to the implementation of the policy to use English to teach Science and Mathematics in all school in 2003, which rendered the Malay Language Smart School courseware inappropriate and the schools had to switch to using the newer teaching courseware produced by the MOE to teach Science and Mathematics in English. Teachers were also not keen on using the courseware which required more preparation time and created more work. They felt they could save precious teaching time, teach more content, and make students understand better by using traditional chalk and talk methods that related directly to examination preparation (MDeC, 2005).

The Smart School Management System (SSMS) was meant to cover nine areas of school management: Financial, Student Affairs, Educational Resources, External Resources, Human Resources, Facilities, School Governance, Security, and Technology. In addition to these management functions, the SSMS was also expected to be integrated with the following systems: Teaching and Learning Materials, Assessment, IT Security Management, Network and System Management, User Support and Help Desk. Consolidating database information across multiple applications to allow access between certain applications and other databases and applications within various divisions in the MOE. SSMS would have act as a common user front-end for access to all Smart School applications. The MOE also wanted portability, flexibility, inter-operability, scalability, usability, and manageability for the SSMS. However, due to technological and financial constraints, the MOE compromised on all the features. This resulted in an integrated SSMS with functionalities that could not be easily decoupled (Azian, 2006). Moreover, the SSMS, which was developed in three releases, was not introduced until a few years into the programmes, as revealed by the MOE/TSS technology review (MDeC, 2005).

The study by local academics found that actual use of the SSMS was minimal to moderate, at 40-50% (MDeC, 2005). This was also supported by the findings from monitoring exercises carried out in the pilot schools by the MOE. In most cases, the Ministry discovered that the schools had not completed the data entry necessary to prime the system, and the staff and students were not using the system correctly (Malaysia, 2001). There were too many systems from other divisions and departments of MOE to be installed and used. Schools found these demands overwhelming. Problems with utilising three of the SSMS components, namely financial management, personnel management and premises, and facilities management were also reported. Most principals and heads of schools rated the success of SSIS as moderate and said the system was not user-friendly and did not simplify routine tasks. Students were not aware that the SSMS was being used in their schools and parents were not well informed about the unique features of their children's 'Smart' school, even though they knew that their children attended a 'Smart School' (MDeC, 2005).

Most schools barely had enough teachers proficient in ICT application, let alone enough teachers proficient to teach science and mathematics (or for that matter, any other subjects) in English as most of them were trained in the Malay medium. In particular, most current primary school teachers also had poor groundings in science and mathematics contents. Surveys (MDeC, 2005; Noraini Idris *et al.*, 2007) consistently showed that a series of short intensive courses were unlikely to leapfrog the teacher's ability to apply ICT in teaching science and mathematics in English. Almost half the teachers surveyed, mentioned that the in-house training for subject teachers was only moderately successful in achieving its objectives. The majority of them rated the in-house training as insufficient to help them apply the SSIS principles (MDeC, 2005). Some of the teachers even thought that it was mandatory for them to use computers in their lessons and had tried using them in ways that were deemed not appropriate. For instance, a language teacher was observed transferring passages from the textbook and then asked the students to read from the computer screens, rather than allowing students to search for information related to the topic on the internet (Malaysia, 2001a).

A number of studies revealed that the level of ICT integration in Malaysian classrooms was unsatisfactory (Lee, 2009; Rosnaini & Mohd Arif, 2008; Lau & Chia, 2008; Rosnaini Mahmud, 2006; Sa'ari, Wong & Roslan, 2005; Ya'acob, Nor & Azman; 2005; Chong *et al.*, 2005; Mohd. Jasmy *et al.*, 2003; Abd. Rahman Daud, 2000; Abdul Razak & Jamaludin, 1998). With heavy teaching loads and a lack of essential ICT skills, teachers were not able to engage in the time-consuming effort of preparing their instructional courseware. Self-paced, self-accessed and self-directed learning could not be carried out due to a shortage of teaching and learning resources. Moreover, most teachers viewed the use of SSIS teaching and learning strategies as time-consuming, especially for large class sizes and they believed they would not be able to "finish the syllabus" since the SSIS strategies did not correspond to the assessment formats used in high stake public national examinations (Azian, 2006).

The benchmarking study by Frost and Sullivan (2005) compared the merits of the SSIS and its components with similar implementations in Australia, Britain, Canada, Ireland, Japan, New Zealand, Singapore and the USA to benchmark their best practices in ICT-mediated education with that of Malaysia. The study noted that no other country had contemplated automating the entire school process as full automation was the practice only in the very affluent residential schools in these advanced countries. They viewed the Malaysian achievement as all the more significant since it had brought such sophisticated advantages within the reach of the average child attending a smart government school. Moreover, most learning courseware in other countries was private-sector-owned and developed for a generic market. In contrast, the Smart School teaching-learning courseware was tailor-made for the Malaysian curriculum (MDeC, 2005). While the Frost and Sulivan study (2005) highly praised the instructional culture propagated in the Malaysian Smart School blueprint as being an "exceptional and visionary" (p.10) model for ICT-mediated education in developing countries, in reality, the implementation of the Malaysian Smart School Project was not as smooth as it had wanted it to be. In view of the questionable quality of the learning courseware delivered and the host of problems with hardware, maintenance, technology obsolescence, and connectivity, ground level SSIS coordinators, subject teachers and management staff did not appear to have been overwhelmed by the changes heralded by SSIS in teaching-learning (MDeC, 2005).

Recently, the MOE was taken to task by the Malaysian auditor-general over RM9.56 million worth of missing equipment for the teaching of Science and Mathematics in English from 812 schools nationwide from 2005 to 2007 (New Straits Times, 2008). According to the Auditor-General's Report 2007, the Education Ministry was allocated RM4.99 billion between 2002 and

2008 to carry out the teaching of Science and Mathematics in English programmes. Of that amount, RM2.21 billion was spent on equipment such as computers, LCD projectors, screens, trolleys, loudspeakers, televisions, and for facilitating training courses. Despite the surplus in funding, no efforts have been made by the ministry to replace the missing equipment. No police reports on the missing or possibly stolen goods were made either. Irregularities in the handling of the funds have also been detected, namely the payment for equipment and the signing of testing and installation certificates prior to the goods being delivered. Despite shoddy installation and the poor condition of the equipment upon arrival, the testing and installation certificates were still signed by the school authorities. The equipment, according to the auditor-general's report, was also not maintained in compliance with standards set out in the 2007 Treasury circulars. Teaching and other computer software were also delivered late to the schools, rendering use of equipment obsolete. No punitive action was applied to the supplier for the delay. Many schools were found to have more laptops than teachers for Science and Mathematics in English, leaving students with excess facilities but no instructors.

A document analysing the use of ICT in education based on the database of 90 ICT projects in education at the UNESCO website (UNESCO, 2003a,b,c; UNESCO 2004a,b) categorized countries into roughly three types: (1) those which are already integrating the use of ICT into the education systems; (2) those which are starting to apply and test various strategies; and (3) those which have just begun and are more concerned with ICT infrastructure and connectivity installation. An analysis of the trends in 17 Asia and the Pacific countries revealed that only 3 countries (Australia, South Korea and Singapore) may be categorized as in the advanced stage of ICT integration into the education system. In terms of connectivity and ICT penetration, almost all classrooms in these 3 countries have been equipped with computers and other ICT infrastructure, have a high student to computer ratio, and a high level of Internet access. As far as the use of ICT in teaching-learning is concerned, these 3 countries have revised their curriculum to ensure that ICT becomes integral nationwide, with e-Learning greatly facilitated by wide access to the Internet connectivity and the delivery of education becoming increasingly online.

Malaysia was placed in the second category together with several neighboring Asean countries. Even though Malaysia has a national ICT policy in education and a master plan to introduce ICT in various aspects of education, from teacher training to teaching and learning it is only in the stage of applying and testing various strategies, and has not yet fully integrated ICT within education. Connectivity and ICT penetration in Malaysia is growing but somewhat slowly, and has yet to reach the level of the more advanced countries. While there have been efforts to integrate the use of ICT in teaching-learning, these efforts have not yet reached systematic nationwide proportions.

PARTNERSHIPS WITH INDUSTRY AND ORGANIZATIONS

Various leading international and local agencies and organizations, notably those international ICT corporations which had invested in Malaysia, also assisted the Smart School initiatives in ICT-related projects by forming partnerships and strategic alliances with the MOE.

Intel Malaysia and MOE jointly spearheaded a project that provides PCs to Malaysian schools to enable students to access information and multimedia resources through the Internet. Intel also set up an *Intel Innovation Centre* at the Regional Centre for Education in Science and Mathematics (SEAMEO-RECSAM) in Penang (MOE/INTEL, 2007).

The Oracle Education Foundation entered into an arrangement with the MOE to enable selected IT coordinators from Malaysian schools to participate in the 21st Century Learning Institute, an online teacher professional development programmes, which allows teachers virtual access to all curricular materials including lessons, practices and additional resources. Teachers can engage in networking with their peers and project learning experts to develop their learning project (Salbiah Ismail, 2008).

Through a joint collaboration between UNDP and Coca Cola Malaysia in partnership with Apple Malaysia and MOE, ICT 'hubs' have been set up in primary schools in semi-urban and rural areas across Peninsular Malaysia. The hubs were equipped with Internet connectivity using Apple hardware technologies (iBooks and iMacs) and software for students and teachers to gain access to and training in ICT and promote e-learning and online communication. Six schools were involved in 2002 (UNDP. 2006; UN Chronicle, 2002).

The Microsoft Partner in Learning (PiL) gave software and grants to teachers to encourage ICT innovation and best practices. Launched in 2003, the initiatives under the PiL programmes have reached more than 101 countries. In Malaysia, 150,000 teachers and 5 million students have benefited from the initiative and up to RM10 million has been invested over the last five years to help improve teaching and learning outcomes in Malaysia. In 2007, the MOE and Microsoft launched the student-focused Generasi-M initiative. It was designed to address the needs of Malaysian students. It provides a platform for them to play a bigger role in enhancing the learning environment in schools, through enhancing communication and collaboration amongst student, and driving deeper dialogue between students and educators (Microsoft, 2007a).

With help from BECTA and the British Council and currently funded by the Ministry of Science and Innovation under the Demonstrator Application Grant (DAG) Scheme, the Malaysian Grid for Learning, MyGfL (E-learning) was developed. It is intended to enhance the formal learning in schools by providing online educational resources (content, information, services, and expertise) to teachers and pupils to use and share.

Other projects include the Penang e-Learning Community project initiated by the state government in 1997 and managed primarily by the Science University of Malaysia. This project spearheaded the development of web presence, web-based services and collaborative web-based tools for the purpose of providing necessary information to the educational community in Penang state. Another state initiative, K-Perak e-Learning Cluster project, was launched in March 2007 to provide professional development for teachers and build e-learning capability in a cluster of five selected schools in the state of Perak. Managed by New Zealand-based Innovation New Zealand Education group, the project offered both face-to-face capacity building and online support. Teaching and learning content is made available to teachers through the project website, which features an interactive forum. Through this capacity building process, teachers are guided to create 'virtual field trips', which emphasised student participation and collaboration between the project schools and schools in New Zealand.

The Prime Education (*Pendidikan Perdana*) project, developed under the DAGS scheme, created a web-based application aimed at developing e-learning and enhancing teacher-student relationships for 800 students and 20 teachers of a secondary school in Kuala Lumpur. Lesson revision programmes were uploaded on to the Internet, where students could access them from home or school. The website allowed students to not only revise lessons, but also to learn new lessons. The site also enabled students and teachers to interact via the Internet in various educational and extra-curricular activities such as the posting of online literary contributions, pen-pal programmes and contests. Activities also included the development of an online examination. A non-government organization, the Malaysian Nature Society, also used its website to promote the appreciation of nature in Malaysia, and educate the students and the public on the responsible uses of natural resources based on principles of sustainability and respect for the natural world.

Through strategic partnerships with different agencies, both local and international, many Malaysian schools not only benefited from gaining or upgrading essential hardware and software, but were also assured of continuous professional expertise and advise on technological strategies, ideas and practices that could enhance teaching, learning and school management. These added initiatives are laudable. Nonetheless, despite getting positive responses from MOE officials, teachers, students, community leaders, and parents with respect to these ICT projects, there was again a lack of systematic effort to evaluate these initiatives to provide educators with evidence-based methods for effectively using ICT to support instruction.

ICT INITIATIVE IN TEACHING SCIENCE AND MATHEMATICS IN ENGLISH

One of the strategic challenges that Malaysia spelled out in its vision was to turn Malaysia into a scientific, progressive and developed nation by 2020. In the quest to achieve a globalised knowledge-based society, it is envisaged that students must learn new skills to meet the demands of a rapid changing world. These skills include a firm grounding in core knowledge subjects, notably science and mathematics, ICT, communication, and self-learning skills as foundations for further development after leaving school.

In the new millennium, the capacity to produce and transmit data using ICT has led to a substantial increase in the application of quantitative methods, scientific analyses, and problem solving skills in many essential jobs. Knowledge of word processing and presentation software to write and present work reports, and use of spreadsheets to conduct financial and other quantitative analyses are fast becoming common entry-level employment requirements. The acquisition of ICT skills is especially tied to science and mathematics, and without exposure to science and mathematics students risk being left out of the job market or attaining a good quality of life. The Malaysian government's emphasis on science and mathematics education and its efforts to use ICT and e-Learning to foster an interest in science and mathematics education is indeed forward-looking. To achieve this aspiration, the Malaysian government made the bold decision to implement the policy of teaching science and mathematics in English at all levels of education.

However, this controversial decision had sparked off wide-spread oppositions from various quarters (Aglionby, 2002; Yoong, 2005; Bernama, 2005; Zulkifli Musa, 2008; The Star, 2008; Suara Keadilan, 2009). Schooled in their mother tongue language, the majority of Malaysian students who had little or no supporting background to learn English are only learning English as a subject formally at the entry level of schooling. Teaching science and mathematics in English at the primary school level implies that one is not merely teaching the subject contents, but also the target English language system as well. It is also pedagogically unsound and unrealistic to expect the students to use English to acquire science and mathematics contents when they do not even have minimum proficiency in English. Science and mathematics subjects are cognitively more demanding and inherently more difficult for most learners, even when studied through their first language or mother tongue (Segalowitz, 1977). They are cognitively high demanding and require learning tasks that involve abstract and formal thought. It would therefore even be more difficult for students to learn science through a second language (English), especially at a very early stage. Most primary and secondary school teachers were also not proficient in English, and most primary school teachers have a poor grasp of science and mathematics contents (Yoong, 2005, 2007). Localised English textbooks, supporting materials, and courseware are also lacking. Given the great diversity of conditions faced and the varying literacy and schooling experiences of the learners in their home languages, a more balanced language literacy programme that gives due consideration to the learner's ability and language background while simultaneously assisting them to acquire basic proficiency in English so that learning science and mathematics would be effective in the long run is a more legitimate solution.

Despite wide-spread oppositions, the government still went ahead with the policy of teaching science and mathematics in English at all levels of education beginning 2003 and a two-prong strategy was adopted. The MOE intensified the short-medium term in-service training for science and mathematics teachers nation-wide to enhance their professional readiness to teach these subjects in English. In addition, a massive investment was launched to "jump start" the use of ICT to support teaching science and mathematics in English in schools across the country. Nevertheless in July 2009, the Malaysian government finally succumbed to public opinion and rescinded the policy of teaching science and mathematics at the primary and secondary school levels, which is to take effect from January 2012. It seems that the policy on the use of ICT to support the teaching of science and mathematics in English will have to be modified accordingly.

Although Malaysian science and mathematics teachers surveyed perceived that they were professionally prepared to teach mathematics and science in English, a large majority of them reported that their pre-service training could not develop their confidence in English speaking. This problem persisted even after the teachers attended the in-service training conducted by the MOE specifically for this purpose. The teachers needed much more training on helping students to overcome difficulties in learning the subjects in English, especially for students who are weak in English or mathematics and science or both (Noraini Idris *et al.*, 2007).

The massive investment to "jump start" the use of ICT to support the teaching of science and mathematics in English included the provision of laptop PCs, LCD projectors, and screens for use with specially designed teaching courseware to teachers of these two subjects. Moreover, they were also offered a critical allowance payment as an incentive. A launching grant of RM5000 to RM15000 was also given to each school to acquire additional reference resources. For schools involved in the Smart School Project pilot tryout, additional facilities like computer laboratories, wireless internet connection and local area network were already provided to assist teachers in their professional tasks.

USE OF ICT IN SCIENCE AND MATHEMATICS EDUCATION

Notwithstanding the weaknesses and teething problems faced by Malaysia in its Smart School initiative, ICT usage in Malaysian education is still impressive, if not laudable by international standards. The development of visionary national ICT policies and the establishment master plans to introduce ICT in education have undoubtedly led to significant improvement in the ICT infrastructure in Malaysian schools, which in turn stimulated and supported the culture of ICT in teaching, learning and school management. The enormous potential of ICT means that in the years to come, teachers and students would have ample opportunities to use these facilities to develop strategies for using ICT in education.

The Frost and Sulivan study (2005) commented on the practice that the Smart School teaching-learning courseware was locally relevant and tailor-made for the Malaysian curriculum while in most other countries learning courseware was private-sector-owned and developed for a generic market. Actually, Malaysia is a small country with limited market potential for private-sector-owned and developed instructional courseware. Since the purchase of available quality courseware developed in the West is expensive and not necessarily completely relevant, developing its own tailor-made learning courseware was the more viable alternative for Malaysia. Even then, the development cost of tailor-made courseware was not cheap. For example, the 1,494 courseware titles created for the Smart School Project tryout cost approximately RM1 million (Bismillah Khatoon, 2007). Adding to the woe was the government's abrupt decision to change the medium of teaching of science and mathematics to English, thus rendering a substantial number of Malay language courseware redundant.

A key issue for concern with any strategy of courseware development lies with the quality of the courseware that has to be developed. For learning courseware to be interesting, interactive, and stimulating, it must contain relevant visual presentation features, including animations that facilitate the understanding of concepts and processes. It may even use simulations and modeling to enhance interactive learning. For science courseware, using virtual environments to simulate experiments that would not otherwise be possible or dangerous to conduct in the laboratory offers new possibilities for data collection and experimentation. Alternatively, digital video may be used to capture experimental processes that cannot or are difficult to be seen in real time, thus allowing more time and opportunity for observation. There are many other possible innovative applications, too.

Quality courseware takes time, effort, expertise, and financial resources to develop, and the effectiveness of its use in improving learning have to be supported by independent research or evaluation before it is put to use. Standard practice of good quality courseware development generally begins with small scale prototyping prior to larger scale tryout. The problem with Malaysia's ICT in education initiatives was that the prototype phase was short-chained. When the government decided to implement the teaching of science and mathematics in English in all schools beginning 2003, the MOE had little choice but to jump straight into large scale implementation phase by producing many science and mathematics learning courseware in English within a short time frame. Outsourcing the courseware preparation through a single vendor (though the vendor is a consortium of companies) was the preferred choice. However, the contract for developing the courseware included only minimal service levels, most of which were restricted to meeting specifications and satisfying user acceptance tests and other technical criteria (Bismillah Khatoon, 2007). Although by contract, the intellectual property rights and copyright of the courseware had been assigned to the government, the vendor is not required to provide post-delivery services, such as updating the courseware and making corrections. The browser-based science and mathematics courseware was produced rather hastily and not independently evaluated. As a result of over-reliance on a single source of supply, deliveries to schools were also not always in time as some companies within the consortium could not produce the courseware on time and according to the quality specified (Bismillah Khatoon, 2007). Independent studies revealed that the pedagogical qualities of the most of the courseware were questionable in relation to the benchmark standards of cognitive loading, interactivity, effectiveness, and user-friendliness, (Ahmad Hanizar *et al.,* 2005).

For example, the science courseware for secondary year 1 students (age group 13+) consisted of eight main topics: physical quantities and their measurements, cell as a unit of life, living things and their classifications, matter, resources on earth, the air around us, sources of energy, and heat. Ahmad Hanizar *et al.* (2005) evaluated the courseware for their pedagogical and communicative configuration, instructional model, cognitive processes and loci of control, learning resources, evaluation, interactive nature of teaching, feedback, help functions and the extent of media utilisation involving texts, images, interactive images, animation and sound. The results revealed that the instructional configuration of the courseware was predominantly in the form of individual instruction rather than in the preferred collaborative learning format. The instructional mode most widely deployed was predominantly information-based, albeit with a low level utilisation of the favoured inquiry-based approach. Interactivity was mostly restricted to answering multiple-choice questions and performing simple activities. The utilisation of online tools and/or consultation with the experts was minimal. The cognitive processes involved low level functioning, especially: information retrieval and memorizing. The use of problem solving, decision making, creation, and invention possibilities was minimal. Nonetheless, the courseware did incorporate a combination of texts, images, interactive images, animation, and sound, to make learning interesting (Ahmad Hanizar *et al.* 2005).

Similarly, the physics courseware for secondary year 4 students (age group 16+) consisted of four browser-based titles involving 110 lessons: Courseware 1 (Linear Motion; Analysing Motion Graphs; Understanding Inertia; Momentum; The Effects of Forces; Impulse and Impulsive Force); Courseware 2: (Gravity; Forces in Equilibrium; Work, Energy, Power and Efficiency; Elasticity); Courseware 3: (Understanding Pressure; Pressure in Liquids; Gas Pressure & Atmospheric Pressure; Pascal's Principle; Archimedes Principle; Bernoulli's Principle); and Courseware 4: (Thermal Equilibrium; Heat Capacity; The Gas Laws). An analysis of the cognitive loading of these lessons showed that the demand was very high. A study of its use in a secondary school found the courseware to be problematic: students encountered cognitive overload when using most of the courseware lessons. Because the courseware was not developed on sound instructional design model, this made learning physics topics (which are highly conceptual) more difficult.

Providing teachers with relevant courseware had its advantages and disadvantages. Because most of the teachers were not proficient in English, let alone teaching these subjects effectively in English, the use of the English courseware was welcomed at the ground level. By and large, presentation technologies involving notebook or stand-alone PC and LCD projectors in tandem with CD-ROM courseware were widely used by teachers to teach science and mathematics in English. However, for teachers who lacked proficiency in science and mathematics and English, it was used to supplant them rather than assist them. Often many teachers used the ICT merely as a means of presentation, replacing the chalkboard or the over-head projectors. Thus, the teacher merely projected the courseware on the screen and the students used the courseware with limited guidelines or teacher supervision (Lee, 2009; Bismillah Khatoon, 2007; Chong, Sharaf & Daniel, 2005).

As the Smart School initiatives are top-down efforts, change management programmes were needed to instill the ownership of the Smart School initiatives among the various stakeholders, notably empowering the teachers and students in developing and managing virtual learning environments for independent and self-directed learning. The effective use of ICT in science and mathematics places considerable pedagogical demands on the teachers, and teachers are probably in the best position to know how a particular application will meet learning objectives. Needless to say, teachers are not only required to be proficient in English, but also be savvy in the use of ICT in teaching-learning. Moreover, teachers with background in ICT training and experience are more likely to implement higher levels of ICT integration in teaching-learning than teachers with little or no ICT training at all. Thus, continuous teacher training in a range of different ICT applications, with time for teachers to develop confidence by exploring them independently, remains a top priority for the Smart School Project to realize its goals, since the ability of teachers in integrating ICT to enhance students' understanding and making learning more effective and meaningful is very crucial.

The practice of outsourcing the development of courseware to large private consortiums, in a way, stifled the local innovative initiatives, notably the teachers and local academics in developing courseware that is most relevant to their classroom situation. In this sense, a significant sum of the large amount of money awarded to private consortiums to develop the courseware could have been used as incentives for ICT-proficient science and mathematics teachers or even university academics, either singly or in strategic alliances to develop tailor-made classroom relevant courseware. It is for this reason that we find it difficult to showcase quality science and mathematics courseware developed by Malaysian school teachers. Nonetheless, there are some exemplary and varied uses of ICT across the science and mathematics curricula that have been exemplified by academics from the local universities. Their impact depends on the context and the ways in which they were used.

One of the simple and yet most important applications of ICT in science and mathematics education is to surf the Internet for multi-media instructional resources, downloading text, audio, graphic, animation and software, and using them to enhance students' learning and understanding of scientific ideas more meaningfully and effectively. With sound pedagogical knowledge, such simple uses of ICT in teaching can have unexpected benefits. Other applications include the use of commercially available software or those developed by institutions of higher learning. The following case study examples illustrate some of these applications.

CASE STUDY 1: USE OF INTERNET RESOURCES TO SUPPORT CHEMISTRY TEACHING

The Chemistry Teaching Method (PGT 211E) course at the University of Science Malaysia (PPIP online, http://ppip.usm.my) required science education major trainee teachers to attend a course involving a micro-teaching session involving the use of ICT. One trainee teacher chose the topic of similarity of chemical properties of the Group 1 elements (alkali metals) of the Periodic Table, focusing on the trend in the reactivity of the elements down the group. The trainee teacher performed a demonstration experiment on the reaction of sodium and potassium with water, highlighting and demonstrating the safety precautions associated with using only a very small piece of sodium and potassium and setting up safety screens prior to putting the metals in a trough of water. Students were asked to observe the vigour of the reactions and take note of the colour change when the pH indicator was added to the resulting solution. After discussing the chemical properties of the reactions, the trainee teacher then showed a video clip he downloaded from the Internet.

The video clip showed the violent reactions of metallic sodium and potassium with water, in which the hydrogen liberated produced subsequent explosions, and even greater reactivity of rubidium and caesium by dropping them into a water-filled bathtub, generating explosions that shattered the bathtub. The students were impressed, as it was uncommon (expensive, and dangerous?) to perform the reactions of rubidium and caesium with water in the laboratory. The Internet resource enabled the teacher to show experiments that would not otherwise be possible. Follow up discussions and interactions led the class to arrive at the expected logical scientific conclusion: the reactivity of Group 1 metals increases down the periodic group.

Actually, the video clip downloaded by the trainee teacher was the alkali metal experiment from Science Abuse (Brainiac, 2004), a British entertainment TV education show. The Brainiac experiment aimed to illustrate the periodic trends in the alkali metal series, but used forged results (Wikipedia, 2008a). The Brainiac experiment purportedly showed that two grams of rubidium and caesium, when dropped into a bathtub filled with water caused large explosions that shattered the bathtub. Similar experiments with caesium or rubidium have been repeated by Gray (2008), MythBusters (Wilkipedia, 2008b), and the Periodic Table of Videos series at University of Nottingham (2008). In none of these cases were the rubidium and caesium reactions extremely violent, or as explosive as the Brainiac episode depicted. The MythBusters could not get the rubidium or ceasium to cause an explosion, even when they used 25 grams of each metal which was over ten times more than the quantity used in the Brainiac experiment. Instead, the chemical reaction caused a brief flame, and the release of hydrogen gas before fizzling out. The Brainiac staff admitted that the explosions had been faked, as very little occurred in the real reaction of rubidium or caesium and water, since the large volume of water over it drowned out the thermal shock wave that should have shattered the bathtub. Instead, the crew set up a bomb in the tub and used that footage to generate the explosion (Goldacre, 2006a; 2006b).

The Chemistry Teaching Method course instructor (Yoong, 2007b) took the opportunity to raise critical issues relating to the faking of the experiment with the students at the end of the micro-teaching session. The video clip was replayed and students were asked to observe the experiment footage carefully to look for visible clues of faking. One student noted that an 'explosives' sign could be seen on the premises. Another student pointed out that no exploding cloud of hydrogen gas, which was expected in a reaction between an alkali metal and water, was visible. The ICT resource actually stimulated discussions among the students. Debates that highlighted ethical issues relating to 'sanitising' or faking data produced by simulations to facilitate or support scientific ideas ensued. There were debates on whether the faking acts which were carried out with good intention were appropriate or not. Some saw the acts as complementing existing resources in filling gaps regarding teaching the 'right' scientific principles where conventional alternatives do not exist. This was especially necessary when eliminating experimental error and increasing visual impact to improve scientific understanding. Other students claimed that it was not authentic and could lead to misconceptions. It is also against the principle of intellectual honesty and can encourage the practice of forging data to fit the 'expected outcomes' when carrying out experiments in the laboratory.

This case study illustrates that pedagogical interactivity can be achieved even by using downloadable Internet resources to facilitate science teaching. This gives the teachers control over their teaching and consequently allows them to teach the topics more effectively.

CASE STUDY 2: LEARNING SCIENCE VIA VIRTUAL FIELD TRIP

The following case study (Salbiah Ismail, 2008) from the K-Perak e-Learning Cluster project provides a good illustration on the use of information resources via the internet in an urban premier primary school. The students participated in a virtual field trip to Karamea, New Zealand in 2005 to learn about the Blue Duck, a protected species. The virtual field trip was meant to help the students learn science in a more relevant and authentic context. A virtual field trip may simply be a photo tour of a famous place, or even extremely detailed and high-tech interactive field trips using video and audio segments with or without sounds and other special effects. Concepts taught in the virtual trip are also reinforced by creating a scrap book or keeping a journal about the trip. Prior to making the actual virtual trip, the teacher prepared the students in order to focus on particular points of interest using the following items:

- *A Diary* – completed by the virtual teacher everyday. Still images were included to explain aspects of the trip. The participating class also received emails summarising the daily activities, all of which were accessed over the Internet.
- *An Audio Conference* – an opportunity for the students to formulate questions based on their understanding of the background information provided before the field trip
- *Schedule* – a schedule of the main activities, location and audio conferencing topics, and topic access times.
- *Questions* – sent before the start of the Field Trip. All participating classes received copies of the questions before the audio conferencing commenced
- *Summary* – sent by the class and displayed on the website, and consisted of activities that helped strengthen the main concepts discussed during the audio conferencing
- *Streaming* – the audio conferencing was conducted using audio conferencing POLYCOM speaker phones and audiostreaming over the Internet (accessed by clicking on a hyperlink in the website).
- *Archive* – all audio conferencing sessions were made available after the activity, by accessing the archived version in the website.
- *Photo Gallery* – all images taken during the Field Trip were made available by password-enabled searches
- *Panorama* – panoramic images on the website gave a wider perspective of the location
- *Video* – a complete list of short video recordings gave insight into the places that were visited virtually, including the individuals who were interviewed, and the topics that were discussed. Each video recording was accompanied by captions.
- *Web Board* - the web board served as the place for sending additional questions to be answered by the virtual teacher or some expert in the field.
- *Ambassadors* - Soft toys were a popular component of the Field Trip experience. The soft toys accompanied the teacher conducting the actual Field Trip in New Zealand, and they were, therefore, able to see, hear, feel and smell everything that the students in Malaysia were not able to do so. Every ambassador had a website and an email address.
- *Competition* – Every Field Trip had a series of competitions and quizzes.

CONCLUSION

In the past, hardware, software and training limitations have tended to reduce the impact of ICT on science education. In recent years, however, hardware costs have fallen, hardware has become more reliable and the ICT skills of teachers have improved. The improvement in ICT infrastructure in many schools and colleges and in the ICT skills of teachers and students has been significant. The Malaysian government has made an ambitious and bold initiative to lay the foundation for ICT applications in education in Malaysia. But these initiatives were costly and fraught with financial and logistic governance problems. Much has been promised to ensure that all schools are equipped with the ICT infrastructure that covers both hardware and courseware and that teachers and administrators are adequately trained to acquire basic ICT proficiency. With the financial crisis looming ahead, Malaysia still faces an uphill battle to move into the advanced stage of fully integrating ICT into the education systems;

The significant improvement in the ICT infrastructure in Malaysian schools is only a first step in a long march towards the implementation of ICT in education. While efforts to integrate the use ICT in teaching-learning have improved as a result of the Smart School initiatives, they have not yet reached widespread proportions in Malaysian schools. There is enormous potential for Malaysian educators to stimulate and support a culture of integrating ICT in teaching, learning and school management. In the years to come, teachers and students would have ample opportunities to use these facilities to develop strategies for using ICT in education. There remains a considerable gap between the aspirations of experts and the realities of the classroom, and a key factor concerns teacher readiness. Findings on the impact of ICT in education in general, and science and mathematics education, in particular, are still largely tentative. Even less clear is the exact nature of the pedagogical principles associated with the use of ICT resources to achieve positive learning outcomes. Researchers and educators should continue to share their effective pedagogy and models of good practice through international collaboration.

REFERENCES

G8 Heads of State. (2000). *G8 Okinawa Communiqué.* Retrieved on December 30, 2006, from http://en.g8russia.ru/g8/history/okinawa2000/4/

Abd. Rahman Daud. (2000). Kefahaman terhadap konsep penggunaan ICT di dalam bilik darjah (Understanding the use of ICT concepts in the classroom). In *Proceedings of the International Conference on Teaching and Learning* (pp. 609-623).

Abdul Razak, H., & Jamaludin, B. (1998). Penggunaan komputer untuk pengajaran dan pembelajaran di sekolah menengah (The use of computer for teaching-learning in secondary schools). [Journal of Education]. *Jurnal Pendidikan*, *23*, 53–64.

Aglionby, J. (2002). *English in schools divides Malaysia: reports on an ethnic revolt over education*. Retrieved September, 26, 2002, from http://guardian.co.uk

Ahmad Hanizar, A. H., Muhammad, Z. M. Z., Wong, S. L., & Hanafi, A. (2005). The Taxonomical Analysis of Science Educational Software in Malaysian Smart Schools. [MOJIT]. *Malaysian Online Journal of Instructional Technology*, *2*(2), 106–113.

APEC, Asian and the Pacific Regional Bureau for Education. (2004). *Integrating ICTS in Education: Lessons Learned: A collective Case Study of Six Asian Countries*. Bangkok, Thailand: UNESCO

Azian, T. S. A. (2006). *Deconstructing Secondary Education: The Malaysian Smart School Initiative*. Paper presented at the 10th SEAMEO INNOTECH International Conference on Learning for Life: Creating Endless Possibilities in Secondary Educatio*n*, Pearl Hall, SEAMEO INNOTECH, Philippines.

Bernama. (2005). *Teach Science, Maths in mother tongue: Chinese movement. Daily Express, Wednesday, October 12, 2005*. Retrieved December 20, 2008, from http://www.dailyexpress.com.my/news.cfm?NewsID=37720

Brainiac: Science Abuse. (2004). *"Alkali Metals". 2004-09-02. No. 1, season 2*. Retrieved from http://video.google.com/videoplay?docid=-2134266654801392897

Chan, F.-M. (2002). *ICT in Malaysian Schools: Policy and Strategies*. Paper presented at a Workshop on the Promotion of ICT in Education to Narrow the Digital Divide, Tokyo Japan. Center for the Research and Support of Educational Practice. Retrieved October 30, 2008, from http://www.comminit.com/redirect.cgi?m=ac050d863535736c87376cac1c064d92

Chan, F.-M. (2006). *Case study one on ICT integration into education in Malaysia: The Malaysian Smart School Project*. Retrieve November 19, 2006, from http://www2.unescobkk.org/education/ict/resources/JFIT/schoolnet/case-studies/Malaysia_ICT.doc

Chong, C. K., Sharaf, H., & Daniel, J. (2005). A Study on the Use of ICT in Mathematics Teaching. [MOJIT]. *Malaysian Online Journal of Instructional Technology*, *2*(3), 43–51.

Chronicle, U. N. (2002). Partnerships: 'E-Learning for Life'. *UN Chronicle Online Edition, 39*(2). Retrieved November 28, 2008, from http://www.un.org/Pubs/chronicle/2002/issue2/0202p22_elearning.html

Fierce Wireless Daily. (2008). *2008 Asia - Telecoms, Mobile and Broadband in Malaysia and Philippines: Executive Summary (Posted July 14, 2008)*. Retrieved on December 20, 2008, from http://www.fiercewireless.com/

Frost & Sullivan. (2005). *Benchmarking of the Smart School integrated solution: Educating hearts and minds*. Malaysia: Multimedia Development Corporation.

Goldacre, B. (2006a, July 15). Sky's limit for big bangs. *The Guardian*. Retrieved November 23, 2008, from http://www.guardian.co.uk/science/2006/jul/15/badscience.uknews

Goldacre, B. (2006b, July 22). Smile while you're faking it. *The Guardian*. Retrieved November 23, 2008, from http://www.guardian.co.uk/science/2006/jul/22/badscience.uknews

Gray, T. (2008). *Alkali Metal Bangs and videos*. Retrieved November 23, 2008, from http://www.theodoregray.com/PeriodicTable/AlkaliBangs/

Intel. (2005). *3 million teachers help students learn to develop 21st century skills*. Retrieved January 2, 2006, from http://www.intel.com/pressroom/archive/releases/20051116edu.htm

Khatoon, B. bt. A. K. (2007). Malaysia's Experience in Training Teachers to Use ICT. In *ICT in Teacher Education: Case Studies from the Asia-Pacific Region* (pp. 10-12). Bangkok, Thailand: UNESCO.

Lau, B. T., & Chia, H. S. (2008). Exploring The Extent of ICT Adoption Among Secondary School Teachers in Malaysia. *International Journal of Computing and ICT Research, 2*(2), 19-36. Retrieved from http://www.ijcir.org/volume2-number2/article3%2019-36.pdf

Lee, J. (2009). Properly Using ICT in the Classroom. *Education in Malaysia Blog*. Retrieved February 3, 2009, from http://educationmalaysia.blogspot.com/2009/02/properly-using-ict-in-classroom.html

Mahathir, M. (1991). *Malaysia: The way forward*. Kuala Lumpur: National Printing Department.

Mahathir, M. (1998). *The Way Forward - Vision 2020*. London: Weidenfield and Nicholson Publications.

Malaysia, M. S. C. (2008). *What is MSC Malaysia / Facts & Figures. National Roll Out - MSC Malaysia Phase 2: MSC Malaysia NEXT LEAP (2004-2010)*. Retrieved December 20, 2008, from http://www.mscmalaysia.my/topic/12066956321774

Malaysia, Government of. (1996). *Seventh Malaysia Plan: 1996-2000*. Putrajaya, Malaysia: The Economic Planning Unit, Prime Minister's Department.

Malaysia, Government of. (2006a). *Ninth Malaysia Plan: 2006-2010*. Putrajaya, Malaysia: The Economic Planning Unit, Prime Minister's Department.

Malaysia, Ministry of Education. (1997a). *The Malaysian Smart School: A Quantum Leap*. Kuala Lumpur, Malaysia: Ministry of Education.

Malaysia, Ministry of Education. (1997b). *The Malaysian Smart School: A Conceptual Blueprint*. Kuala Lumpur, Malaysia: Ministry of Education, 11 July 1997. Retrieved November 28, 2008, from http://www.msc.com.my/smartschool/downloads/blueprint.pdf

Malaysia, Ministry of Education. (1997c). *The Malaysian Smart Schools Implementation Plan*. Kuala Lumpur, Malaysia: Smart School Project Team, Ministry of Education. Retrieved on November 28, 2008 from http://www.ppk.kpm.my/special/sslmPlan.pdf

Malaysia, Ministry of Education. (1997d). *The Smart School Concept Request for Proposals for Teaching and Learning Materials*. Kuala Lumpur, Malaysia: Ministry of Education, Malaysia.

Malaysia, Ministry of Education. (1997e). *The Smart School Concept Request for Proposals for a Smart School Management System*. Kuala Lumpur, Malaysia: Ministry of Education, Malaysia.

Malaysia, Ministry of Education. (1997f). *The Smart School Concept Request for Proposals for a Smart School Assessment System*. Kuala Lumpur, Malaysia: Ministry of Education, Malaysia.

Malaysia, Ministry of Education. (1997g). *The Smart School Concept Request for Proposals for a Smart School Technology Infrastructure*. Kuala Lumpur, Malaysia: Ministry of Education, Malaysia.

Malaysia, Ministry of Education. (1997h). *The Smart School Concept Request for Proposals for Systems Integration*. Kuala Lumpur, Malaysia: Ministry of Education.

Malaysia, Ministry of Education. (2001a). *A report on the collaborative monitoring of the Smart School Pilot Project for 2000*. Kuala Lumpur, Malaysia: Schools Division, Ministry of Education. (Translated title: Original in the Malay language).

Malaysia, Ministry of Education. (2001b). *A report on the collaborative monitoring of the Smart School Pilot Project for 2001*. Kuala Lumpur, Malaysia: Schools Division, Ministry of Education. (Translated title: Original in the Malay language).

Malaysia, Ministry of Education. (2003a). *Educational development plan 2001-2010*. Kuala Lumpur, Malaysia: Ministry of Education.

Malaysia, Ministry of Education. (2003b). *Benchmarking of the Smart School Integrated Solution*. Kuala Lumpur, Malaysia: Educational Technology Division, Ministry of Education.

Malaysia, Ministry of Education. (2003c). *User Acceptance and Effectiveness of the Smart School Integrated Solution*. Kuala Lumpur, Malaysia: Educational Technology Division, Ministry of Education. (Translated title: Original in the Malay language).

Malaysia, Ministry of Education. (2003d). *Education development plan: 2001-2010*. Kuala Lumpur: Ministry of Education.

Malaysia, Ministry of Education. (2006b). *Pelan Induk Pembangunan Pendidikan 2006-2010* (*Blueprint for Education Development 2006-2010*). Putrajaya, Malaysia: Ministry of Education.

MDeC, Multimedia Development Corporation. (2004). *Multimedia Super Corridor Impact Survey 2004: Performance of MSC-Status Companies in phase I (2003)*. Cyberjaya. Multimedia Development Corporation. Retrieved November 28, 2008, from http://cbdd.wsu.edu/edev/Nigeria_ToT/tr510/documents/UNDPAPDIP_ICT_in_Education.pdf

MDeC Smart School Department. (2005). *Malaysian Smart School Roadmap 2005 – 2020: An Educational Odyssey – A consultative paper on the expansion of the smart school initiative to all schools in Malaysia*. Cyberjaya: Multimedia Development Corporation. Retrieved October 30, 2008, from http://www.msc.com.my/smartschool/downloads/roadmap.pdf

Microsoft. (2007a). *Partners in Learning Progress Report 2007. October, 2007*. Retrieved November 28, 2008, from http://download.microsoft.com/download/2/c/5/2c5f1568-368a-45bf-9a1f-7b38935ae7cc/PiL_Report_complete.pdf

Microsoft. (2007b). *Press Release: Ministry of Education and Microsoft launch Generasi-M Kuala Lumpur, December 10, 2007*. Retrieved November 28, 2008, from (http://www.microsoft.com/malaysia/press/archive2007/linkpage4363.mspx

MOE Intel. (2007). *MOE Intel School Adoption Project Phase I: Project Report*. Malaysia, Ministry of Education and Intel Malaysia.

Mohd Jasmy, A. R., Ismail, M. A., & Norsiati, R. (2003). Tahap kesediaan penggunaan perisian kursus di kalangan guru Sains dan Matematik. (The level of readiness in using courseware among science and mathematics teachers). In *Prosiding Konvensyen Teknologi Pendidikan ke 16 (Proceedings of the 16th Education Technology Convention)* (pp. 372-380).

MSC. Multimedia Super Corridor, Malaysia. (1996). *Malaysia's Multimedia Super Corridor (MSC) Background. What Is The Multimedia Super Corridor?* Retrieved April, 1 1996, from http://www.jaring.my/~webmster/msia-new/rnd/msc.html

Nations, U. (2005). *World Summit on the Information Society: Tunis commitment*. New York: United Nations.

New Straits Times. (2008). *Auditor-General's Report: Hauled up over RM9.56 m 'missing' equipment. 30 August 2008*.

Noraini, I., Loh, S. C., Norjoharuddeen, M. N., Ahmad, Z. A. R., & Rahimi, M. S. (2007). The Professional Preparation of Malaysian Teachers in the Implementation of Teaching and Learning of Mathematics and Science in English. *Eurasia Journal of Mathematics, Science & Technology Education, 3*(2), 101-110. Retrieved from http://www.jmste.com

OECD, Organization of Economic Co-operation and Development. (2001). *The well-being of nations: The role of human and social capital*. Paris: OECD.

OECD, Organization of Economic Co-operation and Development. (2006). *Are Students Ready for a Technology-Rich World? What PISA Tells us*. Paris: OECD.

MHS Resources Sdn Bhd. (2005). *Preventive Maintenance Report (PM01) dated 17 Feb. 2005*.

Rosnaini, M. (2006). *Kesediaan teknologi maklumat dan komunikasi asas dalam pendidikan guru-guru sekolah menengah (Basic ICT readiness in the education of secondary school teachers)*. Unpublished doctoral dissertation, Universiti Kebangsaan Malaysia.

Rosnaini, M., & Mohd, A. Hj. I. (2008). Factors Influencing Ict Integration In The Classroom: Implications To Teacher Education. In *Proceedings of the 2008 EABR & TLC Conferences*, Salzburg, Austria.

Sa'ari, J.R., Wong, S.L., & Roslan, S. (2005). In-service Teachers' Views toward Technology and Teaching and their Perceived Competency toward Information Technology. *Journal of Technology, 43*(E), 1-14.

Salbiah, I. (2008). *ICT in the Classroom: A Malaysian Perspective*. Educational Technology Division, Ministry of Education, Malaysia. Retrieved November 28, 2008, from http://www.moe.gov.my/43seameocc/download/MALAYSIA-%20ICT%20and%20School%20Linkages.pdf

Segalowitz, N. (1977). Psychological Perspective on Bilingual Education. In B. Spolsky & R. L. Coopewr (Ed.), *Frontiers of Bilingual Education*. Rowley, MA: Newbury House Publishers.

Sharifah, M., & Lewin, K. M. (Eds.). (1993). *Insights into science education planning and policy priorities in Malaysia*. UNESCO, Paris: International Institute for Education Planning (IIEP).

Suara, K. (2009). *March 7 demo against teaching Science, Maths in English*. Retrieved January 21, 2009, from http://www.suarakeadilan.com/sk/english/2009/01/1960

Teoh, B. (2005). Revisiting Malaysia's Smart School Pilot Project. Monday, October 24, 2005 http://www.it-sideways.com/2005_10_01_archive.html

The Malaysian Smart School Portal. (n.d.) Retrieved from http://www.msc.com.my/smartschool/whatis/index.asp

The Smart School Roadmap 2005-2020: An Educational Odyssey. (n.d.). *A consultative paper on the expansion of the Smart School initiative to all schools in Malaysi October2005*. Retrieved from http://www.csdms.in/gesci/pdf/MALAYSIA-SMARTSCHOOLS-roadmap.pdf

The Star Online. (2008). *Vernacular schools prefer Maths, Science in mother tongue*. Retrieved November 1, 2008 from http://thestar.com.my/news/story.asp?file=/2008/11/1/nation/2431447&sec=nation

UNDP. (2006). *E-Learning for Life (Apple Malaysia)*. United Nations Development Programme Regional Centre Bangkok: Asia-Pacific Development Information Programme. Retrieved November 28, 2008, from http://www.apdip.net/projects/undp/my03/view

UNESCO. (2003a). *ICT Policies of Selected Countries is the Asia-Pacific*. Retrieved November 28, 2008 from http://www.unesco.org

UNESCO. (2003b). *Meta-survey on the use of technologies in education in Asia and the Pacific*. Paris: UNESCO

UNESCO. (2003c). *Trends in the use of ICT in Asia and the Pacific. 29 June 2003*. Retrieved November 30, 2008, from http://www2.unescobkk.org/education/ict/v2/info.asp?id=11012

UNESCO. (2004a). *Integrating ICT into Education: A Collective Case Study of Six Asian Countries - Indonesia, Malaysia, Philippines, Singapore, South Korea, Thailand.* UNESCO Asia and Pacific Regional Bureau for Education. Retrieved November 28, 2008, from http://unesdoc.unesco.org/images/0013/001355/135562e.pdf

UNESCO. (2004b). *The Malaysian Smart School Project. Case Study One on ICT Integration into Education in Malaysia*. UNESCOBKK. Retrieved on November 28, 2008, from http://www2.unescobkk.org

University of Nottingham. *The Periodic Table of Videos*. (2008). Caesium. Retrieved November 23, 2008 from http://www.periodicvideos.com/videos/055.htm

Wikipedia. (2008a). *Brainiac: Science Abuse Alkali metal experiment with forged results*. Retrieved November 20, 2008, from http://en.wikipedia.org/wiki/Brainiac:_Science_Abuse

Wikipedia. (2008b). *Viewer Special Threequel episode of MythBusters.* Retrieved November 23, 2008, from http://en.wikipedia.org/wiki/MythBusters_(2008_season)#Episode_114_.E2.80.93_.22Viewer_Special_Threequel.22

World Bank. (2003). *ICT and MDGs: A World Bank perspective.* Washington, DC: World Bank.

Ya'acob, A., Nor, N., & Azman, H. (2005). Implementation of the Malaysian Smart School: An Investigation of Teaching-Learning Practices and Teacher-Student Readiness. *Internet Journal of e-Language Learning & Teaching, 2*(2), 16-25.

Yoong, S. (2005). *Crossing Linguistic & Cultural Borders: Problems and Challenges in the Teaching Science and Mathematics in English.* Keynote paper presented at TEACH 2005 symposium on the Teaching of Science and Mathematics in English, Universiti Sarawak Malaysia, UNIMAS.

Yoong, S. (2007a). *Smart Schools for Science & Technology: The Malaysian Experience.* Keynote paper presented at the Turkev School of Science & Technology Symposium, Izmir, Turkey.

Yoong, S. (2007b). *Using ICT in Chemistry Teaching: Reactivity of Alkali Metals. Chemistry Teaching Method, PGT 211E: PPIP ONLINE.* Retrieved from http://ppip.usm.my

Zulkifli, M. (2008). Teachers Lack Good English Command. *Skor Career Blog.* Retrieved September 8, 2008, from http://skorcareer.com.my/blog/teachers-lack-good-english-command/2008/09/08/

ENDNOTES

1 The others are Electronic Government, Telemedicine (later renamed Telehealth), Multipurpose Card, Research and Development Clusters, Worldwide Manufacturing Web, and Borderless Marketing

2 RM or *Ringgit Malaysia* is the Malaysian currency equivalent of the dollar: RM1.00 ≅ ε0.22 ≅ US$0.28

Chapter 10
AVU's Experience in Increasing Access to Quality Higher Education through e-Learning in Sub-Saharan Africa

Bakary Diallo
African Virtual University, Headquarters Nairobi, Kenya

Sidiki Traoré
African Virtual University, Regional Office Dakar, Senegal

Therrezinha Fernandes
African Virtual University, Regional Office Dakar, Senegal

ABSTRACT

Universities and other tertiary institutions in developing nations around the world are facing major challenges in meeting the demand for increasing access to higher education (HE): limitations imposed by inadequate funding, poor infrastructure and sometimes lack of political vision, added to the demographic explosion, make it almost impossible for some of these developing nations to ensure access to all to higher education solely through the conventional face-to-face mode. In this context, the Information and Communication Technologies (ICTs) are providing an alternative to face-to-face education. Moreover, they have the potential to significantly increase access to quality higher education, improve management of tertiary institutions, increase access to educational resources through digital libraries and open education resources, foster collaboration and networking between universities, foster collaboration between the private sector and tertiary institutions, enhance sub-regional and regional integration and facilitate the mobility of teachers and graduates. In Sub-Saharan Africa (SSA), the African Virtual University (AVU), a Pan African Inter-Governmental Organization initially launched in Washington in 1997 as a World Bank project, works with a number of countries toward reaching the goal of increasing access to quality

DOI: 10.4018/978-1-60566-690-2.ch010

higher education and training programmes through the use of ICTs. The AVU has been the first-of-its-kind in this regard to serve the Sub-Saharan African countries. In this chapter, the AVU's twelve years experience in delivering and improving access to quality higher distance education throughout Africa will be discussed. The AVU has trained more than 40,000 students since its inception; this is the proof that it is possible to achieve democratization of tertiary education in Africa despite many challenges.

INTRODUCTION

A demand for the democratisation of the access to university has been formulated during these past two decades in developing nations around the world where a great part of the population has limited access to higher education. Universities and other tertiary level institutions in these countries are facing major challenges in meeting the demand: access limitations imposed by inadequate funding, poor infrastructures, sometimes lack of political vision and in addition to the demographic explosion, make it virtually impossible for some of these developing nations to ensure access to higher education to all solely through the conventional face-to-face model of education. These universities are thus faced with recordstudent enrolments that far exceed their existing capacity to accommodate and provide for effective learning.

In Africa, the demand for admission into universities has never been greater: there is a high demand for post-secondary tertiary and non-tertiary education and, as Africa's population continues to grow, the pressure for access to a university education is likely to grow. Akilagpa (2004) describes the challenges facing African higher education as follows:

- Inadequate resources to meet demand,
- Small number of universities per country,
- Small but rapidly increasing number of private universities,
- Low tertiary education expenditure per person, but very high relative to gross domestic product (GDP) per capita,
- Low household incomes affecting the ability of African students to attend higher education,
- Poor infrastructure for teaching, research, and ICT with consequent weak links amongst African Higher Education Institutions (HEI) and the global knowledge system altogether,
- Ageing faculties, lack of incentives to attract younger staff and continued brain drain,
- Inadequate financial and logistical support from governments,
- Weak linkages between academia and the social and productive (private) sectors of the economy.

To mitigate this challenge, an increasing number of universities have launched innovative methods of delivering education by embracing Open, Distance and e-Learning programmes to run alongside the conventional face to face programmes. In SSA,, one of the strategic objective of the the African Virtual University (AVU) is to help a number of countries to overcome obstacles by enhancing the capacity of universities so they can increase access to quality higher education programmes and training through the use of ICTs. The objective is to build capacity and support economic development by leveraging the power of modern telecommunication technologies to provide and/or improve access to quality education for large number of students and professionals learners (including those who otherwise would have been denied access). More so, it aims at bridging the digital divide and knowledge gap between Africa and the rest of the World by dramatically increasing access to global up-to-date educational resources.

In this chapter, a comprehensive and *qualitative* presentation of the AVU's methods of delivering high quality e-learning courses and increasing access to higher education during many years will be made to show how multiple literacy is being promoted.

AN ANALYSIS OF THE AFRICAN VIRTUAL UNIVERSITY MODELS OF DELIVERING E-EDUCATION SINCE ITS INCEPTION

Readers should be advised that a fair amount of the information reported in this chapter has been selected from (Universalia, 2005).

Introduction to The African Virtual University

The African Virtual University (AVU) was initially launched in Washington in 1997 as a World Bank project, with headquarters in Nairobi, Kenya and a Regional Office in Dakar, Senegal. The AVU is a Pan African Inter-Governmental Organization whose mission is to significantly increase access to relevant, affordable, quality and flexible higher education and training through the innovative use of modern information communication technology and innovative methodologies, all the while bridging the digital and knowledge divide between Africa and the rest of the world, as a response to the challenges of African universities. The AVU has been the first-of-its-kind in this regard to serve the Sub-Saharan African countries. It was also established to assist African countries in developing human resources that are critical in the transition of African economies to knowledge-driven economies. The main, but not sole, areas of focus of the AVU's interventions are:

- maximizing different modes of delivery in institutions through the use of ICTs so as to reach a large numbers of students and professionals simultaneously in multiple location,
- developing and disseminating Open Distance and e-Learning (ODeL) academic contents on the African Continent,
- building capacity in African tertiary education institutions through setting-up of state-of-art e-learning centres and training personnel to use ODeL methodologies, and
- managing the delivery of ODeL degree, diploma and certificate programmes in Africa, especially in the area of science and technology.

The AVU targets two key post secondary education student cohorts (Universalia, 2005). The first being students who have completed secondary school but do not access tertiary education for a variety of reasons, including economic reasons or inadequate space. The second cohort include those who have tertiary education and are looking professional development (which may include specific training, short-term courses).

To achieve its mission, AVU partnered with several stakeholder institutions and a network of more than 53 African Partner Institutions (PIs). Throughout its evolution, AVU has also benefited from the support of several international content provider universities, as well as various funding partners including the World Bank, the Canadian International Development Agency (CIDA), Australian Aid (AusAID), the Department for International Development (DfID), the Partnership for Higher Education, the African Development Bank (AfDB) and the United Nations Development Programmes (UNDP).

The AVU has acquired, over its 12 years of its existence, the largest network of Open Distance and e-Learning institutions in Africa: the AVU has expended its reach across Africa to an increas-

Figure 1. The African Virtual University presence in Africa. © 2004 The African Virtual University. Used with permission.

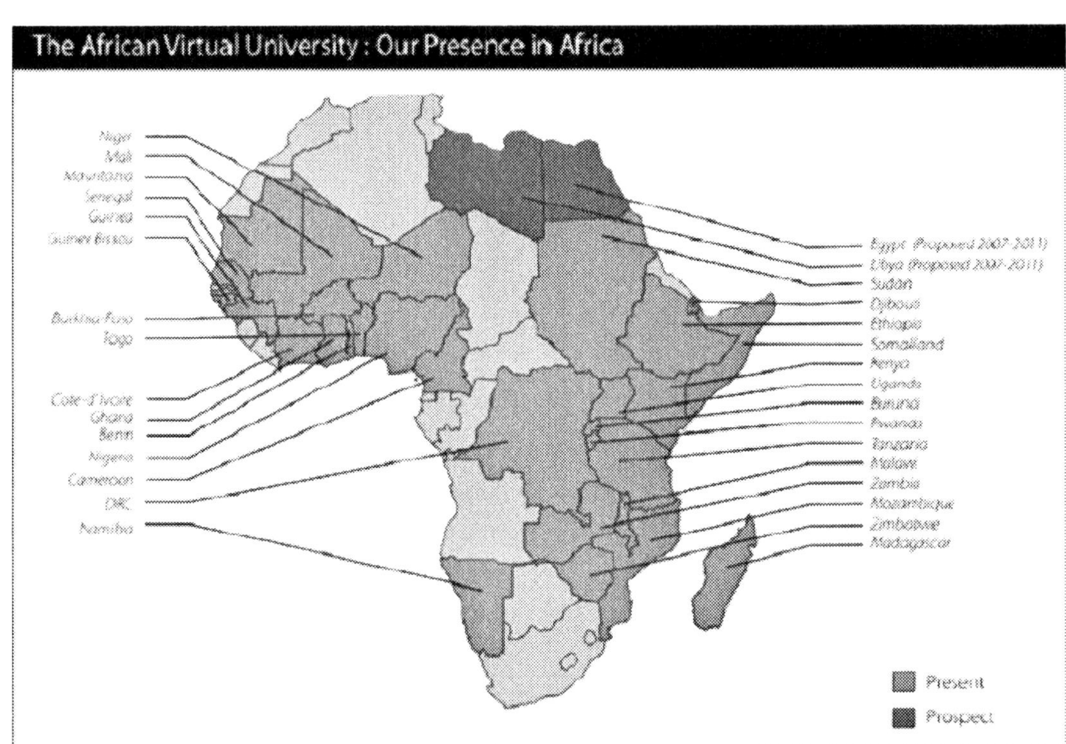

ing number of Francophones, Anglophones and Lusophones tertiary education institutions. It has established more than 53 learning centres (LCs) spread in 27 countries in SSA, the majority located on central campuses of African universities. AVU also have activities in post-conflict societies such as Mozambique and Somalia.

Background History of the AVU

AVU was initiated through a *Pilot Phase* (1997 – 1999) designed to confirm the functional and operational feasibility of using ICTs to enhance and disseminate knowledge on a large scale' (Universalia, 2005, p9) in SSA, to establish the operating structure of the organisation and to provide the foundation for implementation of the operational phase. The project was managed from the World Bank in Washington D.C. The delivery model used during this phase consisted of an integration of satellite and Internet technology to provide one-way video and two-way audio transmission of courses, digital and e-mail services. Course content was obtained from institutions in North America (U.S.A, Canada) and Europe (Ireland, Belgium) in subject matters such as Calculus, Physics, Computer Literacy, Biology, Chemistry and Mathematics and delivered to Learning Centres in AVU's Partner Institutions (PIs). All the courses provided during the pilot phase were free of charge for the AVU's partners..

AVU successfully piloted the delivery model and, altogether, 25 Learning Centres in 15 Anglophone and Francophone African countries participated in the pilot phase. During this stage, AVU:

- delivered in excess of 3000 hours of instructional programmes sourced from leading universities in North America and Europe,
- registered more than 23,000 students in its semester long courses,
- enrolled close to 2500 professionals in executive business seminars,
- provided 1000 personal computers to Learning Centers,
- achieved more than 40 per cent participation of women in AVU's pre-university courses at the most active Learning Centers.

Based on the success of the pilot phase, Vice-chancellors and Rectors of participating African institutions met and resolved to proceed to the Transitional Phase (2000 – 2002) of the AVU, characterised by the transition from the World Bank project to an independent organisation headquartered in Nairobi, Kenya. AVU was registered as a Non-Governmental Organization under African leadership in Washington D.C. on February 24, 2000 and later in Kenya on May 11, 2000. Its management structure comprised a Board of Trustees, which has appointed a Board of Directors, Chief Executive Officer and a management team. The AVU became an Intergovernmental organization on 2003 and is now governed by a Board of Directors.

The *Full Operational Phase* (2002) started in January 2002 and was marked by the beginning of 3 fully accredited degree programmes in Computer Science (Anglophone and Francophone Africa) and Business Studies (Anglophone Africa) and the deployment of an asymmetric Internet, video broadcast data network and Internet-based learning management system (WebCT). Students accessed course content 'delivered from the external content provider university through a mixed delivery mode using on-line, video and print material. They interacted with instructors through Internet-based technologies such as email, chat and discussion forums Universalia, 2005, p5). By June 2004, the AVU had over 1400 students enrolled in its degree and diploma programmes and 1270 students in the Short Professional and Continuing Education programmes, in 13 African institutions in 9 countries.

The External Partner University Model

From the start, the AVU realised the External Partner University (EPU) delivery model (Universalia, 2005). This pedagogical model is based on mixed delivery modes in which content from external university providers (e.g., USA, Australia, Canada and Europe) is made available to students on various university campuses in Africa in AVU's LCs. AVU has contracts with 3 external providers for the delivery of degree/diploma programmes: Royal Melbourne Institute of Technology (RMIT) and Curtin University, both from Australia, and Université Laval from Canada. Short-term courses were offered from North American (U.S.A and Canada) and European (Ireland and Belgium) universities.

AVU was granted non-exclusive licences for the intellectual property rights of the degree and diploma programmes externally delivered. AVU, through some of its PIs (the Lead Partner Universities, see below) was credited the right to locally deliver the content of courses on the African continent. Initially, three Lead Partner Universities (University of Dar Es Salaam of Tanzania for the Computer Science Diploma/ Degree Anglophone, University of Addis Ababa of Ethiopia for the Business Studies Diploma/ Degree Anglophone and Université Gaston Berger of Senegal for the Computer Science Diploma/ Degree Francophone), were expected to take on the delivery of programmes when the contract with the External Partner University expired. The idea was that the pipeline students and newly enrolled students would receive the Lead Partner University' Award (Diploma and Degree). The Lead Partner University model was rejected by

AVU Partner institutions and was abandoned in 2006 when the AVU shifted paradigm.

TECHNOLOGY: OVERVIEW OF THE AVU'S NETWORK INFRASTRUCTURE IN 1997

The AVU technological infrastructure back in 1997 consisted of a broadcast network with the up-link at COMSAT Tele-port in Clarksburg, Maryland, USA and multiple receive-only sites (AVU's learning centres) spread across SSA. This network utilized digital video and audio broadcast over the New Skies Satellite (NSS) 803 in C-Band. Full interaction was achieved through two-way audio utilizing telephone land lines as return links and also through the use of Internet technologies like email, forums and chat boards. This one-to-many nature of the network offers the cheapest way of reaching millions of people simultaneously in real time. *A proposed* minimum requirement to establish an AVU learning centre for running AVU programmes (AVU, 2001) included various technical specifications (eg C-band receive only antennae, internet connection, 64kbs, standby generators, etc) as well as physical room specifications (eg, Classroom, capable of taking a maximum of 50 students comfortably including LCD Projectors and screen and back-up TV monitors mounted so that students could view lectures without straining to see the screen. Computer Laboratory, capable of taking at least 25 students at a time with 25 Pentium PCs and 2 network printers installed and networked and a LAN servers).

Some LCs received material during the pilot phase from the World Bank for them to enroll in AVU's short courses. Along the years, the cost of the satellite broadcast with COMSAT became prohibitive and the Internet based network became more accessible and affordable to PIs. AVU then migrated to Internet based network and so did the LCs.

Pedagogical Approach

A flexible mixed mode delivery approach had been adopted in External Partner Universities, in recognition of the student-centred environment in which AVU operates. This approach aimed to maximize the use of technology in course delivery but it was also responsive to the needs of students for interaction and aware of the problems that may occur with technology from time to time.

The delivery model of the AVU which used a combination of on-line materials, synchronous and asynchronous video broadcasting, pre-packaged learning materials on CD ROMs and DVD, as well as synchronous and asynchronous chat sessions, is presented here. The following modes of delivery were applicable (RMIT, 2002):

- **Satellite TV broadcast and video:** Real time lectures. Some subjects included live broadcasts which provided additional resources in the form of lectures covering difficult material in depth, summary lectures at the conclusion of a module or an expert view of using the course material in an industry environment;
- **Internet:** In the Francophone Degree Programmes, the software 'Interwise' (AT&T, 2008) was used latter to deliver the courses. This technology allowed real time lectures to occur during which students could see the PowerPoint presentations, the videos, and the board where instructors write. The software 'Elluminate' (Elluminate Inc., 2008) is now used to deliver the Francophone Degree Programmes. This mode of delivery has the same features like Interwise. The prime delivery mode and repository of all course material through the learning management system was WebCT;
- **Course facilitators:** Each group of students is supported by an experienced facilitator at the local Learning Center and is

in contact with the lecturer delivering the course outside Africa via email or chat

- **Tutorials:** Classroom based activities which reinforce and explore the concepts covered in the material;
- **Laboratories:** In most courses, two hours of practical were scheduled each week for students to engage in problem solving exercises, tests and quizzes. Additionally, students undertook some private study and practice of concepts being taught;
- **Course discussion forums:** Social and academic interaction between all the students taking the same courses in different countries were provided through forum discussions. These provided an opportunity for learning through discussion with other students and with facilitators and lecturers from the External Partner University;
- **Email:** A form of communication which students used. to ask questions to their instructors;
- **Supplementary CDs:** Supplementary CDs were provided that contained both back up material and additional high bandwidth resources;
- **Videotapes** for students to watch lectures;
- **Textbooks** and supplementary notes.

Change of Paradigm

After several years of testing the External Partner University (EPU) model, the AVU noticed limitations with this model, recognising that its design was not suitable for the African context. Indeed, the EPU model relied on many assumptions that have proven to be inaccurate over the years (Universalia, 2005):

Technology: The EPU model did not take into account the lack of Information Technologies readiness of some African countries and, in many programmes, Internet connections were slow and inadequate. The model was very much technology-driven while some African countries still have limited control of satellite and bandwidth. The mode of delivery was heavily handicapped by its dependence on a large bandwidth. Also, the mode requires class attendance. Some consider this incompatible with virtual learning modes which requires flexibility and individualised learning. Class attendance is also a constraint for any large-scale expansion;

Costs: Another limitation of the EPU model is the high costs associated with the acquisition of rights to the various courses. Although it may well be true that from the point of view of the providers, the courses are genuinely not overpriced, especially given their own quality assurance requirements, they were unaffordable in Africa. The AVU tuition fees were high, given per capita income in the region, and as a result, a very limited number of students enrolled in the Programmes made the agreements between the EPUs and AVU financially unsustainable for the AVU;

Local Ownership: A serious shortcoming of the EPU model was the lack of ownership of course content by the AVU. The AVU acted primarily as a broker of content and this created frustrations in all the universities. The PIs wanted more content ownership of the e-learning programmes offered. In other words, they wanted African content and not necessarily foreign content.

In addition to that, the PIs faced some challenges on their own: inadequate academics and facilities, inadequate learning and teaching resources and connectivity to implement effective distance education programmes. Therefore, it was more than necessary for the AVU to rethink its strategy.

In the Accenture's report (Accenture, 2001) a shift in AVU's role was suggested, from direct service provider of degree programmes to architect and facilitator of an education network. In AVU's 2004-2009 Business Plan (AVU, 2004), a new delivery approach was proposed based on Open Distance and e-Learning (ODeL) methodologies. It was proposed 'that AVU enhances the institutional capacities of partner universities to utilise ODeL delivery approaches to offer their own qualifications. The aim of implementing a flexible, mixed mode methodology is to increase access to *demand-driven* programmes developed with partner universities and reach those located in rural and urban areas' (Universalia 2005, p7). The mission of the AVU then shifted from the EPU model of delivering brokered content to facilitating the ODeL model (see below). The AVU mission was then defined as follow: "The AVU intends to be a leading continental and virtual educational network collaborating with, and supporting African higher education institutions in enhancing their institutional capacities to utilize open, distance and e-learning delivery methodologies so as to dramatically increase access to high quality demand-driven graduate and undergraduate programmes in disciplines critical for Africa's social and economic development." (AVU, 2004, p2)

The Open Distance and E-Learning (ODeL) Model (Universalia, 2005; Diallo & Rasugu, 2008)

At the end of the first 24 months of offering internationally accredited programmes in Africa through the PIs, the shortcomings of this approach became apparent: the approach was costly, difficult to scale, economically unviable, did not lead to significant skill transfer and was putting more pressure on the physical facilities at the PIs. At the same time, it became obvious that the real problem was not the absence of programmes on African campuses but the availability of these programmes in modes other than the traditional classroom based methodologies. The AVU decided to commence work immediately with its partner universities to enhance their capacity to leverage the potential and benefits of ODeL.

The AVU has re-formulated its approach from brokered overseas programmes into content where African universities collaboratively develop and offer their own highly-demand driven programmes through the Open Distance and E-Learning (ODeL) approach. This approach deals with EPU model shortcomings and results in flexibility and adaptability to user's need. It also provides a sustainable model of learning that is well-adapted to the needs of Africa. This will enable programmes to be delivered in a cost effective manner and less resource-intensive.

AVU faces many challenges in course production. It also faces challenges with design for delivery in multiple countries. These countries have varied infrastructures for telecommunications and connectivity and different languages, learner support systems and other issues such as urban vs. rural, etc. But ODeL can be modified to address IT readiness in African countries. It is also demand-driven and custom made to meet the needs of Africa.

The AVU has refined its mission, vision and strategic objectives in 2007 in order to mitigate a serious financial challenge. The RMIT Computer Science programmes in Anglophone Africa did not finish as planned because of lack of funding. This created tension between AVU, RMIT and the concerned Partner Institutions. The externally brokered programmes were all heavily subsidised by donors such as Ausaid, the World Bank and CIDA.

The AVU's overall mission now is "to facilitate the use of effective ODeL methodologies in African Tertiary institutions". Its vision is to be the leading Pan-African Open Distance and eLearning network. The main strategic objectives are (Diallo & Rasugu, 2008):

- To increase access to high quality Open Distance and eLearning (ODeL) resources that are relevant to Africa,
- To enhance the ODeL capacity of African tertiary educational institutions,
- To enhance and sustain a network of Partner Institutions using ODeL,

- To build and sustain partnerships with institutions that can support the African Virtual University Mission,
- To carry out research and evaluation activities on ODeL on the African Continent,
- To build and sustain a committed and effective African Virtual University organization,
- To develop and implement a fundraising strategy in support of all of the above objectives with focus on Governments, the private sector and International Organizations.

AVU's key objective is to expand access to higher education to a greater number of Africans through the capacity enhancement of Partner institutions to develop and deliver ODeL degree/diploma programmes.

AVU'S PROJECTS AND PROGRAMMES

In this section, some of the achievements of AVU will be described. Included in these achievements are the projects that fall under the ODeL approach and the various diploma programmes from the EPU's model. The AVU has a number of significant achievements, which are briefly described below:

The AVU Capacity Enhancement Programmes (Rippon, 2005): Expertise from AVU global partners are utilized to enable the Partner Institutions (PIs) to own the process of curriculum development and implementation so as to increase equitable access to tertiary education. The expected developmental impacts are an increase in the number of courses offered utilizing ODeL methodologies and an improvement in the quality of learning and teaching using mixed mode delivery.

The AVU Bandwidth Consortium (AVU, 2006): Connectivity has continued to be a serious drawback for most of the AVU's Partner Institutions, hampering the delivery of programmes. However, the AVU has helped tackle this situation by negotiating an attractive contract for the supply of VSAT equipment and of satellite bandwidth as a solution to solving the problem of connectivity.

The Digital Library: The Digital Library provides large sets of digital resources for use by AVU's students and facilitators and constitutes a major component of the learner support system. It serves as a gateway to the world's virtual collections of scholarly information contained in vast databases of text and journals in French and English languages. The library targets the "open access" resources that are within the public domain, making it possible for the majority of African students to gain unrestricted access to the collection.

The Open Educational Resources Initiative (AVU, 2006): The Open Educational Resources (OER) Initiative was a two-year project which started in August 2006 and was funded by the William and Flora Hewlett Foundation's Education Programmes. It was a collaboration of the AVU and its partners within and outside Africa to ensure the rational and coordinated expansion of the OER movement in Africa. By examining the initiatives being undertaken and establishing the gaps that exist, the AVU intended to work systematically at creating an enabling environment for OERs to flourish across the continent in support of higher education and training.

The Teacher Education in Sub Saharan Africa (TESSA) Programmes (AVU, 2006): The TESSA Programmes aim is to research and develop high quality 'open content' multimedia resources and course design guidance for teachers and teacher educators working in Sub-Saharan African countries. The Portal build was being managed by the AVU during its inception. The Open University of UK took the lead of the project in 2007.

The Virtual Consortium Project (Diallo & Ragasu, 2008): The Virtual Consortium Project aims to offer accredited demand-driven graduate and undergraduate diploma and certificate programmes that are critical for Africa's socio-economic development. Already, 12 institutions from 10 countries in Africa that are participating in the AfDB/UNDP Multinational Project have signed up for membership. The AVU Virtual Consortium is a result of a formal request from Universities across Africa that is part of the Teacher Education programmes. The AVU will through the consortium coordinate a wide range of online programmes to be offered to students in Africa and around the world, thus offering students several options to pursue their studies. The AVU Virtual Consortium will first build on the existing Teacher Education Consortium and expand progressively to other subjects such as Business Studies, Computer Science, Health Science, Agriculture and Environmental studies, as well as Communities of Practice on eLearning.

Academic Programmes (AVU, 2005): AVU also proposed several online courses designed to increase individuals' knowledge and skills and leading students to a degree or professional certificate in many areas that are critical for the development in Africa. In these programmes, delivered in a distance-learning format, which accommodates multiple learning levels, while creating flexibility to suit individuals' work or life schedules, foreign content (from Canada, USA, Australia, etc.) were or still are brokered and offered to the AVU PIs. Those courses are intended to provide skills for career advancement. The AVU offered and still offers degree, diploma and certificate programmes to students and professionals across Africa. In collaboration with consortium of Africa and international Universities, the AVU has offered the following programmes through its LCs: Degree and Diploma Programmes in Computer Science (Anglophone) The Business Studies Programmes, started in February 2004 and ended in 2007, in partnership with Curtin University of Technology in Australia. It consisted of a 3-year Degree Programmes and a 1-year Diploma Programmes. The Business Studies Programmes ran at four partner institutions in Kenya, Tanzania, Ethiopia and Rwanda. The RMIT Computer Science Programmes started in 2003 in partnership with the Royal Melbourne Institute of Technology (RMIT) University and offered degree and diploma programmes in Computer Science under the following titles: Bachelor of Applied Science (Computer Science) and Diploma of Computer Science. There were nine participating institutions based in Kenya, Ghana, Ethiopia, Tanzania, Rwanda and Namibia. Unfortunately, the programmes did not finish as planned due to lack of funding. As a result, tensions arose between the parties. Degree and Diploma Programmes in Computer Science (Francophone)- The Virtual au Service de l'Afrique Francophone (VISAF) that later became the Programme d'Informatique de l'Université Laval à l'Afrique Francophone (PILAF) project consists in delivering a Canadian bachelor of computer science from Université Laval, Canada, to students from Francophone West Africa. The Programme is funded by the Canadian International Development Agency (CIDA) and managed by the African Virtual University (AVU) in partnership with the Association of Universities and Colleges of Canada (AUCC). The aim of this Programme is to train creative and competent African ICT professionals and Computers Science Engineers. This degree level is still being delivered to two cohorts of students in 9 Partner Institutions in eight countries: Senegal, Mauritania, Niger, Benin, Burundi, Cameroon, Burkina Faso and Mali. This project is a good example of how eLearning can help reduce poverty in Africa because these students are in a very good position to improve their own economic condition and that of their respective country as the majority of graduates intend to stay in their home countries and contribute to the local workforce.

Short, Professional and Continuing Education courses (SPaCE): The AVU also offered a wide range of certificate and professional courses to students and professionals. In partnership with international universities, the AVU delivered programmes in Computer Science, Journalism, Information Technology, and Business Communication.

The Teacher Education Project (Diallo & Wangeci, 2008): The AVU has an on-going Teacher Education (TE) Project for ten African countries developed collaboratively with 12 participating institutions. This is an initiative to develop e-learning compliant teacher education content. This initiative aims at enhancing the capacity of teachers in the use of ICTs in teaching and learning Mathematics and Sciences and developing their capacity to use ICTs in Education; developing and promoting research in teacher education in order to encourage evidence-based decision-making in all aspects of teacher development; and establishing and strengthening relevant partnerships with other teacher education initiatives in Africa and globally. The project covers ten African countries namely: Ethiopia, Kenya, Madagascar, Mozambique, Senegal, Somalia, Tanzania, Uganda, Zambia and Zimbabwe. The Teacher Education programme produced modules on ICT skills and ICT in the classroom to be offered as stand alone qualifications or components in broader programmes run by the PIs. This initiative has also developed 4 full Bachelor of Education (B. Ed) of Math and Science programmes (Mathematics, Physics, Chemistry and Biology). Innovative methods of integrating ICTs in Education will be also part of the B. Ed programmes. In-service teachers receive school-based education courses, while students who are not employed are able to study from their homes, workplaces or resource centres through the ODeL methodologies proposed by the AVU. This allows the programmes to reach participants in hard-to-reach areas where an important segment of them live and work. A total of 73 modules of Mathematics, Sciences, ICT basic skills and ICT integration components, and professional courses have been developed and released as Open Education Resources in June 2008. The programmes have also led to the establishment of a consortium working structure that cuts across the three languages in the participating universities. There has also been the establishment of a community of practice in Teacher Education who form a virtual network of academics working in different countries. As to mainstreaming gender, this initiative aims at encouraging more female students to enrol in the science-based education programmes and raise awareness about gender equality and women's empowerment including gender and HIV/AIDS. The activities under this initiative include: development of gender specific marketing materials for education programmes; convening gender training workshops; development of a gender module; implementing a scholarship scheme for female students.

These efforts show that AVU programs was attractive to students and corresponded to the requirements of the African context. The AVU matched the demands of African partner universities with content provider university courses. The AVU's programs addressed pressing needs in Africa, as well as the challenges of education in Africa. For example, the AVU's computer science and other technology related courses ensured that Africa gradually addresses the connectivity gap that exists with the rest of the world. AVU's courses in journalism are relevant in supporting good governance initiatives involving many African countries. Business courses also help strengthen private sector development in Africa, especially with regard to an entrepreneurship culture. The AVU's courses in business, journalism, and computer science addressed Africa's emerging civil society, private sector and technological sector. Consequently AVU's programmes matched the needs of the African continent.

THE BENEFITS

In this section, several benefits gained from the AVU's projects and programmes by the students and PIs are identified. Through its twelve years of existence, the AVU has been able to train some 40,000 students. This is a major achievement given the many challenges the organization has faced.

The PIs recognise that the major benefit resulting from their collaboration with AVU is that their involvement in delivering AVU programmes enabled their universities to introduce and promote e-learning and multiple literacy on their university campuses (Universalia, 2005). Nowadays, most universities and government officials believe that e-learning is very relevant to address the challenges facing the African tertiary education institutions and the shortage in teacher education. The partner institutions are now anchored in e-learning and in most of them; new initiatives in e-learning are led, with or without AVU's collaboration. Also 75% of students surveyed in (Universalia, 2005) strongly see e-learning as a delivery mode that supports learning and skills development.

There are several considerations from a multiple literacy angle (the development of technologicial, pedagogical, and subject knowledge.)

One key outcome of the AVU Capacity Enhancement Program is the professional development of the PIs personnel through various training offered by the AVU ODeL initiative. They gained a lot of relevant experience in e-learning and most are referred to in their respective institutions. Most of the Learning Center (LC) managers who had also gained much experience throughout the management of the programmes offered at their LCs have capitalized by that very valuable experience and are now considered as the expert in this domain in their institutions, and for some, in the entire country!

On a pedagogical aspect, the traditional teaching methods were challenged in the AVU Partner Institutions by the introduction from the AVU of completely innovative ways of doing things with the support of information communication and technologies. Several professors from the African institutions benefited from training delivered by the External Provider Universities and their idea of teaching were then changed.

As for the technological aspect, the AVU was permitted to correct the lack of IT-readiness of some of the personnel of the LCs and has contributed to the integration of ICTs at their universities. Computers were used for the first time by the personnel of the LCs at some Partner Institutions. Similarly, the Internet connection was established where none existed before. A different world was there for them to grasp. To help them do so, the personnel (mostly the technicians) of the LCs were trained by AVU to be able to work effectively with these new technologies. Most of the technicians were then regarded as very important assets (which caused some recurrent problems at the PIs as the turnover of technicians was high: the majority of the technicians left the LCs for better jobs).

Another very important thing to point out is the contribution of the AVU in reducing the brain-drain from Africa. The AVU's model of delivering courses contributed to reducing the brain-drain of SSA as students do not have to leave their countries to access high quality programmes. As the brain-drain in Africa is a very important matter which needs to be addressed, AVU's approach of education is to be considered as a relevant solution.

Very important also is the capacity of AVU to reach students in conflict or post-conflict zones as in Burundi, Somalia, Democratic Republic of Congo and Mozambique as teachers located in safer zones can still teach to the students via the ICTs. For example, in 1999, AVU's students from Burundi kept attending the AVU classes while the civil war was still going on. Also, the AVU has been playing the role of a Peace Maker especially in post-conflict zones like Somalia. Some seven Somali universities are part of the AVU 53 Universities in its network!

MAIN CHALLENGES ENCOUNTERED (UNIVERSALIA, 2005)

From the start to the present, the AVU's progress has not been smooth. It has dealt with several business plans and that together with the turbulent external context of Africa made it difficult to fully implement these plans. In addition, there has been a high turnover in senior leadership that has to a certain extent had a de-stabilising effect (Universalia, 2005).

On the quality front, technological development levels at African universities and partner countries resulted in delivery constraints. Students and facilitators and LCs managers were prompt to note serious limitations in technologies, hardware, software or access to adequate technological infrastructure, even in the LCs were some material were delivered by the AVU because the material was damaged at some point. These factors were beyond AVU's control or mandate. The ODeL learning architecture is addressing constraints and provides flexibility in delivery modes (Universalia, 2005).

Regarding cost-effectiveness, the AVU's External Partner University delivery model was not cost-effective. 'A cost effective programmes delivery model would have been one that allowed the organisation to maintain a low cost per successful student' (Universalia, 2005, p22). Add to that the use of expensive delivery technologies, the need for considerable facilitator support when the delivery technologies fail, course production, student recruitment, student services, marketing and the operating costs of the AVU and it is very difficult to be cost-effective. Worse, the model made it impossible for AVU to reach its goal to give access to higher education to the poor.

Students often had difficulties paying the high fees of the External Provider University model (mostly for degree and diploma programmes). Fees may be paid by students in irregular small installments. However some students are not accustomed to paying for education. Not surprisingly some LCs identified difficulties in fee collection, accounting practices and revenue sharing (Universalia, 2005). So, student fees collection becomes a difficult challenge for the AVU.

LESSONS LEARNED, RECOMMENDATIONS AND WAY FORWARD

One important lesson learned is that AVU has reconsidered its strategy as a broker of content. It was too expansive and AVU was often seen as a competitor by several African partner universities. The alternative proposed ODeL model has a more cost-effective approach. It also offers flexible delivery modes. These meet the needs of different target groups. The ODeL model could meet many of the criteria of cost-effective programmes. It deals with three main variables that influence cost effectiveness: student enrolment, sophistication of delivery media and face-to-face meetings between students and teachers and facilitators. To deal with the issue of students not being able to pay their fees, a student loan fund for e-courses could be developed. The AVU could create a system where students may be able to register once fees are collected (Universalia, 2005). Another important lesson learned is that it is possible to use ICTs to increase access to higher education, despite the many challenges facing the African continent. In this regard, the experience of the AVU is asset.

CONCLUSION

'The African Virtual University is a vision that speaks to the future of the African continent and the well-being of its people' (Universalia, 2005, p55) by enhancing the capacity of partner institutions to increase equitable access to students at a moderate cost.

Everyone now recognizes the importance of tertiary education and the formation of quality teachers both for meeting the Millennium Development Goals (MDGs) and for overall economic and social development. The African Virtual University's mission is thus relevant and very important in this regard. It responds to some pressing needs in Africa and it responds to the Poverty Reduction Strategy Programmes of the partner countries (Universalia, 2005).

The AVU, as a hub of its network, has a pivotal, catalytic and supportive role to play in the endeavour of Sub-Saharan Africa in regard to the improvement of its higher education and the training of its teachers in terms of the development of multiple literacy, because of the following (Diallo & Ragasu, 2008):

- Its experience in distance education and training and especially in utilizing ICTs in teaching and learning at the tertiary level in Africa;
- Its access to an African and global network of expertise in ODeL methodologies and practices; its ability to marshal technical, financial and human resources for the purpose of enhancing capacity in ODeL methodologies, systems and practices and implementing *e-Learning* programmes (such as teacher education) across its network;
- Its experience in brokering, purchasing, and implementing VSAT systems and managing bandwidth for tertiary institutions in Africa;
- Its experience in developing and setting up institutional, tutor and learner support systems and resources (such as the digital library) for distance education programmes that make use of ICTs in Africa;
- Its experience in developing cross border institutional partnerships (including consortium programmes) at the national, regional and continental levels in Africa;
- Its ability to identify and react to key challenges associated with higher education and training in Africa.

AVU faced challenges in introducing groundbreaking technology and it has struggled to change perceptions and behaviours. But organizations like the AVU are needed to meet complex innovation requirements of developing regions (Universalia, 2005). Since its inception, AVU has made significant efforts and positives changes in its Sub-Saharan Africa partner countries and should be praised for its perseverance.

REFERENCES

ACCENTURE. (2001). *African Virtual University Strategic Review, Final Report*. Boston, USA: Accenture.

Akilagpa, S. (2004). *Challenges Facing African Universities*. Accra, Ghana: Association of African Universities. AT&T. (2008). *AT&T Connect©. Conferencing for the Entreprise*. Retrieved March 2, 2009, from http://www.interwise.com/emc_solution_eroom.html

AVU. (2001). *AVU Equipment Technology*. Nairobi, Kenya: African Virtual University.

AVU. (2004). *The AVU Business Plan (2004 – 2009)*. Nairobi, Kenya: African Virtual University.

AVU. (2005). *Background Information*. Nairobi, Kenya: African Virtual University.

AVU. (2006). *Vice-Chancellors' Meeting Report*. Nairobi, Kenya: African Virtual University.

Diallo, B., & Rasagu, P. (2008). *Elsevier Connect Interview*. Nairobi, Kenya: African Virtual University.

Diallo, B., & Wangeci, C. (2008). *Brief of the AfDB/AVU Multinational Support Project*. Nairobi, Kenya: African Virtual University.

Elluminate Inc. (2008). *Elluminate Live! Participant Quick Reference Guide*. Retrieved March 2, 2009, from http://www.elluminate.com/support/index.jsp

Rippon, N. (2005). *AVU Capacity Enhancement Programmes (ACEP) Phase 1*. Nairobi, Kenya: African Virtual University.

RMIT. (2002). *Facilitators Training Manuel*. Melbourne, Australia: Royal Melbourne of Technology.

UNIVERSALIA. (2005). *Evaluation of the African Virtual University, Volume I, Final Report*. Montréal, Canada: UNIVERSALIA

Chapter 11
Multiple Literacies in the ICT Age:
Implications for Teachers and Teacher Educators, an Australian Perspective

Heather Fehring
RMIT, Australia

ABSTRACT

The exponentially changing world of the Information Age is reflected in the emphasis on multiple literacies and the impact of information communication technology (ICT) in teaching and learning practices in global educational environments. Students' learning, teachers' curricula and teacher education programmes are being adapted to these changing circumstances. The concept of multiple literacies has had a powerful influence on classroom practice. Multimodal and multidimensional curricula have become the standard for students from a very young age to lifelong learners. While discipline-specific literacies such as scientific literacy are widely acknowledged as essential components of a multiple literacies concept, notions of 'information literacy' have taken centre stage in discussions of students' ability to access, retrieve and critically evaluate the information that floods the ICT driven delivery modes of the 21st century. However, it is important to remember that learning is a complex process, and that "Who is looking after our children?" is still an essential question to ask.

INTRODUCTION

Students in the 21st century live in an amazing world dominated by Information Computer Technologies (ICT). However, their education is still driven by a 20th-century, outcome-based philosophy of curriculum design, in which the measure of academic capital is successful achievement of outcomes specified in an educational course or curriculum. In this new world of learning, ICT is perceived to be a way of overcoming issues of equity and achieving success for all. Online learning communities are seen as a major policy focus for many education systems. In Australia, as in other countries, a variety of policies emerged early in the twenty-first century to support and enhance educational opportunities in relation to ICT-based education, as for example: *Maintaining the edge: Strategy overview 2000–04, Backing*

DOI: 10.4018/978-1-60566-690-2.ch011

Australia's ability action plan and *Bridging the digital divide*. These policy initiatives were designed to achieve equity of access to information and communication technologies for all students, regardless of socio-economic status or geographic location. In November 2007 a Commonwealth Labor Party Government came to power in Australia. This new government announced its 'Digital Education Revolution', a programmes aimed at the delivery of "a world-class education system for Australia" (DEEWR., 2009, Overview section) and nothing less than a transformation of "teaching and learning in Australian schools that will prepare students for further education, training and to live and work in a digital world." Accordingly, the Australian Government promised $2 billion to provide for:

- the National Secondary School Computer Fund, providing grants of up to $1 million for schools to assist them to provide for new or upgraded information and communications technology (ICT) for secondary students in years 9–12
- the Fibre Connections to Schools initiative, a contribution of up to $100 million to support the development of fibre-to-the-premises (FTTP) broadband connections to Australian schools
- collaboration with states and territories and Deans of Education to ensure new and continuing teachers have access to training in the use of ICT that enables them to enrich student learning
- $32.6 million over two years to supply students and teachers with online curriculum tools and resources to support the national curriculum and conferencing facilities for specialist subjects such as languages
- the development of online learning and access that will enable parents to participate in their child's education
- $10 million over three years to develop support mechanisms to provide vital assistance for schools in the deployment of ICT provided through the National Secondary School Computer Fund. (DEEWR., 2009, p.13)

These points are but a few examples of government initiatives occurring in Australia and indeed, initiatives are occurring globally. They serve to demonstrate the importance that governments place on the development of educational systems and infrastructure geared towards new generations of ICT users. Naturally, the government's emphasis on ICT and learning is reflected in school classrooms, with curriculum being adapted and created to incorporate the new technologies available. This in turn has changed teacher education institutions programmes which prepare pre-service teachers for the world of schooling.

The "digital revolution" has changed the nature of literacy and impacted not only on the learning capabilities of students, but also, on teaching practices for the global education community. Teachers and teacher educators have had to come to terms with new generations of learners, and this has involved a revised notion of what it means to be literate. The emergent phenomenon multiple literacies involves changing notions of skills, practices, knowledge, and ways of socially and culturally communicating on a global basis (Anstey & Bull, 2006). New understandings have moved into the old space of literacy evolving into the concept of multiple literacies commonly used in the 21st century. The multiple literacies concept not only refers to reading and writing but also accommodates multiple modes of communication, such as the ability to be visually literate in terms of graphics and the many forms of multimedia communication. The concept of multiple literacies embraces other forms of literacy such as scientific literacy (Thomson & De Bortoli, 2008) and information literacy (Hancock, 1999; Lankshear, Snyder, & Green, 2000; Snyder, 1997). The dimensions of literacy have expanded considerably and are being incorporated into the daily routine of primary and secondary school students and teachers throughout the world in a myriad of ways.

As an academic working in pre-service teacher educator, in this chapter, I will deal with some of the challenges facing students and teachers in the Information Age. By exploring what is now involved in the new dimensions encompassing what we used to know as literacy the chapter scaffolds the concept multiple literacies and its importance to teacher educators. The integral relationship of multiple literacies and ICT is explored and together how they impact on educational environments and teaching practices. I also highlight some of the tensions for teachers and teacher educators that must not be forgotten when we are dealing with the diverse range of students' educational and learning needs. The chapter highlights the need for information literacy to become part of curriculum design and implementation in classrooms. The chapter also briefly explores the concept of scientific literacy as it relates to multiple literacies and discusses the changing notions and the contested nature of this term as it relates to teacher education.

BACKGROUND

The generic term 'literacy' often refers to reading, writing, speaking and listening in a written and oral language. However, literacy has been redefined many times over the years. In 1850, to be literate one had to be able to "Write (one's own) name and read aloud a passage from (the) Bible" (Gill, 1993, August 17, p. 13). As defined by UNESCO in 1951, "A person is literate who can, with understanding, both read and write a short simple statement on his everyday life" (UNESCO., 1990, p. 15) . By 2000, the OECD studies of literacy assessment used in the PISA (Programmes for International Student Assessment) research define literacy as:

> *... the ability to understand, use and reflect on written texts in order to achieve one's goals, to develop one's knowledge and potential, and to participate effectively in society. This definition goes beyond the notion that reading literacy means decoding written material and literal comprehension...*
>
> *The focus of PISA is on "reading to learn" rather than "learning to read". Students ... are expected to demonstrate their proficiency in retrieving information, understanding texts at a general level, interpreting them, reflecting on the content and form of texts in relation to their own knowledge of the world, and evaluating and arguing their own point of view.* (OECD., 2001, pp. 21 - 22)

In addition, learners in the Information Age must access, understand, transform and transmit information at an ever increasing rate. *Accessing* information requires identifying and finding printed, oral, and graphic information; *gaining* information requires comprehension, analysis, synthesis, and evaluation; *transforming* information requires writing, speaking, and representing; and *transmitting* information means publishing or disseminating transformed knowledge. The term *multiliteracies* is used to describe these multiple abilities. (Kibby, 2000, p. 380)

Such changing concepts of literacy has engaged students and teachers in the Information Age to become involved in a powerful and challenging reorientation to the creation of learning environments. ICT is leading educational innovation communication possibilities with Wikis, Blogs, Vlogs, eLearning, mlearning (mobile phone learning) and other new forms (still to be created) enabling the transformation of learning interfaces. These examples are illustrations of what is possible to incorporate into the everyday curriculum of primary, secondary and tertiary learning environments. As educators working in the field of teacher education there are many challenging issues facing the implementation of the multiple literacies classrooms of the future. The virtual world of cyberspace and social networking spaces such as Facebook, MySpace and even YouTube have created a new meaning to interactive communication. Teacher educators need to understand the possibilities of the constantly changing world of ICT and adapt curriculum and teaching and learning practices to accommodate what is now possible. However, teachers need to simultaneously challenge the inappropriate usage of ICT when it conflicts with sound teaching and learning principles which encompass the holistic education of children's diverse needs (Convery, 2009).

Context

"Cultural theorists such as Stuart Hall (1996) use the term 'New Times' to refer to the social, economic, political, and cultural changes which characterise the present" (Snyder, 1999, p. 1). This much contested catch phrase quickly became the cliché 'New Times, New Learning'. However, as teachers and teacher educators there is a real need to respond to these changing needs and new challenges. Our first challenge is to understand the nature of these new concepts and what demands such knowledge brings to the teaching and learning curriculum we design and implement. We are faced with a number of questions, such as the following:

- What does the term multiple literacies mean?
- How can we teach for multiple literacies?
- What are the expectations of multiliterate learners?
- What is the impact of ICT on the creation of learning environments?
- The following sections of this chapter address these questions beginning with an exploration of the concept 'what does multiple literacies mean?'

WHAT DOES MULTIPLE LITERACIES MEAN?

There are in a sense two interlocked components of the concept multiple literacies. One refers to the multiple forms of literacy now required to operate in the world of education and work. These concepts of multiple literacies embrace such terms as information literacy and scientific literacy and a host of other discipline-specific literacies. Information literacy is now a skill that library users need in the 21st century. Many major libraries around the world have an information literacy website to assist current users (CQU., 2009; RMIT., 2009a). The dynamic nature of scientific literacy and its changing conception is a reality for students and teachers to embrace (Brown & Krumholz, 2002; Kolsto, 2001; Laugksch, 2000; Murcia, 2005a; Thomson & De Bortoli, 2008).

The other concept of multiple literacies refers to the multimodal and multidimensional aspects that the learning of literacy skills now encompasses (Anstey & Bull, 2006; Healy, 2008). Not only speaking and listening skills and reading and writing in different paper based text genre; but also reading visual images, responding to multimedia texts and interacting, via hyperlinks, with different audiences.

The original term 'Multiiteracies' is attributed to a group of international literacy educators who met in 1994 in New London, Hampshire, USA (New London Group., 1996). This group were acutely aware of the increasing impact of ICT, globalisation and social diversity on the teaching and learning of literacy. Since 1994 the concept has evolved and been used widely around the world to refer to the new demands of literacy required in the Information Age (Anstey, 2002; Cope & Kalantzis, 2000; Unsworth, 2001). Perhaps the best way to address the question 'what does multiple literacies mean' is by providing an example. In Victoria (Australia) the curriculum for Year 11 and 12 is managed by the Victorian Curriculum Advisory Authority (VCAA). Students who sit

Figure 1. Part of the 2006 GAT Writing Task 1 Component (VCAA, 2006). Sourced from the General Achievement Test (2006), (c) Australian Council for Educational Research, illustrated material in the public domain.

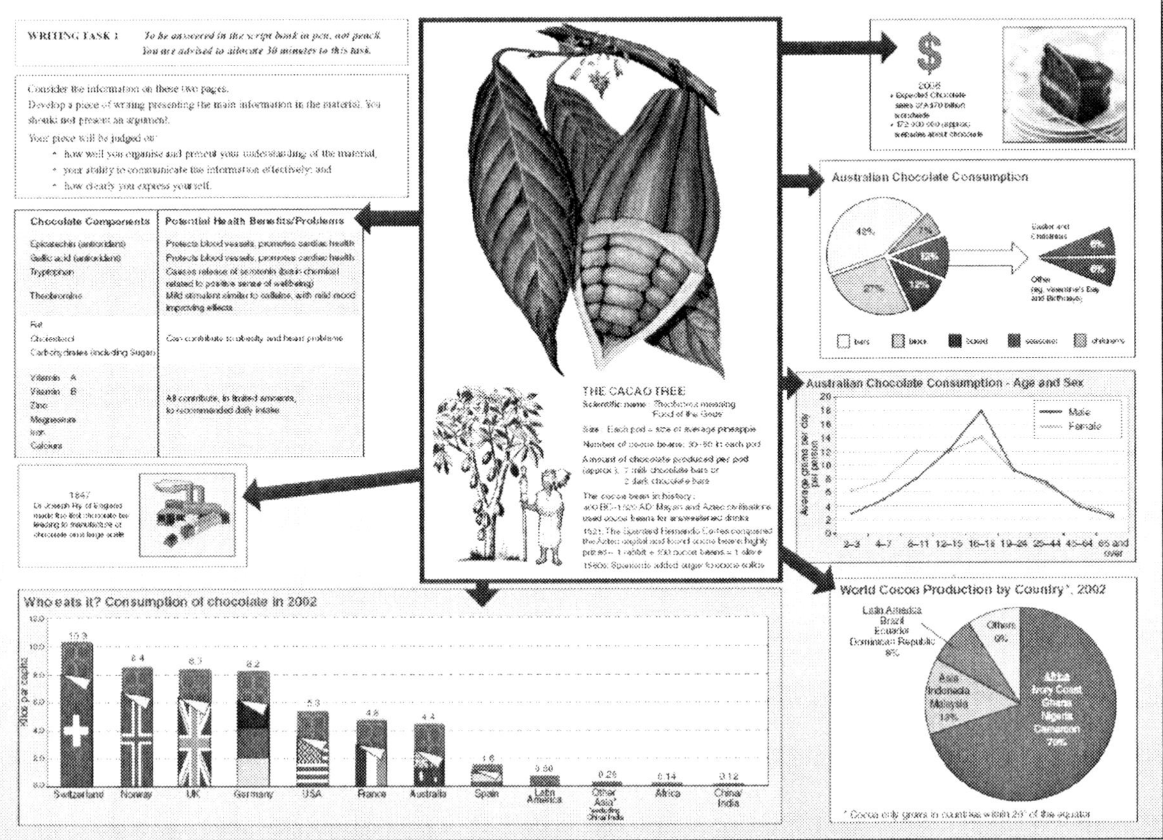

the Victorian Certificate of Education (VCE) examinations must undertake a test know as the GAT (General Achievement Test).

The General Achievement Test (GAT) is a test of general knowledge and skills in: written communication, mathematics, science and technology, humanities, the arts and social sciences. ... How is the GAT used? The VCAA will use students' GAT scores as a basis for: contributing to statistical moderation of School-assessed Coursework; reviewing school assessments in School-assessed Tasks; checking the accuracy of examination marking; calculating the Derived Examination Scores (VCAA., 2009).

The following illustration (Figure 1) is taken from Part 1 of the Writing Task 1 component of the GAT in 2006 (VCAA., 2009). The example was a double-page illustration asking students to: "Develop a piece of writing presenting the main information in the material. You should not present an argument. Your piece will be judged on: how well you organise and present your understanding of the material, your ability to communicate the information effectively; and how clearly you express yourself" (VCAA, 2006, p. 2 - 3).

The format of the GAT Writing Task 1, such as the above example, began in the early 2000s (VCAA, 2005, 2006). Prior to 2000 written passages were provided for students to read and construct a variety of different responses about in a variety of genre. This illustration above is an excellent example of the multidimensional component of multiple literacies (although acknowledging that the test construction requires unimodal responses due to the paper and pencil based construction of the testing situation). The example demonstrates the literacy requirements facing and challenging students of the 21st century. Students must be able to not only read text, graphs, charts and visual representations as part of everyday literacy tasks but also they need to be able to interpret, reconstruct and convey meaning to such diverse material. In internet, or online, communication tasks the literacy skills required extends to live audio visual interactions and multimedia tasks. The multiple literacies skills needed to complete this GAT task are complex and comprehensive. However, in the not so distant past and in fact there still are many literacy assessment procedures which require students to read a passage of text and answer multiple choice questions, matching question and answer responses or write short answer responses. These single dimension assessment tasks are being phased out as the influence of technology exponentially increases the capacity to present multidimensional and multimodal material. Furthermore, as the economic and cultural worlds of work demand such skills students and teachers are responding to this impact of ICT. Such multimodal and multidimensional influences on the notion of what it means to be literate have had correspondingly enormous influences on curriculum design and teaching and learning practices in the classroom. Such influences have encompassed moving from a more silo like orientation of isolated discipline specific design of literacy, mathematics and science subjects to an integrated cross-discipline approach to curriculum design. Cross discipline inquiry learning approaches, multilevel classrooms and the incorporation of ICT equipment into classroom curriculum has facilitated this change. However, such changes have raised a number of issues related to teachers' curriculum design knowledge, the availability of high technology software programmes and the corresponding computers equipment to run such programmes efficiently and effectively and flexible teaching and learning spaces to accommodate such learning environments.

Issues to be Addressed

The following examples illustrate something of the complexity of issues surrounding multiple literacies and ICT in the classroom. Scenarios typical of many Australian schools indicate a number of challenges facing educators in terms of both technological access and curriculum design considerations.

It is not uncommon in any primary school classroom in Australia to see students working on integrated curriculum projects which involve complex and multimodal presentations. For example, a Year 5/6 SOSE (Studies of Society and Environment) and Science Unit[1] incorporating *Rights and Responsibilities* and *Sustainable Conservation* might involve students in the following sequence of activities: In small groups children begin by brain storming and designing a "concept map" of their collective knowledge and of the possible features to be researched. This requires flexible classroom design features.

Students might then research their chosen topic using a variety of resources; Internet and WWW, CDs, text books, TV programmes, relevant organisations and associations and computer programmes. Teachers support and encourage students to analyse, classify, order and synthesise the information researched in support of their argument/s. This requires high level computer equipment and broad band access to be efficient.

Students are expected to critically evaluate the data collected and assemble credible and substantiated information. This process can involve multiple considerations such as linking cause and effect, comparing and contrasting information gathered, questioning the information for bias or prejudice, and identifying possible solutions to the original question asked.

Students would independently select the format for their final project presentation. This involves a twofold decision-making process.

First, there are decisions to be made about whether or not to incorporate a variety of software programmes suitable for the project. Primary school students now have a variety of options to choose from Kid Pix Studio Deluxe; Inspiration (which can create flow charts, diagrams and concept maps see https://www.inspiration.com/); MicroWorlds (to produce multimedia diagrams); KAHooTZ (the Australian Children's Television Foundation see http://www.kahootz.com/kz/ has developed a fantastic online educational programmesme. Students can create interactive communication environments using their own genre and incorporating digital photography, animation, music, quizzes and a variety of other multimedia format) (Janssens, 2002); Publisher (to design a brochure); PowerPoint (which can not only create interesting slide presentation but also has the capacity to incorporate multimedia material to support an argument being formulated); hyperlinks to relevant web sites, Excel and in some case for the more experienced ICT users, Access data bases to create spreadsheets and data charts.

Secondly, what will the final production encompass? Will the project be an interactive Blog, will it be a group web page on a school server, will it be a paper-based report, will it be a video production, will it be a CD-ROM, or will it be a combination of a variety of presentation media? Once again, this requires high level computer equipment and broad band access to be efficient.

This scenario illustrates a dynamic learning environment. However, a balance needs to be taken into consideration when educators are planning curriculum studies and learning environments for students in the Information Age. Not everything can, or should, be embraced without question. Currently, a controversial practice is the use of SMS and text-messaged forms of spelling, truncated and often vowel-less forms, which are seemingly necessitated by modern gadgetry such as the iPhone. Debate over the acceptance of SMS text in other contexts, such as the classroom, has raised much argument. Many challenges face teachers and teacher educators when introducing complex, multiple literacies-based learning environments. As teachers and teacher educators we must recognise the necessity of incorporating critical literacy into all curriculum subjects. We also need to understand the different reading skills required for screen-based versus paper-based texts. A number of researchers have begun investigating this very issue and have identified skills and strategies in reading programmes that are applicable to both modes. In addition, research studies have identified new skills that need to be incorporated into multiple literacies curricula (Bearne, et al., 2007; Lankshear, Snyder, & Green, 2000; Snyder, 1997).

As teacher educators we need to be aware of both the positive and negative aspects of the influence of ITC on learning environments. The positive impact of ICT on students' learning can include the following benefits:

- The revolution in the speed of access to, and quantities of information that can be made available to students (Thomson & De Bortoli, 2008).The writing process becomes a much easier task to draft, revise, refine, edit and publish using the word-processing capabilities of computer technology.
- The power and scope of the features of multimedia computer software programmes such as KidPix, Inspiration and KAHooTZ (Australian Children's Television Foundation) (Hawken, 2008).

The possible negative impact of ICT on students' learning can include:

- The dehumanising of personal interaction and the reduction in face-to face social contact as witnessed by chat-room sessions;
- Although a much debated topic, there is perceived to be an increasing problem of being able to discriminate what is real and what is not in digitally enhanced/generated images (Beavis, 1997). Some digital environments are so realistic that it is almost impossible for the human senses to be able to differentiate real images from fictional images in the multimedia presentations (Anderson & Bushman, 2001; Clark, 1993; Funk, 1993);
- The ability to critically evaluate when assessing online sources of information;
- The extended hours of computer usage and possible health-related issues for the youth of the future;
- The increasing and detrimental effects of "cyber bullying"(Australian Government Net Alert., 2007); and
- The exponential rise in plagiarism and the introduction, in Secondary schooling and Higher Education institutions of the compulsory use of such software programmes as 'Turnitin' (a programmes that identifies uncited use of material in a piece of work).

RESEARCH STUDIES INVESTIGATING MULTIPLE LITERACIES, TECHNOLOGICAL CHANGE AND TEACHERS

Research studies related to students, teachers and multiple literacies is a fairly recent area of investigation in Australia. Some studies research curriculum design and related classroom practices involving multiple literacies (Anstey & Bull, 2006; Healey, 1999; Healy, 2008; M. Kalantzis, Cope, & Fehring, 2002; Neville, 2004; Nixon, 2003). Other studies have investigated the interaction of multiple literacies and ICT learning environments (Hancock, 1999; Hytch, 2008; Jordan, 2008; Lankshear, *et al*., 2000; Snyder, 1997; Walters, 2007; Yelland, 2001). The research raises many issues for Australian teachers and teacher educators faced with rapidly changing curriculum environments, as the following examples will serve to illustrate.

Mary Neville (2004) reports on a project trialled in Queensland which combined an adaptation of the New London Group's 'Multi literacies' framework (New London Group., 1996) called Learning by Design approach (Kalantzis & Cope, 2004) and the Queensland Department of Education curriculum materials such as Literate Futures: Reading (Anstey & Bull, 2004). As Neville states:

The aims of the 2003-2004 project were to:

- Train a group of teachers in Multiple literacies pedagogical practices,
- Offer and encourage multimodal expressions of meaning and process of communication in the learning processes: linguistic, visual, audio, gestural and spatial, and
- Experiment with the *Learning by Design* (Kalantzis and Cope 2002 – 2004) curriculum documentation template. (Neville, 2004, p. 20)

Twenty primary and secondary teachers from 19 schools worked with staff from The Gap Cluster Learning and Development Centre (Literacy) based in Stafford, Queensland. The teachers planned and implemented very successful multiple literacies units of work with their classes. These curriculum units involved the students in the knowledge processes – experiencing, conceptualising, analysing and applying covering a range of genre and the creation of multimedia projects such as PowerPoint presentations and Video documentaries.

Marlene Walters (2002-2007) undertook her doctoral research investigating how learning could be enhanced through utilising information and communication technologies (ICT) and by incorporating the explicit teaching of thinking skills across the curriculum using an inquiry learning approach with children in a Year One/Two classroom in Victoria, Australia. The research design was of a qualitative nature and incorporated a constructivist (naturalistic) inquiry process, case study and action research approaches and qualitative methods of data collection. Her classroom based research demonstrated that computer technology supported learning and was useful in developing the higher-order skills of critical thinking, analysis and inquiry by engaging students in authentic, complex tasks within collaborative learning contexts. The technology, together with thinking skills, allowed the students in her classroom to work at their own pace and encouraged them to take initiative and learn independently. During the implementation of this research, patterns of behaviour displayed by the participating students emerged. These major and observable changes were identified as stages of progress: discovering and engaging, demonstrating, analysing, and synthesising, signify the stages the participating students displayed. Walters described the stages as follows:

- **Discovering and Engaging:** students knew very little about thinking or ICT, but they were interested in both and began to participate and engage in related activities. All the participating students went through this stage and beyond.
- **Demonstrating:** students were beginning to use both thinking skills and ICT in their daily routines, but were still in the stage of trial and error and consolidating their understandings. All the participating students went through this stage and beyond.
- **Analysing:** students were using both thinking skills and ICT continuously. These had become intrinsic to their daily practices and were used as a matter of course. Most of the participating students were working within this stage by the end of 2002.
- **Synthesising:** students had established skills and were using them to explore and go further. Some of the participating students were operating within this stage by the end (Walters, 2007; Walters & Fehring, 2008, p. 18)

Robyn Henderson (Henderson, 2004) took quite a different approach in her research conducted with families and the exploration of the influences of multiple literacies (Henderson, 2004). As she states in her opening paragraph:

It's not unusual for a multiliteracies approach to literacy teaching to be equated with the literacies associated with computers and other information communication technologies. Although literacy educators need to take account of these emerging technoliteracies, a multiliteracies approach should encompass much more than that. What can sometimes be forgotten are the diverse social, cultural and literate practices of homes and communities - the literacy backgrounds that seem to often play a determining role in whether children are successful or not in school settings. (Henderson, 2004, p. 11)

Henderson's study involved children from itinerant and bilingual families. She asked the questions "Why is it that difference is not always visible to teachers? If recognising difference is one of the challenges of a multiple literacies approach, how can we work to address that challenge?" (Henderson, 2004, p. 11). She identifies some simple strategies for teachers to adopt when assessing the recognition of difference. These strategies include: *'Use a wide lens'* (Hill, Comber, Louden, Rivalland, & Reid, 1998) which refers to teachers considering sociocultural practices in both children's home and community contexts. Keeping an open mind and looking through multiple lenses before making assumptions about children's capabilities. Question assumptions one makes about children and their families. Identify students' strengths rather than focusing on deficiencies. Where possible talk to students and their families about their expectations, needs and aspirations. Henderson argues that " ... multiple literacies pedagogy incorporates critical framing, helping to show students that all literate practices are framed by social and cultural contexts" (Henderson, 2004, p. 13) is an important strategy. This research highlights the knowledge teachers need to know in relation to the sociocultural and linguistic diversity of students when planning multiple literacies classroom curriculum.

The research studies reported above highlight some of the understandings that as teachers and teacher educators we need to reassess in relation to multiple literacies and classroom practices. However, there are other, discipline related literacies, which embrace aspects of the concept of muliliteracies that also need to be examined and incorporated into our re-conceptualisation of teaching and learning environments encompassing the technological driven Information Age. Two very crucial discipline-related literacies that are particularly important are the areas of Information Literacy and Scientific Literacy.

INFORMATION LITERACY

Information literacy is a phenomenon of the late 20th century and is a term first coined by Paul Zurkowski in a 1974 report *The information services environment, relationships and priorities* ED 100391 (Bundy, 2004, p. 45). The concept now refers to a person who is able to:

- recognize the need for information;
- determine the extent of information needed;
- access information efficiently;
- critically evaluate information and its sources;
- classify, store, manipulate and redraft information collected or generated;
- incorporate selected information into their knowledge base;
- use information effectively to learn, create new knowledge, solve problems and make decisions;
- understand economic, legal, social, political and cultural issues in the use of information;
- access and use information ethically and legally;
- use information and knowledge for participative citizenship and social responsibility; and
- experience information literacy as part of independent learning and lifelong learning. (Bundy, 2004, pp. 3 - 4)

This all-encompassing concept is a reaction to the abundance of information now available from a huge variety of sources, which "raises questions about authenticity, validity, and reliability" (Bundy, 2004, p. 3). Access to information does not guarantee that critical understanding and meaningful learning will necessarily occur. Rather, students and teachers alike need to enter into new and challenging exploratory modes of learning (Chung & Neuman, 2007).

The impact on changes in practice related to the incorporation of multiple literacies has had far-reaching effects in not only primary classrooms but also, secondary and higher education institutions. The importance of understanding the concept Information Literacy has also impacted on classroom curriculum and library practices. As for example, a simple and elementary concept of challenging the source of the information that our students soak up as if by osmosis through internet searches presented by nameless individuals in data bases transported through hyperspace from a virtual reality are now built into many school programmes. The impact of information literacy has also changed the role of librarians. Teachers and librarians are now working together as curriculum design colleagues rather than independent educators and resource providers (Asselin & Lee, 2002; Herring, 2006; Kuhlthua, 2004). Li Wang (2006) reports an interesting study undertaken by Information Skills Librarians at the University of Auckland. The librarians have explored using socio-cultural learning theories employing learner-centred and collaborative learning approaches to redesign the information literacy courses offered by the library staff. The study demonstrates the following results:

> *... student-centred and activity-based approach reduces the time teacher-librarians spent delivering lectures and increases the time spent interacting with students whilst they are doing learning tasks. Therefore, the role of the teacher-librarians shifts from an authority to a facilitator. In the community-of-learners environment, knowledge is created in community instead of being transmitted. Students are encouraged to take initiatives and to express themselves and they are responsible for their own learning. Their lifelong learning skills therefore are developed.* (Wang, 2006, p. 8)

Bilal and Kirby (2002) undertook an interesting comparative study that "examined the success and information seeking behaviours of seventh-grade science students and graduate students in information science in using Yahooligans! Web search engine/directory" (Bilal & Kirby, 2002, p. 649) to a fact-finding task. This mixed methods study design involved 14 seventh-graders students and nine graduate students. The study investigated cognitive, affective and physical behaviours and examined searching and browsing moves; backtracking and looping moves; screen scrolling; target location and deviation moves and the time undertaken to finish the tasks required. As would be expected, the study found that the more experienced graduate students were more successful in finding the target answers and were more effective and efficient in using Yahooligans! However, the study did find that the graduate students and the seventh-graders shared common information-seeking behaviours. Furthermore, the study "... revealed that Yahooligans! had several limitations that affected children and graduate students' information seeking. System designers should improve the structure of keyword searching and provide intelligent interfaces that support children and adults' information seeking" (Bilal & Kirby, 2002, p. 667). What this study highlights is the fact that there is a need, not only for hardware designer, but also instructional designers, to be constantly reappraising the information architectural features of human and technological interfaces.

Knowledge of the meta-language of any data source or informational retrieval system, whether it is information literacy or scientific literacy, is imperative to the success of all teaching and learning activities. Research studies exploring the physical design features of the interface between the learner and the machine is another area that needs constant investigation. Furthermore, such research needs to explore the needs of all categories of learners; children, youth, adults, parents, workers, and the disabled.

Studies such as the ones reported in this chapter have led librarians to develop tools to assist students to be critically literate when it comes to using databases involving multiple literacies sources (CQU., 2009; RMIT., 2009a, 2009b, 2009c). Table 1 below is but one example of a checklist developed by librarians to facilitate students' abilities to critically evaluate information obtained from web-based online resources.

Many information management researchers have investigated what strategies children use to explore the internet, the vocabulary necessary to understand and successfully retrieve information and the issues faced by the students accessing online materials and exploring possible solutions to maximising successful learning (Bilal, 2005; Bilal & Kirby, 2002; Dresang, 2005; Herring, 2006; Madden, Ford, & Miller, 2007; Madden, Ford, Miller, & Levy, 2006; Wilson, 2006). These studies are often explorative and raise many ques-

Table 1. Checklist to evaluate WWW sites, developed by the RMIT University library personnel for use by Foundation Study students (year 11 and 12 secondary school)

Authority
Ask yourself who puts information on the internet? The answer is *anyone*.
When assessing a website can you easily identify who the author is? Is contact information included?
If you can identify the author you need to ask if they have the credentials and/or experience to write on the topic? Are they an expert in the field?
What is the authority and credentials of any contributing or sponsoring organisation?
Accuracy
Ask yourself what kind of universal standards exist for web publishing? The answer is that there are no universal standards.
You need to assess whether the information is of a suitable standard to use within a university essay or project.
Look to see if the information is consistent with other information you have found. This is particularly important with websites like Wikipedia. Is the information reliable and error free? Are there contradictions or gaps in the information? Don't rely on just one source.
Does the information provide citations? Are the conclusions based on evidence?
Currency
Ask yourself whether it is important for information on your topic to be up-to-date?
This is particularly important for science topics. Would it be useful to have information on a topic like genetic engineering from a website that was last updated in 2000? On the other hand, it might not be as important for a topic on an art movement from the last century.
Some indication of currency can be found by looking at when the webpage was created or last updated. This is often found at the bottom of the page.
You might also look to see if the website has a latest news section. Dead links are also an indication that the webpage is no longer updated.
Purpose
Ask yourself if the website has a bias or commercial interest? Is the purpose of the website to persuade, inform, parody, sell?
Are you going to go to a cigarette company website to find information on the health effects of smoking?
Going to an opposition political party website to find information on the prime minister may give you useful and interesting information, but be aware that the purpose of the website is probably to persuade you vote for an alternative prime minister.
Does the author's bias negatively affect the information being presented? Is opinion being represented as fact?
Information about the purpose can often be found in the "About" section of a website. Are the values/goals of any contributing organisation made clear? Check links to other sites.
Coverage
Ask yourself whether the information is relevant to your topic?
Does it offer something you don't know already? Is the information intended to be comprehensive or specialised? Is it well-written and is the style appropriate for the information?
What country/region does the information cover?
Intended audience
Ask yourself who is the intended audience of the information?
How would a website on the importance of washing your hands differ if intended for medical practitioners or for school children?
Does it contain jargon that only a specialised audience can understand? Is it pitched at a level useful for your purposes?

(RMIT., 2009b)

tions about design issues related to Information Architecture and the internet and its importance as one component of the information space. However, as such studies build on the techno-picture being constructed about user use we will gain a greater awareness about the teaching and learning requirements of both students and teachers. Curriculum design and implementation related to the multitude of avenues of access to information that ICT offers is and will continue to reflect this new knowledge.

Educating students about how to obtain information and developing strategies to critically evaluate online data and data sources has become a fundamental part of the education process and especially in relation to multiple literacies. As for example, Gordon, (2002) conducted a study with tenth-grade students and web-based searching and recommended the use of concept maps as a strategy to facilitate the conceptual skills critical to online searching procedures (Gordon, 2000). The principle of introducing critically inquiry processes now begins in Primary schools and continues with learners throughout their lives. Pre-service teachers need to become knowledgeable about how to, not only teach themselves information literacy skills, but also to be able to transfer such knowledge to the new generation of learners in our schools.

The second component of multiple literacies that is particularly important in this chapter is the area of Scientific Literacy. This element of multiple literacies is explored in the next section.

SCIENTIFIC LITERACY

Scientific literacy is not a new concept. It was documented in the literature in the late 1950s (Laugksch, 2000). However, the concept has evolved to encompass the broader scoiocultural issues facing society. Scientific literacy is also not an uncontested concept. The quest for science education reform parallels educational reform in other disciplines which have challenged the theory and practice debates for decades.

In the United Kingdom two significant reports (House of Lords Select Committee on Science and Technology., 2000; Millar & Osborne, 1998) lead the way in changing the thinking about the concept of and place of science in English schools. Traditionally, school science had been perceived as preparing future students for science related careers. A view which educationists increasingly felt did not reflect the reality of the politically changing global environment. The new UK reports reflected that science curriculum should enhance scientific literacy for all students and that science in schools must equip all students for science for citizenship. This change in thinking about the nature and conceptualisation of scientific literacy influenced the subsequent English National Curriculum implemented in 2006 (Burden, 2005).

The American Association for the Advancement of Science undertook two very significant explorations of what constituted a science-literate American: *Science for All Americans* or Project 2061(American Association for the Advancement of Science., 1989) and *Benchmarks for Science Literacy* (American Association for the Advancement of Science., 1993). These studies were based on a belief that the 21st century world would be shaped by modern science and mathematics and technological knowledge would be of fundamental importance for future generations of Americans. Alternatively, a more controversial conception of scientific literacy was forwarded by Deboer as a "broad and functional understanding of science for general education purposes and not preparation for specific scientific and technical careers" (Deboer, 2000, p. 594). This is not a universally accepted definition of the term but illustrates the contested nature of this concept. Bybee developed a hierarchical framework for scientific literacy which encompassed four dimensions:

1. Nominal
2. Functional
3. Conceptual and procedural
4. Multidimensional (Bybee, 1997;. Bybee, Powell, & Trowbridge, 2008).

The nominal and functional dimensions of his framework relate to the literacy aspects of vocabulary. This refers to understanding the meta-language of the discipline; being able to read, write, and enter into science discourses with meaning. These are everyday life applications of scientific literacy. The third dimension of his framework relates to understanding conceptual schemes and processes; the notion of scientific inquiry, scientific reasoning and problem solving and the application of such knowledge to real-life situations. The fourth and highest dimension, the multidimensional level of scientific literacy, relates to "... philosophical, historical, and social dimensions of science and technology. Here students develop some understanding and appreciation of science and technology as they have been and are part of the culture. Students begin to make connections within scientific disciplines, between science and technology, and between science and technology and the larger issues of social challenges (Bybee, et al., 2008, p. 91 - 92). If a researcher chooses only one dimension from this framework and uses this to define and then construct tests to measure scientific literacy the results may be quite different from the researcher who chooses an alternative definition of scientific literacy as the foundation of an assessment programmes. This situation is common to many of the large-scale research studies whether it is measuring multiple literacies, scientific literacy or mathematical literacy.

In the OECD Programme for International Student Assessment (PISA) testing programmes (OECD., 2006; Thomson & De Bortoli, 2008, 2009) the following definition of scientific literacy was used:

... an individual's scientific knowledge and use of that knowledge to identify questions, to acquire new knowledge, to explain scientific phenomena, and to draw evidence-based conclusions about science-related issues, understanding of the characteristic features of science as a form of human knowledge and enquiry, awareness of how science and technology shape our material, intellectual, and cultural environments, and willingness to engage in science-related issues, and with the issues of science, as a reflective citizen. (OECD., 2006, p. 12)

The emphasis in this OECD definition is more related to young adults' knowledge and process in the discipline area of science. This definition is more directed towards future study and work as citizens adapting to rapidly changing societal changes The OECD began the PISA project in 1997 and began testing from 2000. It has continued to assess 15 year old students' reading literacy, mathematical literacy and scientific literacy in OECD member countries every 3 years (Thomson & De Bortoli, 2008). Scientific literacy only became a major domain for the first time in the Programme for International Student Assessment (PISA) in 2006 (Thomson & De Bortoli, 2008, 2009). However, the primary focus of the OECD PISA testing project is on assisting governments to maximise the productive capacities of their educational systems. Questions guiding the development of PISA are the following:

- How well are young adults prepared to meet the challenges of the future? What skills do they possess that will facilitate their capacity to adapt to rapid societal change?
- Are some ways of organising schools and school learning more effective than others?
- What influence does the quality of school resources have on student outcomes?
- What educational structures and practices maximise the opportunities of students from disadvantaged backgrounds? How equitable is education provision for students from all backgrounds?
  (Thomson & De Bortoli, 2008, p. i)

PISA regularly surveys the knowledge and skills of an international sample of 15-year-old students (2000, 2003, 2006). In 2006, 57 countries and 400,000 students participated. In Australia, this involved 356 schools and 14,170 students (Thomson & De Bortoli, 2009).

As the PISA example demonstrates there is a political and operational definition of scientific literacy which does not necessarily resonate with the entire scientific community of scholars. Lau (2009) and Yin (2008) have undertaken interesting analyses of the changing definitions of scientific literacy in relation to the PISA testing programmes 2000, 2003 and 2006 and the results reported in relation to Hong Kong students (Lau, 2009; Yin, 2008). However, the driving force for reform in educational practices has always resulted in contested theoretical, philosophical, ideologies and applied practices.

As the PISA project continued and the dynamic nature of the concept of scientific literacy evolved what PISA demonstrates is that scientific literacy has taken on a new level of significance in the Information Age. Thinking scientifically, understanding and applying scientific knowledge and being able to critically evaluate the scientific information relayed by the media have emerged as part of the multiliteracies and ICT suite of curricula concerns (Jarman & McClune, 2007). PISA assesses three interrelated stands of scientific literacy: scientific knowledge or concepts, scientific processes (as for example, i) identifying scientific issues, ii) explaining phenomena scientifically, and iii) using scientific evidence) and scientific situations and context referring to the application and use of knowledge (Thomson & De Bortoli, 2008). The reason why scientific literacy is so important to the multiple literacies curriculum of the future is because "... globalisation and computerisation are changing the labour markets and societies, and a different set of skills will be needed by those entering such markets.... Students preparing for the work force of the future will need to be able to solve problems for which there is no clear solutions, and be able to communicate their ideas effectively" (Thomson & De Bortoli, 2008, p. 243). The results of the PISA testing of scientific literacy provide evidence for policy makers in relation to the provision of educational resources relevant to the 21st century. In addition, the data provides information for teachers and teacher educators in relation to designing culturally relevant curricula which reflects the demands of modern society.

There is a growing body of research studies that interpret scientific literacy to be more than knowledge of scientific terms and procedures. The current view of scientific literacy includes a greater emphasis on science's interaction with society, the role that science has in shaping the decisions, values and actions that people take in their lives. This broader view includes an understanding of the nature of science, values critical questioning of the assumed tenets of scientific theory and emphasises the importance of creativity and imagination (Kolsto, 2001; Laugksch, 2000; Murcia, 2005a). This view of scientific literacy sits comfortably with the idea of the significance of sociocultural practices within the underpinning theoretical constructs of multiple literacies. Murcia (2005a) a teacher educator, describes a contemporary framework of scientific literacy in the following terms:

Scientific literacy can be thought of as a blend of these three knowledge dimensions:

- Nature of science;
- Interaction of science and society and;
- Enduring and important scientific terms and concepts.

Scientific literacy is clearly about KNOWING but it is also about A WAY OF THINKING and ACTING.

Being scientifically literate requires the confidence, interest and or disposition to use or put into action a blend of these knowledge dimensions for engaging with science in context. As such, it requires the ability to:

- *Use science as a tool for inquiry or discovery;*
- *Use science for learning, informing or contributing to problem solving; and*
- *Critically reflect on the use or role of science in context.*(Murcia, 2005a, p. 5)

This framework for scientific literacy requires a reorientation to the teaching of science beginning in the Primary classroom of our schools. This in turn has ramifications for teacher educators in the preparation of pre-service teachers. Murcia (2005a) has undertaken work in the area of how to engage students in teaching and learning practices so that such practices contribute to scientific literacy. She suggests classroom strategies like student- centred learning and real world examples of science interacting with society would enhance student interactions with the science curriculum. As for example, the simple use of newspapers articles about advances in science would facilitate students' experiences of the relevance of science to everyday life (Murcia, 2005b). Kolsto (2001) examined the notion of the science dimension of controversial socioscientific issues as a way of developing teaching models to promote science education for citizenship. He developed a matrix of eight content–transcending topics which he believes would assist science educators in this process (Kolsto, 2001). Susan Rodrigues has investigated the criteria necessary for the notion of pupil appropriate context in science to enhance learning (Rodrigues, 2006). These studies point towards the development of what has been called authentic or real life teaching and learning practices. Integral to the teaching and learning practices in the classroom of course is the need to reflect and reconsider teacher education institutions and their preparation of teachers for science.

Multiple literacies, information literacy and scientific literacy are having a significant impact on the provision of future educational policies, procedures, practices and classroom curriculum implementation. The implications for teacher education is another essential component to be considered. The following section explores some of the ramifications.

IMPLICATIONS FOR TEACHER EDUCATION

Students and teachers have embarked upon a new journey of learning which as educators we need to understand so we can create best practice for all future learners. In creating best practice educators need to not only understand the technological capabilities of the ICT innovations, but also, the meta-cognitive literacy expectations and teaching and learning implications. The teachers' role as a transmitter of knowledge has changed to a facilitator and creator of independent learning environments. The diagram in Figure 2 illustrates the complex learning environments that teachers function within. The educational system in which a teacher operates is context dependent and this fact needs to be taken into consideration by teachers as they develop 21st century student-centred learning environments. In addition, discipline specific knowledge must be integrated with interdisciplinary understandings and high level communication skills to produce the learning environments which will cater for the needs of all stakeholders of the future.

Take for example, this simple illustration of a Primary school teacher designing curriculum activities for 8-year-olds as she takes on the role of a facilitator and creator of independent learning environment for her students.

A Year 3 Primary school teacher introducing a topic called Location. By introducing the students to the functions of an interactive white board and *Google Earth* the students can explore a bird's eye view of the world drilling down from Australia, Victoria, Melbourne and finally to their own school and classroom.

Figure 2. The creation of student-centred 21st century learning environments

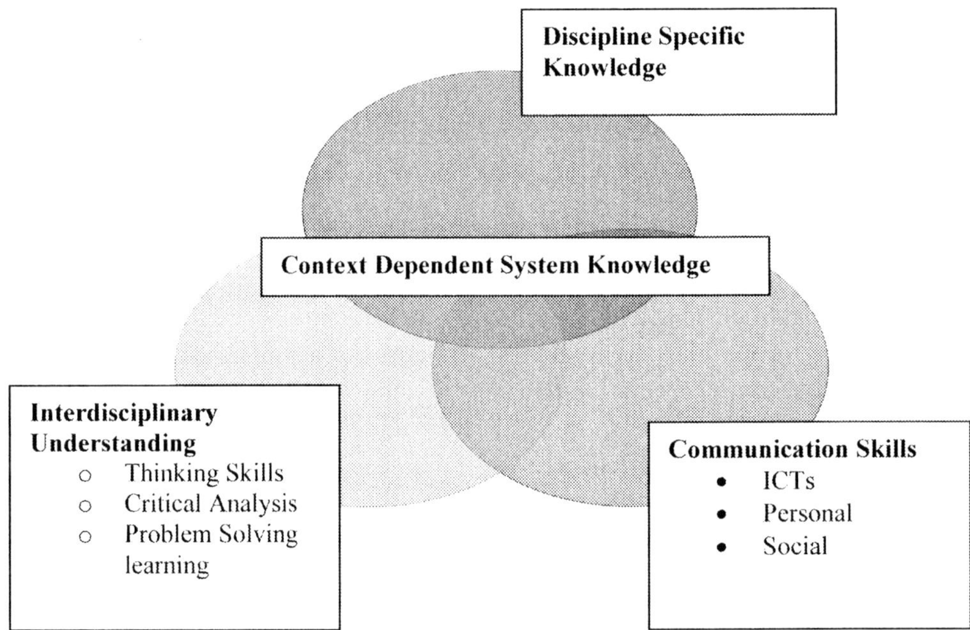

The educational system's policy and procedures provides the ICT equipment required to enable this activity to happen. The students explore different discipline knowledge through interdisciplinary understanding and inquiry processes developing their thinking skills, critical inquiry capabilities and communications skills.

There is a massive mind set change in knowledge for teacher educators and teachers and in the provision of new learning environments equipped with technology to implement this delivery of education. As Australian is, so are governments all around the globe, coming to terms with the technological demands that new learning environments require in the 21st Century (Australian Information & Communications Technology in Education Committee (AICTEC), 2008; Curriculum Corporation for MCEETYA, 2008; MCEETYA and MCVTE, June, 2008a, June, 2008b). Such government policies are coming to terms with the provision of the ICT hardware necessary to permit the creating of new learning environments. However, teachers and teacher education are coming to terms with the impact on students' learning and the creation of best practice in the classrooms of the future. Considerable work has been undertaken in the area of teacher education and teacher educators by Susan Rodigues (Rodrigues, 2005). She has investigated the requirements of continuing professional development for teachers related to science and ICT (Rodrigues, Marks, & Steel, 2003) and the need for the constant professional renewal of teacher educators who must respond to, and reposition themselves, to the changing needs of the multiple science communities of users (Nixon, Cope, McNally, Rodrigues, & Stephen, 2000). This valuable work needs to continue so we can draw together ICT, multiple literacies, information literacy and scientific literacy to develop best practice for the future. However, there is also a need to revisit the role of teachers' duty of care in relation to children and the creation of learning environments in the 21st century. Such issues are raised and addressed in the next section.

THE NEW GENERATION OF E-LEARNERS: WHO IS THINKING OF THE CHILDREN?

Young e-Learners of the 21st Century are currently four to five years old. They are the post-Generation Y babies, the Generation Y2 or Z users of the future. [2] Such students are supplied with computers in every classroom, some schools even require every student to have a personal laptop and the wireless internet is now a reality rather than just a dream. They are sitting at computers for hours on-end playing games that utilize the insidious cognitive learning strategies devised by psychologists involving 'Operant Conditioning'. They are subtly and subliminally encouraged to believe that killing and decapitating small cute flying creatures have nothing to do with reality. At six years of age they are accessing information through human-computer interaction (HCI) (McKay, 2008; Preece, 1994), hyper linking through virtual cyber space and integrating multimedia texts of unknown origins to find material for a school project. They sit in non-ergonomically designed furniture motionless for hours concentrating on bright colourful screens designed to capture, conquer and suppress the physical active that is required to develop growing bodies. Aliza Rothman (2003) reported on a simple survey conducted by America Online/Digital Marketing Services, Inc. An Online survey conducted in Opinion Place among more than 2,000 children aged 7 – 12 years and parents of children aged 7 – 12 years, nearly half of the children (46%) go on line at least four times a week and nearly 20% go online every day"(Rothman, 2003). What would be the results of such a survey conducted in 2010? This is not to imply, in any way, that the integration of ICT into educational classroom environments needs to be halted. However, as teacher educators I believe it is our responsibility to be continually vigilant about the use and inappropriate use of advances in communication technologies. As teacher educators we need to keep in mind how we can use ICT for the enhancement of students' learning rather than just the incorporation of high-tech programmes and equipment because it is available (Convery, 2009). Alternatively, as teacher educators we need to be aware of change and adapt teaching and learning to meet the changing needs of society and the changing work force.

As I observe the keyboard of my new laptop which has a Vista operating system and Word 2007 installed I am reminded of the complex relationship between new inventions and change. An interesting historical fact in relation to the design of the computer keyboard still exists for our children today. The QWERTY keyboard is an antiquated phenomenon. The name QWERTY, "... (reflects the first row of keys on the top alphabetic row on the left hand side of the keyboard). Christopher Sholes invented a type writer in 1886 that arranged the keys in such an order that prevented frequently used key-mechanisms from colliding when struck, according to popular myth. Since then, this style of keyboard arrangement has been challenged from time to time. Because colliding keys are no longer an issue; other keyboard configurations have emerged. Increasing the speed of typing saw an advent of a DVORK keyboard (named after the inventors, Dvork and Dealy, in the 1930s). ...

Although the newer designs make more sense for easy use, they have not been widely accepted and we continue to use a QWERTY design" (McKay, 2008, p. 9).

Who is looking after the children of the future? Do computer designers, manufacturers and instructional designers consider the complex interaction of learning styles of human beings? Whose responsibility is it to design, develop and mass-produce web-based educational system (WBES) that actually take into consideration the physical, cognitive, educational and social needs of the new generation of e-Learners? As teacher educators it is our responsibility to continue to ask the questions about the best interests of the students we teach and the educational environments we create to maximise learning.

We need to consider the increasing necessity for the new generation of HCI users to be not only multiliterate, but also, critically literate (Fehring & Green, 2001) in order to be empowered as learners. The one-size fits all model cannot be sustained when we take into consideration the diversity of students' needs. In schools of the 21st Century, Howard Gardner's theory of Multiple Intelligences (Gardner, 1983, 1993a, 1993b, 1998, 1999) has important educational consequences for all design features in educational programmes. Gardner's Multiple Intelligences theory relates to the following identifiable abilities:

Linguistic
Logical-mathematical
Spatial
Musical
Bodily-kinesthetic
Interpersonal
Intrapersonal
Naturalist.

The singular text-based curriculum has vanished and has been replaced by curricula incorporating diverse content, processes and flexible delivery. All teaching and learning environments need to capitalize on these theories of learning and communicating. The multimodal presentation of information through e-Learning possibilities needs to be capitalized for all learners but not at the cost of other equally important learning environments. Awareness by educators of the need for a student to experience learning environments in a range of diverse situations to maximise the growth of the whole student is still very important.

ISSUES RELATED TO YOUNG E-LEARNERS OF THE FUTURE

Kindergartens in the 21st Century are introducing computers into their play environment. Computer equipment (chairs, desks, computer screens, laptops) that has standard adult and industry designed specifications which may not have had children's needs specifically in the design brief. Standard keyboards are not designed for hand spans of 3 - 4 year old children. Such equipment has not been ergonomically designed for the physical stature of kindergarten nor primary school children. We are already witnessing young children developing inefficient keyboard skills because we are not teaching touch-typing skills to these young learners until after inappropriate habits have already been established. We have young learners spending hours a day on computers both for pleasure and for schooling. Young children email each other regularly and for extended periods of time, they are surfing the Internet for information to complete project work, they are involved in extended courses designed for schools to overcome the disadvantage of scare resources. Perhaps more significantly they are involved in educational games where they create personalities and relationships, raise children and control the lives of the participants in a simulated environment. We are already starting to count the costs in terms of inappropriate chat room contacts (Australian Government Net Alert., 2007), social isolation and poorly developed interpersonal skills (Funk, 1993), physical issues related to poor posture and eyesight related issues for children with vision problems. There is a tenuous, but growing, link that the ability for some online computer users to differentiate the line between reality and the virtual environment can be dangerously fine and in some cases related to dysfunctional behaviour (Anderson & Bushman, 2001; Clark, 1993; Funk, 1993). This is not to imply that there is an automatic need for censorship or the banning of products that flood the markets related to e-Learning, computer games, use of the Internet or email communication. But it is a word of caution to keep in mind the importance to remember the complex interaction of human, learning and technology in curriculum design issues.

There is a gender issues re-emerging in the 21st Century related to the ICT industry. In the 20th Century we had gender issues related to the content of texts, gender biased practices in our schools and the workforce and gender biased occupations that had to be challenged. In the 21st Century we are beginning to see a similar situation emerging but this time the context is ICT. For example, there is a growing awareness that the content of some of the current computer games is gender biased. Computer games are often dominated by violence, aggression and competition: traits that would historically appear to be more male characteristics. The questions that are beginning to emerge "Is there a gender bias in the language, logic, multimedia material selected and the types of illustrations used to represent situations in curriculum programmes?" "What affect does this have on developing gender relationship, especially of young children?"

CURRICULUM CONSIDERATIONS AND THE HUMAN ELEMENT

The holistic nature of educational experiences has taken a back seat to the outcomes based content driven instructional designed curriculum. The dominance of current technological capabilities seems to have temporarily overridden the need to consider the student as a social being in a global environment that in actual fact demands more social interaction and people skills than ever before.

Instructional designers involved in eLearning need to work in partnerships with a range of educational professionals to produce the very best teaching and learning environments possible. As educators we need to revisit and give serious consideration to such issues as diversity in learning styles, gender-related curriculum and the human and societal consequences and costs of social isolation. The notion that teaching and learning is a social activity requiring the interaction of human beings is still fundamentally important to the creation of effective learning environments. Interactive participation by learners is a facet of eLearning capabilities that is still in its embryonic stage. Techniques are still being developed for 'realtime' interaction. However, unless, as educators, we keep this fundamental learning tenet a priority for all of us involved in the development of educational material and programmes it may well be displaced for an economically rationalized alternative such as online chat sessions.

The development of reusable learning objects (RLOs) that are universally generic and therefore cannot be context specific must now address the issues of gender bias, inappropriate meta-tagging and stereotyped and outdated concepts. For example, there is tendency for instructional designers to build into curriculum courses for children cartoon-based characters, sitcoms and comedy in the belief that this will appeal to primary school students and students in the Middle Years. Such situations can be very sexist in their orientation, out of touch with the reality of the psychology of the audience and unrelated to the complex concepts that are being covered in the curriculum. It is the combination of these three above mentioned factors (learning *styles*, social communication and gender issues) that needs to be re-examined in relation to the current status of multiple literacies and e-Learning in the ICT world. There are a number of simple features that can be built into our online learning environments (Snyder, 2000). For example, there is a need for all of us involved with education to embed critical literacy skills in our curriculum programmes and courses. Teaching students how to interrogate, analyse and evaluate information obtained from apparently authentic internet sources is imperative. This has been referred to in the earlier section on information literacy skills. In addition, a core component of every course or programmes could be an offline cooperative learning group, peer group, or small group task. Such an activity can contribute to the social interaction between humans that is a fundamental aspect of any workforce requirement. Have you ever seen a job description that does not ask for "good interpersonal skills"? This attribute begins in kindergarten, develops in primary and secondary schools and is a desirable skill in the workforce. It is not developed in an isolated fully online course with the occasional asynchronous or synchronous chat session between students and/or teachers. Future global reconciliation between nations may well also depend on interpersonal skills of the highest order.

IN SUMMARY

My proposition in this chapter is simple, as teacher educators we need to constantly revisit, rethink and reconceptualise teaching and learning practices as they relate to notion of multiple literacies, the impact of ICT, and the notions of information literacy and scientific literacy. We need to consciously keep in mind the necessity of maintaining a balance in the curriculum designed and the method of delivery of our teaching practices. Flexible delivery, mixed mode and multimodal dimensions of curricula need to be embedded into our educational programmes from the youngest to lifelong learners. This is the responsibility of everyone involved in the chain of planning, designing, production and delivery of education programmes, not just teachers and teacher educators. However, as teachers and teacher educators we are specifically charged to consider the best interests of all interested parties in the education process. When we are using multimodal and multidimensional learning interfaces as part of multiple literacies programmes we need to consider the diversity of the backgrounds of our students; the relevance of our teaching and learning practices, and how to maximise the benefits of ICT in the learning process.

REFERENCES

American Association for the Advancement of Science. (1989). *Project 2061 - Science for all Americans*. Retrieved March 18, 2009, from http://www.aaas.org/

American Association for the Advancement of Science. (1993). *Benchmarks for science literacy*. New York: Oxford University Press.

Anderson, C. A., & Bushman, B. J. (2001). Effects of violent games on aggressive behavior, aggressive cognition, aggressive affect, physiological arousal, and prosocial behavior: A meta-analytical review of the scientific literature. *Psychological Science*, *12*, 353–359. doi:10.1111/1467-9280.00366

Anstey, M. (2002). *Literate futures: Reading*. Coorparoo, DC., Queensland: State of Queensland, Department of Education.

Anstey, M., & Bull, G. (2004). *Literate futures. Professional development. The teaching of reading for a multiliterate world. Years 7 - 12*. Brisbane, Queensland: Literate Futures, Curriculum Strategy Branch, Department of Education, The State of Queensland.

Anstey, M., & Bull, G. (2006). *Teaching and learning multiliteracies. Changing times, changing literacies*. Kensington Gardens, South Australia: Australia Literacy Educators' Association in conjunction with the International Reading Association, Newark DE.

Asselin, M. M., & Lee, E. A. (2002). 'I wish someone had taught me': Information literacy in a teacher education programmes. *Teacher Librarian*, *30*(2), 10–17.

Australian Government Net Alert. (2007). *Cyber bullying. Information sheet*. Retrieved January 2, 2009, from http://www.netalert.gov.au

Australian Information & Communications Technology in Education Committee (AICTEC). (2008). *Research Report: Interoperability standards across the Australian education & training sector*. Retrieved October 7, 2008, from http://www.aictec.edu.au

Bearne, E., Clark, C., Johnson, A., Manford, P., Mottram, M., & Wolstencroft, H. (2007). *Reading on screen. Research report*. Leicester, UK: United Kingdom Literacy Association.

Beavis, C. (1997). Computer games, culture and curriculum. In I. Synder (Ed.), *Page to screen. Taking literacy into the electronic era* (pp. 234-255). St Leonards, NSW: Allen and Unwin.

Bilal, D. (2005). Children's information seeking and the design of digital interfaces in the affective paradigm. *Library Trends, 54*(2), 197–208. doi:10.1353/lib.2006.0013

Bilal, D., & Kirby, J. (2002). Differences and similarities in information seeking: Children and adults as web users. *Information Processing & Management, 38*, 649–670. doi:10.1016/S0306-4573(01)00057-7

Brown, C., & Krumholz, L. R. (2002). Integrating information literacy into the science curriculum. *College & Research Libraries, 63*(2), 111–123.

Bundy, A. (Ed.). (2004). *Australian and New Zealand information literacy framework. Principles, standards and practice*. Adelaide: Australian and New Zealand Institute for Information Literacy (ANIIL).

Burden, J. (2005). The new science curriculum at key stage 4. *Education in Science Education, 213*(10-12).

Bybee, R. W. (1997). *Achieving scientific literacy: From purposes to practices*. Portsmouth, NH: Heinemann.

Bybee, R. W., Powell, J. C., & Trowbridge, L. W. (2008). *Teaching secondary school science. Strategies for developing scientific literacy* (9th ed.). Columbus, OH: Pearson Education/Merrill Prentice Hall.

Chung, J. S., & Neuman, D. (2007). High school students' information seeking and use for class projects. *Journal of the American Society for Information Science and Technology, 58*(10), 1503–1517. doi:10.1002/asi.20637

Clark, C. S. (1993). TV Violence. *CQ Researcher, 3*(12), 167–187.

Convery, A. (2009). The pedagogy of the impressed: How teachers become victims of technological vision. *Teachers and Teaching, 15*(1), 25-41. Retrieved March 1, 2009, from http://dx.doi.org/10.1080/13540600802661303

Cope, B., & Kalantzis, M. (Eds.). (2000). *Multiliteracies. Literacy learning and the design of social futures*. London: Routledge.

CQU. (2009). *Information literacy at CQU Library*. Retrieved January 5, 2009, from http://www.library,cqu.edu.au Curriculum Corporation for MCEETYA. (2008). *Digital education- Making change happen. Learning in an ONLINE world*. Retrieved October 7, 2008, from http://www.icttaskforce.edna.edu.au/icttaskforce/go

Deboer, G. E. (2000). Scientific literacy: Another look at its historical and contemporary meanings and its relationship to science education reform. *Journal of Research in Science Teaching, 37*(6), 582–601. doi:10.1002/1098-2736(200008)37:6<582::AID-TEA5>3.0.CO;2-L

DEEWR. (2009). *Digital Education Revolution. Australian Government Policy*. Retrieved January 6, 2009, from http://www.digitaleducationrevolution.gov.au

Dresang, E. T. (2005). The information-seeking behavior of youth in the digital environment. *Library Trends*, *54*(2), 178–196. doi:10.1353/lib.2006.0015

Fehring, H., & Green, P. (Eds.). (2001). *Critical literacy. A collection of articles from the Australian Literacy Educators' Association*. Newark, DE: International Reading Association.

Funk, J. B. (1993). Reevaluating the impact of video games. *Clinical Pediatrics*, *32*(2), 86–90. doi:10.1177/000992289303200205

Gardner, H. (1983). *Frames of mind. The theory of multiple intelligence (Tenth Anniversary Edition)*. New York: Basic books.

Gardner, H. (1993a). *Creating minds*. New York: Basic Books.

Gardner, H. (1993b). *Multiple Intelligences: The theory in practice*. New York: Basic Books.

Gardner, H. (1998). A multiplicity of intelligences. *Scientific America presents: Exploring intelligences, 94*(4), 18-23.

Gardner, H. (1999). *Intelligences reframed: Multiple intelligences for the 21st century*. New York: Basic Books.

Gill, M. (1993, August 17). Literacy - always a slippery customer for wordsmiths. *Age*, 13.

Gordon, C. A. (2000). The effects of concept mapping on the searching behavior of tenth-grade students. *School Library Media Research, 3*. Retrieved January 15, 2009, from http://www.ala.org/ala/mgrps/divs/aasl/aaslpubsandjournals/slmrb/slmrcontents/volume32000/mapping.cfm

Hancock, J. (Ed.). (1999). *Teaching literacy using information technology. A collection of articles from the Australian Literacy Educators' Association*. Carlton South, Victoria, Australia: ALEA in conjunction with the IRA Newark, DE.

Hawken, S. (2008). Lockie Leonard multi-literacy DVD-ROM and resource package. Australian Children's Television Foundation. *Literacy Learning: The Middle Years*, *16*(3), 67–68.

Healey, A. (1999). *Children reading in a post-typographic age: Two case studies*. Unpublished doctoral dissertation, Queensland University of Technology, Brisbane.

Healy, A. (Ed.). (2008). *Multiliteracies and diversity in education. New pedagogies for expanding landscapes*. South Melbourne, Australia: Oxford.

Henderson, R. (2004). Recognising difference: One of the challenges of using a multiliteracies approach. *Practically Primary*, *9*(2), 11–14.

Herring, J. E. (2006). A critical investigation of students' and teachers' views of the use of information literacy skills. *School Library Media Research, 9*. Retrieved January 15, 2009, from http://www.ala.org/ala/mgrps/divs/aasl/aaslpubsandjournals/slmrb/slmrcontents/volume9/informationliteracy.cfm

Hill, S., Comber, B., Louden, W., Rivalland, J., & Reid, J. (1998). *100 children go to school: Connections and disconnections in literacy development in the year prior to school and the first year of school*. Canberra, ACT: Commonwealth Department of Employment, Education, Training and Youth Affairs.

House of Lords Select Committee on Science and Technology. (2000). *Science and society*. London: HMSO.

Hytch, T. (2008). Transforming pedagogies with new technologies. The power of protest. *Engineers Australia, 43*(2), 47–50.

Janssens, J. (2002). Kahootzing around in Year 3. *Practically Primary, 7*(3), 10–11.

Jarman, R., & McClune, B. (2007). *Developing scientific literacy. Using news media in the classroom*. Maidenhead, Berkshire, England: Open University Press, McGraw-Hill Education.

Jordan, K. (2008). 'But it doesn't count, sir' - A conversation about using electronic discussion in VCE English. *Engineers Australia, 43*(2), 59–62.

Kalantzis, M., & Cope, B. (2004). *Learning by design*. Altona, Victoria: Common Ground Publishing.

Kalantzis, M., Cope, B., & Fehring, H. (2002). Multiliteracies: Teaching and learning in the new communications environment. *PEN, 133*(March).

Kibby, M. W. (2000). What will be the demands on literacy in the workplace in the next millennium? *Reading Research Quarterly, 35*(3), 380–381.

Kolsto, S. D. (2001). Scientific literacy for citizenship: Tools for dealing with the science dimension of controversial socioscientific issues. *Science Education, 84*(3), 291–310. doi:10.1002/sce.1011

Kuhlthua, C. C. (2004). *Seeking meaning: A process approach to library and information services* (2nd ed.). Westport, CT: Libraries Unlimited.

Lankshear, C., Snyder, I., & Green, B. (2000). *Teachers and techno-literacy: Managing literacy, technology and learning in schools*. St. Leonards, NSW: Allen & Unwin.

Lankshear, C., Snyder, I., & Green, B. (2000). *Teachers and techno-literacy. Managing literacy, technology and learning in schools*. St Leonards, Australia: Allen & Unwin.

Lau, K.-C. (2009). A critical examination of PISA's assessment on scientific literacy [Published Preprint Online]. *International Journal of Science and Mathematics Education*, 1-28. Retrieved March 18, 2009, from http://springerlink.com.ezproxy.lib.rmit.edu/content/x6620h1156810417/fulltext.pdf

Laugksch, R. C. (2000). Scientific literacy: A conceptual overview. *Science Education, 84*(1), 71–94. doi:10.1002/(SICI)1098-237X(200001)84:1<71::AID-SCE6>3.0.CO;2-C

Madden, A. D., Ford, N. J., & Miller, D. (2007). Information resources used by children at an English secondary school. Perceived and actual level of usefulness. *The Journal of Documentation, 63*(3), 340–358. doi:10.1108/00220410710743289

Madden, A. D., Ford, N. J., Miller, D., & Levy, P. (2006). Children's use of the internet for information-seeking. What strategies do they use, and what factors affect their performance? *The Journal of Documentation, 62*(6), 744–761. doi:10.1108/00220410610714958

MCEETYA and MCVTE. (2008a, June). *Charter of principles for cross-sectoral collaboration and interoperability across the Australian education and training sector*. Retrieved October 7, 2008, from http://www.aictec.edu.au

MCEETYA and MCVTE. (2008b, June). *Joint Ministerial statement on information and communications technologies in Australian education and training: 2008 -2011*. Retrieved October 8, 2008, from http://www.aictec.edu.au

McKay, E. (2008). *The human-dimensions of human-computer interaction. Balancing the HCI equation* (Vol. 3). Nieuwe Hemweg, Amsterdam: IOS Press.

Millar, R., & Osborne, J. (Eds.). (1998). *Beyond 2000: Science education for the future*. London: Kings College.

Murcia, K. (2005a). *Science for the 21st Century: Teaching for scientific literacy in the primary classroom*. Paper presented at the CONASTA54 University of Melbourne.

Murcia, K. (2005b). Science in the newspaper: A strategy for developing scientific literacy. *Teaching Science*, *51*(1), 40–42.

Neville, M. (2004). Brisbane multiliteracies project. *Practically Primary*, *9*(2), 20–22.

New London Group. (1996). A pedagogy of multiliteracies: Designing social futures. *Harvard Educational Review*, *66*(1), 60–92.

Nixon, H. (2003). New research literacies for contemporary research into literacy and new media? *Reading Research Quarterly*, *38*(3), 407–413.

Nixon, J., Cope, P., McNally, J., Rodrigues, S., & Stephen, C. (2000). University-based initial teacher education: Institutional re-positioning and professional renewal. *International Studies in Sociology of Education, 10*(3), 243-261. Retrieved January 4, 2009, from http://dx.doi.org/10.1080/09620210000200062

OECD. (2001). *Knowledge and skills for life: First results from the OECD Programmesme for International Student Assessment (PISA) 2000*. Retrieved December,17, 2008, from http://cat/lib.rmit.edu.au/vwebv/searchBasic

OECD. (2006). *Assessing scientific, reading and mathematical literacy*. Paris: OECD.

Preece, J. (1994). *Human-computer interaction*. Harlow, UK: Addison-Wesley.

RMIT. (2009a). *Information literacy and RMIT University library*. Retrieved January 5, 2009, from http://www.rmit.edu.au/library

RMIT. (2009b). *Library skills for foundation studies*. Retrieved January 8, 2009, from http://rmit.libguides.com/content.php?pid=6688&sid=42212#106369

RMIT. (2009c). *Library research skills for education students*. Retrieved January 9, 2009, from http://rmit.libguides.com/content.php?pid=4587

Rodrigues, S. (Ed.). (2005). *International models of teacher professional development in science education: Changes influenced by politics, pedagogy and innovation*. New York: Nova Academic Press.

Rodrigues, S. (2006). *Pupil-appropriate contexts in science lessons: The relationship between themes, purpose and dialogue*. Retrieved January 2, 2009, from http://dx.doi.org/10.1080/02635140600811544

Rodrigues, S., Marks, A., & Steel, P. (2003). Developing science and ICT pedagogical content knowldge: A model of continuing professional development. *Innovations in Education and Teaching International, 40*(4), 386-394. Retrieved January 21, 2009, from http://dx.doi.org/10.1080/1470329032000128413

Rothman, A. (2003). *America online 'youth wired' survey finds kids are online an average of four days a week; nearly 20% go online every day. Survey published 28th September 2003*. Retrieved January 10, 2009, from http://corp.aol.com/press

Snyder, I. (Ed.). (1997). *Page to screen. Taking literacy into the electronic era*. St Leonards, NSW: Allen & Unwin.

Snyder, I. (1999). Using information technology in language and literacy education: An introduction. In J. Hancock (Ed.), *Teaching literacy using information technology. A collection of articles from the Australian Literacy Educators' Association* (pp. 11-30). Carlton South, Victoria, Australia: ALEA.

Thomson, S., & De Bortoli, L. (2008). *Exploring scientific literacy: How Australia measures up. The PISA 2006 survey of students' scientific reading and mathematical literacy skills*. Camberwell, Victoria, Australia: Australian Council for Educational Research Ltd.

Thomson, S., & De Bortoli, L. (2009). PISA in brief from Australia's perspective. Highlights from the full Australian report: Exploring scientific literacy: How Australia measures up. *The PISA 2006 assessment of students' scientific, reading and mathematical literacy skills*. Retrieved January 10, 2009, from http://www.acer.edu.au/ozpisa/reports.html

UNESCO. (1990). *Compendium of statistics on illiteracy - 1990 Edition* (Statistical Reports and Studies No. 31). Paris: UNESCO, Division of Statistics on Education, Office of statistics.

Unsworth, L. (2001). *Teaching multiliteracies across the curriculum: Changing contexts of text and image in classroom practice*. Buckingham, UK: Open University Press.

VCAA. (2005). *General Achievement Test (GAT) 2005*. Retrieved January 3, 2009, from http://www.vcaa.vic.edu/vce/exams/gat/gat.html

VCAA. (2006). *General Achievement Test (GAT) 2006*. Retrieved January 3, 2009, from http://www.vcaa.vic.edu.au/vce/exams/gat/gat.html

VCAA. (2009). *General Achievement Test (GAT)*. Retrieved January 5, 2009, from http://www.vcaa.vic.edu.au/vce/exams/gat/aboutgat.html

Victorian Curriculum Assessment Authority (VCAA). (2005a). *Victorian Essential Learning Standards (VELS). Revised 10th December 2005*.

Victorian Curriculum Assessment Authority (VCAA). (2005b). *Victorian Essential Learning Standards. Revised Edition 10th December 2005*. Retrieved January 2, 2009, from http://vels.vcaa.vic.edu.au/links/standards.html#5

Walters, M. (2007). *An Investigation of Adopting, Adapting and Integrating of Information and Communication Technology (ICT) and Incorporating the Explicit Teaching of Thinking Skills Across the Curriculum*. Unpublished doctoral dissertation, RMIT University, Melbourne.

Walters, M., & Fehring, H. (2008). *Integrating ICT and thinking skills - have schools moved on?* Unpublished journal article. RMIT University.

Wang, L. (2006). *Information literacy courses- a shift from a teacher-centred to a collaborative learning environment*. Paper presented at the 4th International Lifelong Learning Conference: Partners, Pathways, and Pedagogies, Yeppoon, Queensland.

Wilson, T. D. (2006). 60 years of the best in information research. On user studies and information needs. *The Journal of Documentation, 62*(6), 658–670. doi:10.1108/00220410610714895

Yelland, N. (2001). *Teaching and learning with information and communication technologies (ICT) for numeracy in the early childhood and primary years of schooling*. Canberra: Department of Education, Training and Youth Affairs.

Yin, D. Y. (2008). *Hong Kong students' performance in scientific literacy. HKPISA Centre*. Retrieved March 10, 2009, from http://www.fed.cuhk.edu.hk/~hkpisa/events/2006/events2006_20071210.htm

ENDNOTES

1 The current Victorian Education system uses a curriculum framework called VELS (Victorian Essential Learning Standards). This framework incorporates six levels of achievement covering Strands, Domains and Dimensions of learning (Victorian Curriculum Assessment Authority (VCAA), 2005a, 2005b) . In Australia a National Curriculum framework is being designed and trialling will commence in 2010 www.ncb.org.au

2 Generation X users are considered to be those born between 1982 – 1995 Generation Y those born post 1995.

Chapter 12
Online Assessment:
Informing Practice in Tertiary Science Education

David Walker
University of Dundee, Scotland

ABSTRACT

This chapter reports the findings of a study which investigated the learning processes, expectations and perceptions underpinning undergraduate science student's use of online assessment. The aims of the study were to: review the way students interact with online assessments within the context of their studies; examine how this interaction was shaped by their preparation and prior experience with online tools; explore the prominence of test anxiety and the extent to which the student's anxieties influenced their interaction with the technology and the learning approach employed; and analyse the student's perspectives on the use of online assessment in higher education and, in particular, in their discipline-specific courses.

BACKGROUND

Rapid developments in the field of e-learning are altering the educational landscape, with evidence emerging that e-learning tools and techniques are being mainstreamed at all levels of higher education (Macdonald, 2004). Proponents of e-learning point to the enhancements in information technologies and the new methods of learning, teaching and assessment as reasons behind this expansion in use. Critics argue that the development of e-learning in higher education has largely been driven by senior managers' misplaced desire to achieve the efficiencies and cost savings promised by educational technology (Weller, 2004). Online assessment is becoming particularly prevalent across the sector, with studies locating the majority of activity within science-based disciplines (Bull & McKenna, 2000; Bull & McKenna, 2004). The growing prominence of online assessment among the science community indicates a need for science educators to demonstrate awareness of the practical and pedagogical issues surrounding the use of formative and summative online assessment. Specifically, these issues should

DOI: 10.4018/978-1-60566-690-2.ch012

be considered from the perspective of the end user; the students who engage with online assessment in the course of their studies. Research focussing on the student experience of e-learning however, has largely been neglected (Creanor, Trinder, Gowan & Howells, 2006), while research into student's experiences of online assessment is limited (Walker, Topping & Rodrigues, 2008). Research focusing on the relationship between assessment and student approaches to learning has long been identified as an area demanding further investigation (Hofer & Pintrich, 1997) with effect, experiences and expectations of online assessment pinpointed as areas prime for exploration (Northcote, 2003).

METHOD

Students drawn from four course modules in Biology, Chemistry, Medicine and Nursing at the University of Dundee formed the cases within the sample. A mixed-methods research design drew on both quantitative and qualitative techniques to facilitate in-depth exploration of student experiences of online assessment. The research design was influenced to an extent by methodological approaches use by others (see for example, Creanor *et al.* 2006; Cassady and Johnson, 2002) to support an interpretative phenomenological analysis approach. Methods involved the collection of objective and quantifiable data through screen capture of student completion of a formative online assessment task, and interpretative analysis of interview data to gain an insight into the students' experiences of these tasks. The interpretative results were subjectively negotiated and emerged through a process of interpretation which sought to establish the meaning that specific experiences or events held for individuals within the sample (Smith & Osborn, 2003). In order to cast some light on the impact of online assessment on student levels of cognitive test anxiety, a modified version of the Cognitive Test Anxiety Scale created by Cassady and Johnson (2002) was utilised with the limited alterations allowing comparison of results with previous findings. Importantly, the instruments selected and the approaches adopted have been adapted to provide as rich an insight as possible into the intentions behind student actions during online assessment events and the impact of online assessment on their learning.

Participants

Four science-based modules made up the sample:

- two first year modules in biology and chemistry from the School of Life Sciences
- one first year module from the School of Medicine
- one second year module from the School of Nursing and Midwifery

Table 1 below presents an indication of the students who participated in the study. The gender imbalance within the sample is to an extent representative of the gender balance of the students enrolled on the modules with females making up approximately 60% of the students studying on the Medicine, Biology and Chemistry modules rising to 86% in the Nursing module.

Procedures and Analysis

Data was gathered from those participants who consented to their participation in this study via the following methods:

- A paper-based questionnaire presented to participants at an initial face-to-face meeting.
- Participant's completion of a formative online assessment recorded using screen-capture software.
- A modified version of Cassady and Johnson's (2002) Cognitive Test Anxiety

Table 1. Participant breakdown (n=15)

Module	Number of Students	Number Male	Number Female
Nursing	3	-	3
Medicine	4	1	3
Biology	4	1	3
Chemistry	4	1	3

Scale administered online seven days prior to participants undertaking a summative online assessment.

- Semi-structured interviews conducted with participants immediately after their completion of a summative online assessment.

The data collection methods were administered in three separate phases. At the initial face-to-face meeting students were first asked to complete the Student Experiences of Online Assessment Questionnaire, which gathered quantitative demographic and background information about the participants plus information relating to their prior experience and perceptions of online assessment. During the same initial meeting the participants were then asked to complete a formative online assessment available within their module which was screen captured using Camtasia Studio™ software.

The second phase of data collection occurred seven days prior to the participants' completion of a summative online assessment. A modified version of Cassady and Johnson's (2002) Cognitive Test Anxiety Scale was developed for online delivery using Questionmark Perception™ and presented to the students via Blackboard™.

The third and final phase of data collection occurred immediately following the participants' completion of a summative online assessment. At the end of each respective module examination, participants were asked to remain behind and take part in a semi-structured interview. The final interview schedule consisted of 29 questions based around the following five topics: Post-Assessment Reaction, Test Anxiety, Preparation, Exam Strategies and Perceptions of Online Assessment. Interviews were recorded digitally.

The screen captured video recordings of the formative online assessments completed by each student were observed in detail with all key actions and times recorded and compiled into activity lists. An activity list is the core output from the process of task analysis, a point acknowledged by Diaper (2004) who stated that "...an activity list is a fundamental (perhaps the fundamental) representation of a task in task analysis" (p. 40). The data contained within each individual activity list was subsequently compiled in Excel and subjected to quantitative analysis to provide descriptive statistical data relating to student use of the formative online assessments by group, question type, question outcome and by gender.

Analysis of the modified Cognitive Test Anxiety Scale was conducted using the methodology established by Cassady and Johnson (2002). Anxiety scores were calculated on the basis of anxiety level for the student group as a whole and also on the basis of module. The data was imported into SPSS™ to facilitate analysis of the internal consistency of the items within the instrument with a value for Cronbach's Alpha calculated.

Each interview conducted with students following their completion of a summative online assessment was recorded and transcribed verbatim. Transcripts were read individually and subsequently coded with each transcript treated as a single unit of analysis.

THE STUDENT EXPERIENCE OF ONLINE ASSESSMENT

Data from the Student Experience of Online Assessment questionnaire is presented, as it provides background information relating to the participants involved in the study and an insight into their attitudes towards online assessment. Quantitative task analysis data emerging from student interaction with formative online assessments is then reported. Finally, themes emanating following interpretative analysis of student experiences of summative online assessments drawn from semi-structured interviews is presented.

Student Experiences of Online Assessment

Each student was asked to complete the Student Experiences of Online Assessment questionnaire. A total of fifteen students drawn from four modules across three Schools (two Colleges) participated in this third stage of the research. Participants were predominantly female (n=12) with males (n=3) only making up 20% of the sample. The mean age of the students within the sample was 26.4 years and the median age was 25 years. The data revealed that over half the sample (53%) had completed sixteen or more online assessments while 47% had completed less than six online assessments since starting university. Closer analysis of the data however revealed that of those students who had completed sixteen or more online assessments since joining university, all were registered within the School of Life Sciences.

Respondents were asked to consider on a scale of one to five (where one equaled problem solving and five equaled factual recall), the type of skills they believed were assessed by the online assessments presented within their modules. Table 2 below reveals that responses were tightly bunched together with the majority (53%) indicating that they perceived the skills assessed to lie somewhere between problem solving and factual recall. However, a significant minority (47%) indicated that they felt the skills assessed were more closely associated with factual recall.

To determine the level of support for use of online assessment within modules students were asked to indicate their level of agreement with the statement: 'I believe that online assessment is an appropriate method of assessing my subject.' The majority of respondents were again found to either agree or strongly agree with the statement (67%) with the majority of those respondents (60%) from the School of Life Sciences. A significant minority (33%) from across the three Schools showed no preference for or against online assessment as a method of assessment within their subject.

In response to the statement 'I would like to see online assessment used in other modules within my course', the data shows that the majority (53%) neither agreed nor disagreed with the statement however, seven out of the eight respondents who selected that option were registered in the School of Life Sciences where online assessment use is already extensive. Students from the School of Medicine were found to be the least enthusiastic about greater use of online assessment within their course, whereas all three students from the School of Nursing & Midwifery either agreed or strongly agreed with the statement, indicating that they would support wider use of online assessment. A measurement of the extent to which support for wider use of online assessment varied together with perceptions of online assessment as an appropriate method of assessment within a specific subject produced a positive correlation coefficient of 0.50.

To gain an insight into whether test mode influenced exam preparation, respondents were asked to indicate their level of agreement in relation to the statement: 'I would prepare as extensively for online exams as I prepare for paper-based exams.' The majority of students (73%) agreed or strongly agreed with this view. A significant minority (27%) were found to disagree with the statement. Notably, all students from the School

Table 2. Skills assessed via online assessment (n=15)

Skills Assessed by Online Assessment	Total No. of Respondents	% of Respondents
1 – Problem Solving	0	0
2	0	0
3	8	53
4	7	47
5 – Factual Recall	0	0

Standard Deviation = 0.52

Table 3. Online assessment as an appropriate assessment method for subject (n=15)

Scale	Total No. of Respondents	SLS	SM	SNM	% of Respondents
Strongly Agree	2	1	0	1	13
Agree	8	5	2	1	53
Neither Agree nor Disagree	5	2	2	1	33
Disagree	0	0	0	0	0
Strongly Disagree	0	0	0	0	0

Note: SLS = School of Life Sciences; SM = School of Medicine; SNM = School of Nursing & Midwifery

of Medicine indicated that they would prepare as extensively for an online exam as a paper-based equivalent, while 75% of respondents from the School of Life Sciences also either agreed or strongly agreed with the statement. An analysis of the extent to which preparation for online exams varied with perceptions of the skills assessed by online assessments produced a correlation coefficient of 0.32, suggesting that the two variables are moderately positively associated.

Findings relating to the statement: 'I would not be concerned about the prospect of completing an online exam in one of my modules' revealed that the majority of students (73.33%) indicated that they either agreed or strongly agreed with the statement suggesting that they do no view the prospect of completing an online exam as threatening or potentially anxiety inducing. Importantly however two students disagreed with the statement, suggesting that they would be concerned about completing an online exam. It must be noted however that this finding does not reveal whether or not these concerns would similarly manifest for different modes of assessment. A correlation analysis between measures relating to clarity of guidance given on how to complete online assessments and low levels of anxiety in relation to the prospect of completing an online examination produced a moderate positive correlation coefficient of 0.29, suggesting that lower anxiety levels are associated with the provision of clear guidance.

IMPACT OF FORMATIVE ONLINE ASSESSMENT ON STUDENT LEARNING

An objective of the research was to explore student interaction with online assessment software. To facilitate such an analysis, each student was invited to complete a formative online assessment which was active within their specific module of study. The completion of each assessment was screen-captured and subsequently subjected to a process of task analysis to provide statistical data to illuminate patterns of use, task dependencies and to identify any specific issues relating to the task that require further consideration.

Formative Online Assessment – Task Analysis Data

Each screen-captured video recording of the formative online assessments was observed in detail with all key actions and times recorded and compiled into activity lists for each student. The data contained within each individual activity list was compiled and subjected to quantitative analysis to provide descriptive statistical data relating to student use of the formative online assessments by group, question type, question outcome and by gender.

Analysis of the data revealed that the mean time spent answering questions within the assessments across all students was 28.76 seconds. On the basis of gender, males (mean = 35.91 seconds) were found to spend more time on questions than females (mean = 26.98 seconds). However, it must be noted that the standard deviation of the mean time spent on questions by males was high (20.22 seconds) and that males only represented 20% of the overall sample.

Further analysis revealed that the mean time spent on question feedback by all respondents was 7.24 seconds. Little variation was noted on the basis of gender with males spending on average 7.30 seconds on question feedback and females spending 7.24 seconds. Analysis of the data on the basis of module revealed that Medical students spent the longest time answering questions (mean = 40.05 seconds) closely followed by Biology students (mean = 37.72 seconds) with Nursing students found to spend the least time answering questions (mean = 11.43 seconds). Analysis of mean times spent reading question feedback on the basis of module again revealed that Medical students spent the longest amount of time reviewing feedback (mean = 10.72 seconds) with Chemistry students found to spend the least amount of time (mean = 4.93 seconds). However, this data is not normally distributed and, as such, median values have been calculated. The data revealed a median value of 5 seconds for Medical students, 3 seconds for Biology students, 8 seconds for Nursing students and 4 seconds for Chemistry students.

In contrast to the findings suggested by the mean times, the median values revealed that Nursing students spent the longest time reviewing question feedback while Chemistry students spent the least amount of time. Closer inspection of the data however, revealed that of the question feedback available to be reviewed, only feedback for 65 of a possible 80 questions (81.25%) was reviewed by students who completed the Medical assessments while feedback for only 10 out of a possible 60 questions (16.67%) were reviewed by students who completed the Nursing assessments.

Table 5 presents descriptive statistics on question completion times by question type.

Analysis of the data revealed that students spent the largest amount of time completing fill in the blank questions (mean 50.78 seconds) and multiple response questions (mean = 40.39 seconds). Students were found to spend the least time completing multiple choice format questions including standard multiple choice questions (mean = 13.90 seconds), yes/no questions (mean = 14.25 seconds) and true/false questions (mean = 9.40 seconds). On the basis of this data fill in the blank questions appear to take the most time to complete however, when median values are calculated, the value for fill in the blanks questions falls to 32.50 seconds, behind pull down list question types, multiple response question types and drag and drop question types with median time values of 34 seconds, 37 seconds and 44.50 seconds respectively. Table 5: Descriptive statistics on question completion times by question type (n=240).

Table 4. Time on formative assessment questions and feedback by group (n=15 standard deviation in parenthesis)

Group	N	Number of Questions	Total Time on Questions (Sec)	Mean Time on Questions (Sec)	Total Question Feedback Reviewed	Total Time on Feedback (Sec)	Mean Time on Feedback (Sec)
All Respondents	15	240	7014	28.76 (13.29)	175	1296	7.24 (5.34)
Male	3	45	1758	35.91 (20.22)	45	340	7.30 (1.73)
Female	12	195	5256	26.98 (11.54)	130	956	7.22 (5.98)
Chemistry	4	40	861	21.52 (17.29)	40	197	4.93 (4.01)
Biology	4	60	2263	37.72 (26.56)	60	324	5.40 (6.03)
Medicine	4	80	3204	40.05 (25.03)	65	697	10.72 (13.66)
Nursing	3	60	686	11.43 (5.97)	10	78	7.80 (4.57)

Table 5.

Question Type	N	Minimum Time on Question Type	Maximum Time on Question Type	Mean Time on Question Type	Median Time on Questions Type	Standard Deviation
Drag and Drop	6	15	57	38.50	44.50	18.51
Fill in Blanks	18	13	149	50.78	32.50	40.04
Multiple Choice	84	3	69	13.90	11.00	9.40
Multiple Response	38	9	83	40.39	37.00	18.18
Numeric	8	5	88	31.88	26.00	26.53
Pull Down List	72	5	122	38.33	34.00	24.30
True/False	10	4	16	9.40	9.50	3.75
Yes/No	4	4	26	14.25	13.50	9.03

Descriptive statistics on the time each student spent on questions on the basis of question outcome is presented in Table 6 below. Analysis of mean times found that students spent the largest amount of time on questions they did not manage to answer (mean = 49.50 seconds) however, this related to only 2 questions. Of those questions the students did attempt, the largest amount of time spent completing questions is shown to be for questions the students only managed to partially answer correctly (mean=46.94 seconds). Furthermore, the data shows that students spent longer answering questions they answered incorrectly (mean=38.05 seconds) than those they answered correctly (mean=23.99 seconds). It must be noted that this data is not normally distributed however a similar pattern emerged when median values were calculated on the basis of question outcome.

Inspection of the data revealed that fill in the blank, multiple response and numeric question types were those answered least successfully by the students with mean values of 2.56, 2.89 and 3.00 calculated respectively. Free text fill in the blank questions were found to be the only question type that was unanswered. True/false questions are found to be the question type most successfully answered by students (mean = 3.80). Notably, despite taking almost three times longer to answer

Table 6. Descriptive statistics on time on questions by outcome (n=240)

Question Outcome	N	Minimum	Maximum	Mean	Median	Standard Deviation
Correct	163	3	122	23.99	17.00	20.20
Incorrect	58	7	149	38.05	32.50	26.30
Partially Correct	17	13	121	46.94	33.00	32.47
Unanswered	2	15	84	49.50	49.50	48.79

than multiple choice questions, pull down list questions are answered more successfully.

The data shows that females were marginally more successful at answering questions than males with mean values of 3.43 and 3.36 calculated respectively. Females were the only group to have left questions unanswered. It must again be noted that males represented only 20% of the student sample with data only available for 45 questions (18.75%) as opposed to 195 questions (81.25%) for females.

IMPACT OF SUMMATIVE ONLINE ASSESSMENT ON STUDENT LEARNING AND ANXIETY

Having explored in detail student interaction with online assessment software in a formative context, it was decided to investigate whether such approaches differed in the context of summative assessments where the personal salience of the task is higher. This analysis also facilitated exploration of the prevalence of test anxiety during summative examinations and identification of the methods students adopted to manage their anxiety levels. Each participant was invited to complete the modified Cassady and Johnson (2002) Cognitive Test Anxiety Scale one week before completing their summative examination, and to participate in a semi-structured interview immediately after completion of their respective examinations. The results obtained are presented as follows:

Descriptive statistics are presented to indicate levels of cognitive test anxiety among the students in the week preceding the examinations.

Qualitative data that emerged following interpretative analysis of post-examination interview transcripts will be presented thematically, illustrated with evidence grounded within the transcripts.

Summative Online Assessment – Cognitive Test Anxiety

The Cognitive Test Anxiety Scale was made available online within each participant's respective Blackboard™ module one week prior to their examination. A total of 14 students completed the questionnaire with only one student from the School of Life Sciences failing to complete before the examination date. Responses on the scale are divided across three levels of anxiety. Low anxiety scores range from 27 to 61, moderate anxiety score range from 62 to 71 and high test anxiety scores fall in the range from 72 to 108 (maximum score possible). Table 7 reveals that the majority of the participants fell into the low anxiety range with only three participants identified in the high anxiety category. The mean anxiety score on the scale was 64.71 (standard deviation = 16.88) - falling within the moderate anxiety category. Total scale internal consistency for the 27 items was high (α= 0.948) which compared favorably with Cassady and Johnson's (2002) original findings (α= 0.91).

Table 7. Cognitive Test Anxiety Scale scores by grouping (n=14)

Score Range	Minimum Score	Maximum Score	No. of Respondents	% of Respondents
27 to 61	43	61	7	50.00
62 to 71	62	68	4	28.57
72 to 108	75	108	3	21.43

The Cognitive Test Anxiety Scale Data indicated that the majority of students in the sample would experience low levels of cognitive anxiety in online examinations.

EXPERIENCES OF SUMMATIVE ONLINE ASSESSMENTS

All fifteen students who engaged in the earlier parts of the study participated in a semi-structured interview following completion of a summative online assessment within their respective modules. Each interview was conducted immediately after the exam had ended in order to maximise the richness of the data. The data obtained in the summative interviews was subjected to a rigorous process of interpretative phenomenological analysis with analysis occurring within the pre-established themes outlined in the interview schedule. Data to emerge from analysis of the five main super-ordinate themes, illustrated with examples grounded within the transcripts, are presented as follows: a) Post-Assessment Reaction; b) Test Anxiety; c) Preparation for Summative Online Examinations; d) Exam Strategies and e) Perceptions of Online Assessment.

Post-Assessment Reaction

Analysis of the student's immediate post-assessment reaction revealed a range of themes relating to their experiences, perceptions and expectations of the event. Reflections on the overall experience were found to be mixed. A significant minority of the students were happy with their performance in the exam and confident that a positive outcome would be achieved.

Among a small group of students however, the dominant reaction was found to be one of relief and closure with the climax of the examination portrayed as a point of liberation, as the following comment from a Medical student illustrates:

'Just a big one of relief from having finished the exam really. Em just like you don't have to study tomorrow or later on, just that kind of think. Like I've totally put it out my mind now.'

The sense of relief is clear but the stress inducing nature of the summative experience is implied if not explicitly stated in the extract above. The student appears to be suppressing her memories of the task in order to release the tension. This tension, manifested in concern about the outcome of the examination and an inability to make a judgment on how well they had performed, was also evident in the analysis of two other transcripts:

"I can't quite think about it properly, it's, it's a bit of a blur. I just eh I wasn't that happy with it so just, I think I'm feeling a bit nervous about when the results come out."

"I think it's very difficult to judge in those circumstances how you've done. I think they're very deceiving and I've felt lulled into a false sense of security on many occasions with them and this one, I think this is one of the most difficult ones that we've ever done really as a, as a group."

In the first extract the Medical student is either unwilling or unable to reflect on what she found to be a challenging experience. Disappointment in her own performance appears to heighten her anxiety about the outcome of the examination. In the second extract, the Life Science student adopts a less emotional stance, refusing to be drawn on how she feels she has performed. The student's concerns are revealed however by her assessment of the difficultly of the examination ('one of the most difficult ones that we've ever done').

Analysis of what each student felt they had personally gained from completing the online examination revealed two main themes with regard to the notion of multiple literacy. Firstly, the majority of students signaled that the exam had performed a diagnostic role in respect of their subject it had helped them to identify their strengths and any gaps in their knowledge. Furthermore, a number of the participants commented how it would help them to improve their revision methods for future online examinations. Secondly, a small number of students commented that completion of the exam had in itself given them valuable exam experience, experience which was of particular value to the Nursing students who were undertaking their first online examination. The following extracts, both from Nursing students, exemplify the way experience of the examination facilitated these outcomes:

'...I think it's made me aware of areas that I don't know as well as I thought I'd might have known. Things that I thought I knew but when it comes down to it I didn't really know them as well as I thought I had but on the other side I did know a lot of things as well so it kind of reinforced what I have taken in, what I have learned, through the semester.'

'Hmm... well I've had experience of doing – that's the first online exam that I've done - so that's given me experience as well of doing that and its also by doing it shown me areas that I could look on, like look into better.'

Test Anxiety

Interviewee's were invited to discuss their thoughts relating to test anxiety. Two thirds of the students described experiencing varying degrees of nervousness prior to the start of their online exam. This anxiety was attributed by a number of the interviewees to difficulties experienced in previous examinations: 'I was quite nervous because the last few exams hadn't gone so well'; '...I knew my in-course work eh certainly didn't meet up to the forty percent'; with a small number of students simply attributing their nervousness to their personality traits: 'I felt pretty nervous but then I always do.'

A range of issues emerged through analysis of each student's views on the factors that influenced their anxiety levels during their exam and the thoughts entering their mind. Issues identified were grouped into two areas: exam specific factors and environmental factors. A small majority of students commented that aspects of the assessment or question format that were found to differ from those previously encountered in their formative assessments or which differed from their expectations was a strong inducer of anxiety. The following comment from a Medical student reveals her frustration at not being provided with what she regarded as proper guidance in order to tailor her preparation for the question formats contained within the examination:

'...the way they formatted the haematology questions were a bit different and there was a lot more kind of case scenarios focusing on like results from blood films and things which hadn't really been the way we'd been taught.'

Environmental influences, particularly the ability to view the monitors of other students undertaking the same examination and the temperature of the room, also emerged as having a strong influence on the degree of anxiety experienced during online examinations. Analysis of the transcripts revealed that a significant minority of students were greatly concerned at not being seen to be looking away from their own monitor during the exam, so as not to be considered to be cheating. In addition, a number of students commented that they sought out specific positions within the IT suite to minimise the stress from those around them. The strongest theme to emerge from analysis of the transcripts centered round the need to maintain a strong focus on the questions and avoid the potential distractions round about them.

Exploration of whether online assessment was less stressful than paper-based examinations again elicited a range of responses. A small majority of those interviewed agreed that the format was less stressful than paper exams, pointing to the higher chance of success and friendlier atmosphere in the IT suites as reasons for taking this view. Written papers were viewed as more physically demanding and more difficult to time manage. Alternative viewpoints were expressed by a smaller number of students, all studying Life Sciences, who either viewed online assessment to be as stressful as paper exams; or more stressful than paper-exams due to the potential internal conflict that can arise when unable to separate two potential answers to a particular question. A Life Science student who had dyslexia, and as such was entitled to extra time in the online examination, commented that she found the format far less stressful than a paper exam because she wasn't as concerned about the standard of her writing:

'Because my writing, my writing is atrocious. My spelling is very bad. Obviously I know you would get compensated for that in a written exam I just feel it's slightly better having it on the computer because then you don't have to worry about your writing and things like that.'

The comment above is revealing and points to a potentially significant advantage of online assessment over paper-based examinations. The online format liberates the student from the difficulties imposed by her dyslexia and enabled her to focus on the assessment – in a sense increasing the validity of the assessment.

Preparation for Summative Online Examinations

The majority described undertaking extensive preparation before their examination. This preparation was characterised by extensive reading of lecture notes and textbooks, use of online resources and detailed note making over a sustained period of time. Formative online assessments figured strongly in the preparation methods outlined by a small majority of the students interviewed, including but not exclusively those students who had adopted the extensive pre-assessment preparation above. When asked to describe their preferred revision methods and the role played by formative online assessment as part of their exam preparation, a range of views again emerged. Preferred revision methods were found to be split between reading lecture notes and textbooks and the more active process of note making and producing mind maps. The vast majority of the students interviewed expressed themselves satisfied with the adequacy of their preparations before their online examinations. A third of the interviewees stated that they would not change their approach prior to future online examinations however a small majority indicated that they would either broaden or refine their revision approach or commence their revision at an earlier stage.

Exam Strategies

A clear split was evident among interviewees regarding the approach they adopted to completing the task in the summative assessment. A significant number of those interviewed described how they tackled each question in sequence, bypassing difficult questions which they returned to having attempted all questions within the assessment. This approach is illustrated in the comments from a Medical student and Nursing student below:

"Go through the questions in sequential order trying to answer as best as possible, if I lose concentration on any of the questions I don't tend to try and start it again I just say ok start afresh go to the next one and come back to that one later."

"Em... to work my way through from one to forty, not to jump about. Eh, to read the question carefully and if I'm not understanding it I will read it again because there was a couple I wasn't too sure about".

The above comments illustrate that these students take a conscious, linear approach geared to maximise the chance of success and ensure that what is being asked within each question is clearly understood. When not initially understood, or if concentration has been lost, the question is passed over and returned to at a later stage to be reconsidered. An alternative approach, geared to minimise levels of anxiety, was described by another group of students. This approach involved identifying easier questions that could be answered successfully first with the objective of boosting confidence before returning to tackle the more difficult questions at the end. The thought process behind this approach was clearly explained by one Life Science student as the following quote illustrates:

"I always try and stay calm, go through the exam, look for the easy questions first, start off on a high, build up your confidence, get to the ones that I can't do if I really can't do them I'll just leave them and come back to them".

The reverse of this approach was described by a single student, again from the School of Life Sciences, who outlined how she answered the questions in reverse order believing the harder questions to be located at the end of the assessment. Notably, the approach was again about building confidence and minimising the effect of anxiety to ensure clarity of thinking:

"I like to start at the end and work back to the first one figuring that the hard questions tend to be put towards the end of the exam and I'd rather do these ones first, even if I can't do them, then get to the easier ones so I've built up my confidence so that I can gradually do more and more of them and then go back to the harder ones because then I find that I'm a lot calmer and able to think about them rationally."

In line with the approaches outlined above, a significant minority of interviewees indicated that reviewing answers prior to submission was one of the most important elements of their approach during the exam. Common influences on the approach taken during summative online examinations were difficult to identify within the transcripts. The type of questions within the assessment, clarity of question wording, number of distracters and question weighting were all mentioned by individual students. The time/question ratio was mentioned as a key factor which dictated whether or not there was scope to revise answers prior to submission. When the respondents were asked explicitly to describe whether their approach differed between an online and written examination a variety of answers again emerged. While two students indicated that they would adopt the same approach on paper as they would online, more than a third of those interviewed indicated that their approach on paper would differ. One Life Science student indicated that greater planning would be required in a written exam to ensure their answer was in an acceptable format. Additional comments mentioned that there was less flexibility in the order that questions could be answered in a written exam but that there was more scope to explain answers and gain marks. A small number of students indicated that they would adopt a more detailed approach to their revision for a written exam, believing that more in-depth knowledge is required.

The majority of students, when asked to reflect on whether they would modify their exam strategy if able to take the same examination again, indicated that they would not alter their approach or would make only minor adjustments. Only two students explicitly stated that they would change the approach they had taken, with one student outlining that she would have allocated more time to the more difficult calculation questions within the examination as opposed to the prompted question types. Those students who indicated they would largely maintain the same approach but with minor adjustments all mentioned allocating more time at the end of the assessment to review answers.

Perceptions of Online Assessment

When asked to explain the reasons they believed online assessment was used within their module two main themes emerged: firstly, the efficiency benefits to staff from automatic marking and, secondly, the capacity of the software to effectively test knowledge and be used to identify areas of student weakness. Just under half of the interviewees cited automatic marking within their responses, identifying the efficiency gains offered by the software as the key reason behind the use of online assessment within their module. In contrast however, one-third of those interviewed indicated that they believed that online assessment was used because it was effective at testing a wide knowledge base and provided both academic staff and students with an indication of areas of weakness. A small number of students from the School of Life Sciences indicated that they thought online assessment was used in their course because it was suited to their subject.

A variety of responses were identified in relation to the question of what type of learning was tested by online assessment. The retention of information emerged as the dominant type of learning perceived to be assessed, with a smaller number of students indicating that they perceived the assessments to be assessing problem solving skills. Where memorisation was viewed as the main type of learning assessed there was recognition that higher levels of learning could be tested if the question was suitably constructed.

Three students explicitly stated that they did not believe that online assessments could assess what they regarded as deeper forms of learning and instead simply assessed ability to effectively employ various exam strategies appropriate to the format. A diverse range of responses emerged when participants were asked to consider the advantages and disadvantages of the online assessment format. The efficiency benefits offered by automatic marking were viewed positively by a small number of students who welcomed the fact that they received their exam marks sooner when their examination was presented online as opposed to paper. Further advantages of the online format, as expressed by a third of the interviewees, included not having to write therefore enabling greater scope to think, and feeling less stressed in an online exam whereas as a written exam was described as being more intense: '…a lot more frantic.' In terms of disadvantages, the precision required to answers fill in the blank or numerical questions was again criticised by a small number of students, a factor which may have influenced the selection of multiple choice questions by the majority of those interviewed as their preferred question format within online examinations.

A clear split emerged between students who favoured greater use of online assessment within their modules and those that expressed that this would be undesirable. Notably, all the students from the School of Nursing, where online assessment is used limitedly, stated that they would be in favour or wider use. In contrast, the majority of students from the School of Medicine and a half of those interviewed from the School of Life Sciences, where online assessment is already used extensively for both formative and summative assessment purposes, expressed that use could not be extended as it was near saturation. Furthermore, some students questioned whether there was an over reliance on the online format at the expense of a more varied assessment strategy. When asked to consider whether they perceived online assessment to be less stressful than other assessment formats, the majority expressed that they thought the format to be as stressful as paper-based tests, with just over a third stating that they felt the format to be less stressful. Those who felt that it was as stressful appeared to take this view because they believed that, irrespective of format, the same amount of studying was required and the importance of the event was not diminished.

DISCUSSION

The investigation reported explored the student experience of online assessment and the way students approached online assessment tasks in order to: gain an insight into the preparation they undertook; evaluate the anxieties they experienced; identify any techniques they employed; and explore their perceptions of the assessment mode and their views on it's appropriateness as a mechanism by which to assess learning. Importantly, the research sought to identify commonalities and differences in on-task strategies in the context of formative and summative assessment tasks.

The Student Experience of Online Assessment questionnaire provided initial background information on the fifteen students who participated in the study, and also provided an early insight into their perceptions of online assessment tasks. The data revealed that all the students within the sample perceived the skills to be assessed by online assessment to lie somewhere between factual recall and problem solving. Further findings of note included the apparent relationship that exists between perceptions of skills assessed by online assessment and extent of preparation for online examinations. In addition, a similarly positive relationship was identified in relation to desire for wider use of online assessment and perceptions of online assessment as an appropriate method by which to assess the subject. Another interesting finding to emerge related to test anxiety with more than two thirds of the students (73%) indicating that the prospect of an online exam would not be a cause of anxiety, with a positive relationship identified between low anxiety and clarity of guidance and support provided within modules (correlation coefficient 0.29).

The results of the analysis of the data obtained from each student's completion of a screen-captured formative online assessment provided task analysis data highlighting patterns of use. Analysis of the mean times students spent answering questions revealed that Medical students spent the most time on each question (40.05 seconds) with Nursing students spending the least time (mean = 11.43 seconds). The variation in mean times may be attributable to the more varied range of question types within the assessment completed by the Medical students, with the assessment completed by the Nursing students comprising entirely of multiple choice questions. Analysis of question completion times by question type would appear to support this view with the median completion time for multiple choice questions found to be 11 seconds whereas median times for drag and drop, multiple response, pull down list and fill in the blank questions found to be 44.50 seconds, 37 seconds, 34 seconds and 32.50 seconds respectively. Analysis of time on questions by outcome revealed that the largest median time (49.50 seconds) was spent on questions that were subsequently left unanswered with the two unanswered questions found to be fill in the blank type, providing initial evidence of difficulty with this form of unprompted questions. Further analysis found that of those questions the students did answer, higher median times were identified for questions answered partially correctly (33 seconds) or incorrectly (32.50 seconds) than those questions answered correctly (17 seconds). This finding suggests that more thinking time is invested in questions the students find challenging than those answered correctly where confidence or guesswork results in a more prompt answer selection. Analysis of question outcome by question type again suggested that unprompted questions were the most challenging question types encountered by the students. However, it must be noted that these question types accounted for only 10.8% of the questions completed by the students within the sample.

Analysis of the quantitative data gathered via the Cognitive Test Anxiety Scale administered prior to the participants' completion of a summative examination combined with analysis of the qualitative data gathered via post-examination interviews revealed four areas for discussion. Firstly, levels and determinants of test anxiety in the context of online examinations were of note as were the preparation methods undertaken in the build up to an online examination. In addition, a wider range of exam strategies were identified in the descriptions of the approach to the task given by interviewees for the summative assessments than those used during formative assessments. Finally, more detailed exploration of the differences in student perceptions of online assessment as an assessment mode is required in a context in which the degree of personal significance is greater.

A key finding to emerge from the analysis related to the effect of environmental (contextual) factors on levels of anxiety during the exam and in the measures students employed to combat this. Factors such as previous exam performance and performance in in-course assessments were mentioned in individual cases, but the effect of being able to observe others within the IT suite emerged as the strongest inducer of anxiety. Specifically, students were concerned about the prospect of looking up from their monitor in case their behaviour was interpreted as an attempt to cheat. The ability to observe others within the room was also identified as a potential distraction which could divert attention from the task.

The majority of the students in this research were found to regard online assessment as less stressful than written examinations. A notable finding which corresponded with similar findings reported by Ricketts and Wilks (2002), and which requires further research, was the apparent benefits of the online format for dyslexic students. Limited evidence within this research suggests that dyslexic students find the online format less stressful due to the absence of concerns about the quality of their spelling.

Analysis of the findings relating to the strategies employed during online examinations revealed that the majority of the students adopted a linear approach which involved answering each question in sequence and returning to questions flagged as requiring attention at the end. A notable finding however was that a significant minority of students were adopting an approach geared at controlling their anxiety levels during the exam. A key finding to emerge related to the effect of time on the techniques students employed while completing their online examination, with time to question ratio a key factor in whether students reviewed their answers prior to submission.

Academics setting online assessment may wish to consider affording greater time during online examinations to facilitate opportunities for reflection and revision of answers. Pilot testing of assessments may provide an indication of appropriate time limits required for successful completion. An important finding to emerge from the analysis involved whether differences in approach would be adopted online as opposed to a paper exam. Although the majority of students indicated they would not alter their approach were they to be able to take the same exam again, a significant minority outlined that their approach during an online examination differed to that adopted during a written examination, a finding signifying a variation on the 'test-mode effect', a concept referring to instances where different results are recorded on identical tests presented in different formats (Clariana & Wallace, 2002).

The final area to produce findings of note surrounded student perceptions of online assessment when elicited following completion of their summative assessment. The majority of students were found to perceive that efficiency gains through automatic marking were the major reason behind online assessments use as a summative assessment method within their module. Only in a small number of cases did pedagogical reasons such as the capacity for both academics and students to identify topics of weakness or suitability for assessing particular subjects emerge.

CONCLUSION

The investigation examined in detail the student experience of online assessment in both formative and summative contexts. The study looked closely at: student approaches to online assessment tasks; the levels and influences of test anxiety experienced in relation to this particular assessment mode; and perceptions of the value and validity of online assessment as an assessment method.

The results presented suggest that when appropriately designed and aligned with course learning outcomes, online assessment can represent a powerful means by which to engage, consolidated and enhance student learning. Investigation of the student experience of formative and summative online assessment revealed evidence of the important diagnostic role performed by formative online assessments within student learning. Task analysis data combined with the findings that emerged from the qualitative interviews revealed that students primarily adopted linear approaches to the completion of assessment tasks, answering questions in order using largely deductive analytical techniques. An important finding to emerge from the data (relating to both formative and summative assessments) was the influence of assessment design on student learning approaches and levels of anxiety. Issues emerged relating to the use of unprompted question types, clarity of instructions and question wording and the transparency and flexibility of marking. Analysis of the data revealed that unprompted question types, for example fill in the blank questions, were the most challenging question type faced by students.

A key finding which supporting the results of previous research conducted by Butler (2003) emerged, revealing concern among a section of students about potential over-reliance on online assessment within their modules and a preference in some cases for a more varied range of assessment methods. Crucially, while the results of this research indicate that online assessment does have the potential to enhance the effectiveness of student learning, effective assessment design and the need for transparency in all aspects of the assessment process are shown to be important determinants of the quality of learning engendered.

Implications for Science Educators

Academic staff who currently incorporate online assessment within their courses or who are evaluating the potential of web-based assessment methods may find it useful to consider the following:

This study indicates that student's perceptions of the skills assessed by online assessment are related to the nature of the learning approach they adopt. It is therefore advisable that use of online assessment be mapped to course learning outcomes and/or the skills the assessments are believed to assess.

Issues were identified in this study with regard to use of unprompted question types, specifically free text questions, and the precise nature of the answers often required. In summative assessments, it is recommended that academics moderate responses to free text questions to ensure that students are not unfairly penalised for typographical/punctuation errors. In order to alleviate anxiety, it is important that students be made aware that this moderation will take place.

Care should be taken in the development of drag and drop and hotspot question types, with clear areas marked to indicate where students are expected to locate markers. If provision of marked areas is deemed inappropriate, consideration should be given to the size of the 'hot' area to ensure the scoring is not unduly sensitive.

Specific consideration should be directed at the time required to answer different question types with this study indicating that drag and drop, multiple response and fill in the blank type questions took approximately 3-4 times longer to complete than multiple choice questions.

This study identified that students also made use of feedback provided for questions answered correctly as a means to extend knowledge and understanding and reinforce learning. If feedback is provided, students should be informed that this will be available and given appropriate guidance on how to access this feedback.

Academics may wish to rethink the value of trailing summative questions within formative assessments. While inclusion of these questions may motivate students to engage with formative assessments, the findings of this research indicate that students may adopt lower-level learning approaches geared at memorisation. This strategy may encourage students to allocate disproportionate amounts of pre-exam revision time to the formative assessments and increase anxiety levels through a perceived need to review all questions.

Academics may wish to consider the use of privacy screens during online examinations (if available) to reduce the potential impact on student anxiety levels caused by being overlooked or being perceived to be cheating when looking away from monitors.

REFERENCES

Bull, J., & McKenna, C. (2000). CAA Centre update. In *Proceedings of the 4th International Computer Assisted Assessment Conference*, Loughborough. Retrieved January 26, 2007, from http://www.caaconference.com/pastConferences/2000/proceedings/jbull.pdf

Bull, J., & McKenna, C. (2004). *Blueprint for Computer-Assisted Assessment*. London: RoutledgeFalmer.

Butler, D. (2003). *The impact of computer-based testing on student attitudes and behaviour. The Technology Source. January/February*. Retrieved November 7, 2006, from http://ts.mivu.org/default.asp?show=article&id=1013

Cassady, J. C., & Johnson, R. E. (2002). Cognitive test anxiety and academic performance. *Contemporary Educational Psychology, 27*(2), 270–295. doi:10.1006/ceps.2001.1094

Clariana, R., & Wallace, P. (2002). Paper-based versus computer-based assessment: Key factors associated with the test mode effect. *British Journal of Educational Technology, 33*(5), 593–602. doi:10.1111/1467-8535.00294

Creanor, L., Trinder, K., Gowan, D., & Howells, C. (2006). *The learner experience of e-learning*. Retrieved April 6, 2007, from http://www.jisc.ac.uk/uploaded_documents/LEX%20Final%20Report_August06.pdf

Diaper, D., & Stanton, N. (Eds.). (2004). *The Handbook of Task Analysis for Human-Computer Interaction*. London: Lawrence Erlbaum Associates.

Hofer, B. K., & Pintrich, P. R. (1997). The development of epistemological theories: Beliefs about knowledge and knowing and their relation to learning. *Review of Educational Research, 67*(1), 88–140.

Macdonald, J. (2004). Developing competent e-learners: The role of assessment. *Assessment & Evaluation in Higher Education, 29*(2), 215–226. doi:10.1080/0260293042000188483

Northcote, M. (2003). Online assessment in higher education: The influence of pedagogy on the construction of students' epistemologies. *Issues in Educational Research, 13*(1), 66–84.

Ricketts, C., & Wilks, S. J. (2002). Improving student performance through computer-based assessment: insights from recent research. *Assessment & Evaluation in Higher Education, 27*(5), 475–479. doi:10.1080/0260293022000009348

Smith, J. A., & Osborn, M. (2003). Interpretative Phenomenological Analysis. In: J. A. Smith (Ed.). *Qualitative Psychology: A Practical Guide to Research Methods*. London: Sage Publications.

Walker, D. J., Topping, K., & Rodrigues, S. (2008). Student reflections on formative e-assessment: expectations and perceptions. *Learning, Media and Technology, 33*(3), 221–234. doi:10.1080/17439880802324178

Weller, M. (2004). Learning objects and the e-learning cost dilemma. *Open Learning, 19*(3), 293–302. doi:10.1080/0268051042000280147

Chapter 13
Enhancing Scientific Literacy through the Industry Site Visit

Jari Lavonen
University of Helsinki, Finland

Antti Laherto
University of Helsinki, Finland

Anni Loukomies
University of Helsinki, Finland

Kalle Juuti
University of Helsinki, Finland

Minkee Kim
University of Helsinki, Finland

Jarkko Lampiselkä
University of Helsinki, Finland

Veijo Meisalo
University of Helsinki, Finland

ABSTRACT

Citizens in contemporary societies are encountering more and more issues that are somehow related to science and technology. Therefore, science and technology education plays an important role in providing students with the knowledge and the competences they need in their life. The research and development project discussed in this chapter focuses specifically on scientific literacy. It is considered as a crucial element of multiple literacies required in modern life. These proficiencies are often referred to in terms such as information literacy, media literacy, environmental literacy, political literacy, computer literacy etc. (see e.g., Lankshear & Knobel, 2003). In order to enhance student scientific literacy the authors introduce a model of industry site visit for lower secondary school science education as a form of out-of-school learning. The potential of the site visit and other learning activities connected to it are discussed in the frameworks of scientific literacy, motivation and interest. The site visit and the activities, such as the use of ICT in reading and writing, are scrutinised with regard to the specified

DOI: 10.4018/978-1-60566-690-2.ch013

educational goals. The analysis of the motivational aspects of the site visit is based on self-determination theory. Self-determined learning could occur when an activity at a site is considered by a learner to be interesting, enjoyable, or personally valuable. Furthermore, the site visit offers role models which are critical for students' choice of advanced studies and careers in science. Some empirical results on both cognitive and affective learning outcomes, as well as challenges that were encountered are presented on the basis of first design and evaluation cycle.

INTRODUCTION: ASPECTS OF SCIENTIFIC LITERACY

Recently, an international comparative study on scientific literacy in OECD countries (OECD, 2007) raised interest in promoting scientific literacy in many countries. We begin by discussing the importance of scientific literacy and specifying our perspective on the concept.

Scientific literacy is a worldwide concern, and improving it constitutes a major objective of formal as well as informal science education (McEneaney, 2003). The rough descriptions 'scientific literacy stands for what the general public ought to know about science' (Durant, 1993) or 'commonly implies an appreciation of the nature, aims, and the general limitations of science' (Jenkins, 1994) do not explain a lot about the actual meaning of the concept, but much less how to measure it. Indeed, the concept of scientific literacy has developed into an umbrella term covering virtually everything regarding science education (see e.g., Shamos, 1995; Roberts, 2007). This ambiguity is due to the diversity of underlying reasons for promoting scientific literacy as being beneficial at both societal and personal levels (Jenkins, 1997; Sjøberg, 1997; Laugksch, 2000; McEneaney, 2003). There exists a substantial and diverse literature related to the concept of scientific literacy (for conceptual overviews see e.g. Jenkins, 1997; Laugksch, 2000; Roberts, 2007), which does not lack attempts to *concretise* meaning for the concept. It appears that the only thing in common in these conceptualisations is the (obvious) notion that one has to have some grasp of scientific content knowledge in order to be scientifically literate.

Accordingly, the perspective on scientific literacy employed in this chapter recognises content knowledge as an indispensable part of being scientifically literate. Even so, our focus here is not on the knowledge *of* science, but rather on the knowledge *about* science. By the latter we mean an understanding of science as an enterprise, procedural knowledge, the nature and limitations of science and the use of scientific knowledge in industry, instead of 'pure' scientific knowledge itself (cf. Ryder, 2001). In terms suggested by Roberts (2007) in his recent review of literature on scientific literacy, our viewpoint in this chapter is closer to his definition of Vision II (starting with situations, proceeding to science that is relevant in them) than to Vision I (looking inward to science itself). This kind of functional and contextualised scientific literacy emphasises personal relevance for citizens, enabling them to make informed independent decisions on personal and communal issues that have a scientific or technological base. The chosen conception is in line with the PISA definition of scientific literacy (OECD, 2007), insights of the STS-movement and the Socio-Scientific Issues (SSI) framework (Zeidler *et al.*, 2005), and recent recommendations for European science education policies (Osborne & Dillon, 2008).

Our interpretation brings scientific literacy close to other elements of *multiple literacies* that are needed to get along in modern society (see above). Especially, this Vision II -type (Roberts, 2007) conception of scientific literacy is intimately linked with *technological literacy*. Many scholars, e.g., Shamos (1995) and Sjøberg (1997), have argued that the so-called socio-scientific issues discussed in reference to scientific literacy are, in fact, primarily based on technology. Thus, grasp of technology (in its widest sense) and its relations to society should be considered as the cornerstone of scientific literacy as well. While literature on technological literacy is still tenuous and disconnected – despite some large-scale efforts to solidify the concept (e.g., ITEA, 2000) – the unified concept of *scientific and technological literacy* (STL) has been used increasingly in the past decades (Jenkins, 1997) in order to highlight the interconnections of the two. In this chapter, we consider functional scientific literacy as such to involve not only science but technology as well.

In line with the goal of enhancing students' knowledge *about* science, our approach to scientific literacy emphasises learning how scientific and technological knowledge is applied in industry and in enterprises. This choice of emphasis is based on two educational goals: we are aiming to facilitate students' entry to working life, and to enable students' further study in science and technology. These goals are based on the economic argument for scientific literacy (Sjøberg, 1997): after all, one rationale for enhancing scientific literacy is the need to ensure a steady supply of scientists, engineers and other science-related professionals (Jenkins, 1997; Laugksch, 2000; McEneaney, 2003). Accordingly, our site visit model aims to contribute to students' interest and attitudes to science. Scientific literacy is commonly regarded to involve also affective factors: e.g., the latest PISA framework (OECD, 2007) provides an expanded interpretation of scientific literacy compared to the earlier frameworks, and includes also students' attitudes as an important element of the concept.

Before presenting our practical methods for approaching the above-mentioned educational goals, we will point out one more aspect of scientific literacy that resonates well with the instructional design of our site visit model. Namely, the *active learning through reading and writing* methods presented later in this chapter supports the often neglected *fundamental sense of scientific literacy*. Norris and Phillips (2003) argue for the importance of this fundamental sense, 'reading and writing when the context is science', while noticing that conceptions of scientific literacy generally tend to be connected to the derived sense, 'being knowledgeable, learned, and educated in science'. They stress that it is not enough to know the meanings of scientific concepts, but to be scientifically literate (in the fundamental sense) one has to understand their interconnections through reading, writing, speaking, listening, and representing (Norris & Phillips, 2003).

As one solution for enhancing the kind of broad scientific literacy presented above, we have developed a model of the industry site visit for lower secondary school science education. We will analyse the potential of how scientific literacy could be enhanced during an industry site visit and the onsite learning activities, such as inquiry activities and article writing, in accordance with the visit. The site visit and activities are called *industry site visit*. The visit model has been developed for the International Materials Science-project (SAS6-CT-2006-042942-Material Science). The model has been developed to help the students become familiar with materials (e.g., paper, metal and plastics) around us; the properties and behaviour of the materials; their microscopic models describing their properties and behaviour; and use of (raw) materials in constructions/manufacturing products. Moreover, the learning activities developed to be used within the visit could help students to become familiar with the careers in material science enterprises and to learn science process skills and, moreover, to increase students' motivation and interests in material science.

The aim of this chapter is to analyse the industry site visit as an environment to enhance students' scientific literacy. However, we will analyse the site visit from a narrow perspective, referring to the knowledge about the materials industry and careers in industry. Then we will analyse the potential of reading and writing activities that could accompany a site visit. Finally, we take another look at motivational and interest issues regarding the site visit. In the discussion section, we will reflect on our experiences of conducting industry site visits.

THE INDUSTRY SITE VISIT MODEL

Development of the Site Visit through the Design-Based Research Approach

As a methodological framework, the Design-Based Research approach (DBR) (Design-Based Research Collective, 2003) has been followed in the design of an industry site visit. It is a general framework for design, development, implementation and evaluation of learning activities. DBR uses a pragmatic frame and emphasises three aspects: 1) a design process is essentially iterative starting from the recognition of the need to change praxis, 2) it generates a widely usable artefact, like learning activities, teaching modules or environments, 3) and it provides educational knowledge for more intelligible praxis.

The design procedure contains four main phases: 1) needs assessment; 2) theoretical problem analysis and definition of the objectives for the module; 3) design and production of the module; and 4) the evaluation of the module. During the design and production phase of the industry site visit model, there were several cycles: designing of a prototype, evaluation of the prototype in real classroom settings, revising three times the objectives based on the evaluation, and re-designing. This phase together with needs assessment has been the empirical problem analysis of the project (Edelson, 2002).

Theoretical problem analysis including our review of relevant literature was the starting point of the design. First we analysed an industry site visit as an out-of-school learning activity. Moreover, we followed the research on motivation and interest in module design. We also analysed learning of the basics of the materials, their properties and structure through classification, modelling, concept mapping, and Predict-Observe-Explain (POE) activities.

Evaluation of the module employed multiple methods as Bell, Hoadley and Linn (2004) have proposed. The main tool was coordinated communication between teachers and researchers (designers). Throughout the communication, researchers helped teachers to articulate their experiences gained whilst teaching. Moreover, multiple methods such as video recordings, student and teacher interviews, and questionnaires to evaluate the outcomes of the visit were used. We report here some theoretical and empirical problems considering the design of an industry site visit. Some preliminary results are reported elsewhere (Loukomies, Lavonen, & Juuti, 2009).

Industry Site Visit as an Out-Of-School Learning Activity

The industry site visit belongs to the growing area of developmental research conducted in out-of-school settings such as museums, science centres, planetariums or astronomical observatories, university laboratories, special exhibitions, workplaces including industry and events, as well as tours and field trips (e.g., Griffin, 2004; Martin, 2004; Falk & Storksdieck, 2005; Braund, Reiss, Tunnicliffe, & Moussouri, 2008). Often the emphasis has been on analysing the characteristics of learning environments and the effects of an environment on students' learning, interest in science, socialisation, personal development and responsibility and, moreover, on stimulating further studies or even a career related with science and technology (Braund & Reiss, 2004; 2006). For example, access to new learning materials and to 'real' science and technology or authentic practical work at a university or in an industrial laboratory can have an effect on both learning and interest (Braund & Reiss, 2004). While much of the related research and development effort has focused on learning outdoors, at museums or science centres or in university laboratories, we are interested here in out-of-school activities in an industry environment.

An industry related site visit uses a school-industry partnership or school-industry/business links. Out-of-school visits make it easier to discuss cross-curricular perspectives, awaken and deepen interest in sciences and show how humans apply sciences (Langsford, 2002). Moreover, the students meet positive and diverse role models of scientist or workers in industry (Bruce & Bruce, 2000). The importance of school-industry partnerships has been recognised also by education policymakers. For example, the New Zealand technology curriculum suggests that schools should seek to develop links with industry as a means of providing real-world examples of technology practice (Brunton & Coll, 2005). From the industry side Parvin and Stephenson (2004) mention three main products from the school-industry partnership: 1) recruitment, 2) contributing to the community, and 3) enhancing reputation. Although the framework or models for industrial site visits have typically been targeted at big business, also small scale local industry or small enterprises are possible sites.

There has been a long tradition of site visits in Finland, involving visits to science centres and museums, and to industry sites. Site visits are highly recommended in the National Core Curriculum that now also includes entrepreneurship education as an inter-curricular issue (FCCS, 1994; NCCBE, 2004). Education for the world of work and technology has been since 1979 a theme of Physics and Chemistry promoted by the curriculum of the upper secondary school. Furthermore, spokesmen of Finnish industry and industry interest groups believe that industry visits can provide students with experiences that are relevant from the point of view of student interest and motivation, and knowledge about science and technology and science and technology related careers. According to the training and labour survey by the Finnish Employers' Confederation, nearly half of all the companies, from every branch of industry and every part of the country, are involved in arranging work experience for students. According to Erkkila's (1996) survey industry site visits have been the most common method of enterprise (entrepreneurship) education. Many of the industrial sites which are open to visits have developed visitor programmes or student materials corresponding to school subjects. A common characteristic among all programmes and study materials is that they guide students to orientate themselves about the place before the visit, gather information at the visit place, and work with the new information afterwards. A typical site visit follows five phases: planning of the visit; organising preliminary activities in small groups; visiting and working in small groups; presenting reports; evaluating visits (Kuitunen & Meisalo, 1988, p. 146). The site visit model, developed for the Materials Science project is presented in Table 1.

The industry site visit model is flexible and allows wide variation when implemented in secondary schools. For example, visits could be organised in connection with a wide range of school subjects. Teachers also allow each student group to follow an individual time schedule while working at the industrial plant, whilst other student groups have each visited a different site. Visits could also be arranged in the form of two-day school camps. Lastly, students themselves could contribute in various ways to the dissemination and acceptability of the idea of industry site visits.

The site visit is analysed below from the perspective of student motivation and careers in science and technology. Table 2 describes how the students were guided to work as reporters and became familiar with the materials used/produced in the site and professions in the site.

Based on the findings of our first design and evaluation cycle, the students ($N = 21$) learned

Table 1. Phases of the site visit according to our model

1. Advance planning by teachers (0.5 – 2 hours):- Preliminary planning on a general level by teachers. Science teachers prepare students to study common materials and a student counsellor informs them about careers, occupations and relevant education while a Mother Tongue teacher teaches them about the process of writing and preparing for the role of a reporter, choosing the site to be visited and informing the school management of the plans to organise a visit to get formal permission, when needed
2. Teacher preparatory site visit (2 – 3 hours):-Find a contact person at the plant, co-planning with the contact person at the plant (discussion about the preliminary goals of the visit dealing with the materials science and technology contents and occupations, description of student skills and abilities)
3. Advance preparation with students (1 – 2 hours):- Form student groups for activities, present preliminary goals for the visit, everyone involved must be made aware of the nature and purpose of the visit (students should be informed of its relation to and implications for their recent, current or impending studies).
4. Practical preparations for the visit (0.5 – 2 hours)- Details of the visit, including the date, time, venue and programmes details must be given to all involved, including the participating students, accompanying teachers and non-teachers, and any staff at the site to be visited who will receive/work with the students. It must also be ensured that everyone involved is aware of contingency plans for unforeseen circumstances, first-aid facilities, emergency telephone numbers.
5. The site visit (2 – 4 hours):- sightseeing around the plant, activities in a small group, different topics as agreed with students
6. Activities after the visit (1 – 3 hours)- reports, discussion, displays, article publication.
7. Evaluation and feedback with teachers and site representatives
8. Collecting ideas for planning future site visits (15 – 30 minutes).

Table 2. Syntax of the site visit analysed in this chapter

Preparation of the visit
Preparing for the site visit and working after the visit requires the students to adopt the role of a reporter. Before the site visit the students find out what reporters do and how to write an article. An article is usually divided into different parts with subtitles and the most fundamental content of it is often summarised in the first paragraph. The article has a matter-of-fact style like news, but unlike news, the article should also include the writer's own opinions and assessments.
When working as a reporter, students first get acquainted with the subject, then plan the perspective of their article and then find a specific focus for their topic, if needed. After getting familiar with the home page of the visit site, the students choose the perspective for their article (for example materials used in the production, quality control, different occupations and the education needed for these occupations or where the products of the site are used). Students determine the aim of their article and what they want to tell their readers.
Site visit
During the site visit students who work as reporters make notes about what they see and hear. They also interview people who are essential to the subject of their article. Students ask questions which they planned beforehand and which are relevant to the subject of their article. The answers to these questions should be written down carefully and recorded if possible.
Students make general notes during the site visit. If students are allowed to take photos, they photograph subjects that are relevant to the topic they are going to write about. Students should note what they have photographed. These notes can be edited so that they explain the pictures in the final article. If it is forbidden to take photos, students find out if they can use pictures from the company web site or artefacts the company produces. In their article students are supposed to consider and introduce their topic in an extensive and multi-faceted way.
Working after the visit
After the site visit students write an article that is based on their notes and interviews. The technique of writing this article is similar to process writing, because a writer lets his or her group members read his or her drafts and elaborates on them further after the feedback given to him/her. After the articles have been presented and discussed in the classroom, they are possibly published to wider audiences. Finally, the students' learning outcomes and the whole industry site visit are evaluated.

during the visit about working life and professions, and how physics and chemistry are applied in practice. They did not learn much about physics and/or chemistry related issues during the visit. In general, the students evaluated the visit relevant to their interests and indicated that they were willing to have similar visits in the future.

Engage Students in Active Learning through Reading and Writing

Within the industry site visit, several types of texts, for example course books, encyclopaedias, commercial materials, brochures, and web-based texts, could be used as sources of information. Once a student understands the meaning of a given text, the reading activates the knowledge that the student already has regarding the subject and then initiates the learning process. This leads into integrating previous knowledge and new information into new knowledge construction. This process of constructing knowledge could be supported by writing or concept mapping activities, in that previous knowledge affects students' reading and writing, and other engagement. Moreover, topics, contexts, and discussions with peers affect students' engagements in science education.

Reading represents an active process in which the student constructs new knowledge by reconstructing a given text. At first, when glancing over a text, the student creates the 'first interpretation' that keeps being reinterpreted on subsequent readings. In addition, both reading and writing involve creating and modifying meanings. For instance, Tynjälä (1999) states that learning by reading, creating meanings, may be facilitated by carrying out writing exercises and discussions. The exercises that support learning may include the following (see also Baker, 1991): activating students' previous views and knowledge, comparing students' previous views and knowledge with the information featured in the text, dissecting the views presented in the text, applying the general principles presented in the text to imaginary practical settings, activating student learning in a small group (discussion, reflection), voicing critical opinions, and writing a summary. The most efficient strategies involve looking for the main points, organising the contents, taking notes, creating mind maps, drafting summaries, and anticipating what the text states next. Examples of strategies for active reading in the different stages of reading are presented in Table 3.

During the writing process, students develop reflective thinking skills. Therefore, a teacher should consider that creating a written piece of work is a whole process which involves students in gathering ideas—through reading, organising ideas into sentences and paragraphs, and they need to be put into a logical order, paragraphs need a main idea and supporting points with a few details, drafting, and editing.

Table 3. Strategies for active reading

Preparing for reading
Preparing for reading involves activating background knowledge which means thinking about what one already knows about the subject. This information is jotted down on a piece of paper in the form of a bullet point list or by sketching a mind map. At the same time those types of questions are generated which can be answered by reading the text. This preparing for reading makes the reading process easier and sets goals for the reader, which in turn help the reader to focus on the subject to be dealt with.
Connecting previous knowledge with new information
After reading, the active reader combines her previous knowledge with new learnt information. This combining can be facilitated either by writing about one's own thoughts after reading or answering questions generated during the reading process.
Taking notes
An active reader takes notes while reading by writing down the key words or creating a mind map on the most crucial issues in the text. She orders the contents by, for instance, recognising, classifying, comparing, and evaluating new information. In addition, she redefines the questions posed at the beginning and evaluates her work.
Posing questions
When students read independently, they can be asked to pose questions that come to mind when reading a given text. These questions voice points that students have not understood. The questions can be collected for all to consider.

Rivard (1994) has pointed out several factors that are crucial when trying to develop learning by writing. These factors involve, for instance, the requirements set for students by the writing exercise, the learning atmosphere in class and the students' meta-cognitive knowledge and skills. Thus such writing exercises that facilitate learning require students to reprocess, question, interpret and synthesise issues and principles already learnt. In contrast, traditional exercises only orient students to represent previous information, copy ideas directly from a source, such as, for instance, a course book or a website. Copying involves moral and ethical problems: It is essential that students learn how to indicate the sources of used information by reference lists etc.

Reading and writing activities presented in Table 4 have been designed to support active learning within the site visit. These activities characteristically result in report texts that outline the results of the given projects.

Learning which is based mainly on reading and writing faces new challenges when learners look up information on the Internet. For instance, reading may entail copying web-based information onto a notepad and writing in turn may entail pasting this information on the document-in-progress. In such cases, learners neither process information nor understand the meanings of new concepts - not to mention integrating these concepts within their own existing knowledge structure. Thereby learning presupposes processing the available information by, for instance, reading and writing. In addition, mind-mapping and knowledge structuring serve as effective means by which to process information (Bentley & Watts, 1989). In fact, by such a processing of information learners learn the skills and thinking necessary for fruitful information processing to take place. Moreover, when processing information in small groups, learners practise collaboration and communication skills as well. These skills are a pre-requisite for professionals serving in various fields of expertise (Tynjälä, 1999).

Based on our first design and evaluation cycle the students learned quite well during their reading and writing activities, how to write an article. Several articles fulfilled the criteria of an article as a text type. There were all the required parts in them except the subtitles. Students had different perspectives in their articles. The most comprehensive articles dealt with the international relations of the enterprise and job descriptions of the employees in the company. On account of writing the articles students' conception of the professions in the technology industry is presumptuously quite extensive. Students met representatives of both genders in occupations of the technology indus-

Table 4. Reading and writing activities supporting active learning

Journals and blogs. A basic method for writing a story is to connect the story line with some sort of tangible action, and both journals and blogs fulfil this function.
Studies based on interviews. The previous example drews from students' own actions as the source of information. Interviews are used to acquire information about the actions of other people. Use an MP3 player or a tape recorder, or take notes while interviewing. Report your results in class or publish.
Reports on branches of material science industry. When writing, information can be drawn from various course books and specialised publications, newspapers, magazines, the Internet the home page of the company to be visited, as well as websites of various organisations, newspaper databases, and home pages of magazines and journals serve as excellent sources of information.
Manuals. Our world abounds in various types of manuals. Once you create a manual for another person, you at the same time learn the topic in question. You could pay special attention to visualisation and the layout.
Booklets. The idea of a booklet is identical to the manual discussed above. You could pay careful attention to the booklet's structure, foreword, headings, contents, visualisations and layout."

try. By interviewing these people in order to get data for their articles, students created a personal relationship with science related occupations.

Motivating Students through an Industry Site Visit

Students' interest towards science and technology and interest in careers related to those fields have been intensively researched since the 1960s. Science in general is quite interesting for students, but most students, especially girls, do not find school science and technology or careers and occupations in those fields interesting (Woolnough, 1996; Jones, Howe, & Rua, 2000; Osborne, Simon, & Collins, 2003; EU, 2004; EU, 2005). According to international ROSE (Schreiner & Sjøberg, 2005) and SAS surveys (Sjøberg, 2000) the social position of scientists and engineers has weakened in developed countries and their lifestyle appears unattractive to students: working hard and alone in a laboratory. Science and technology occupations are perceived as being under-estimated in relation to the workload. Furthermore, several students had stereotypical views of science and technology occupations, like 'I want to work with people and scientists do not do that', 'I think engineering is a man's job', and 'I want to be a nurse so I do not need science'. However, discovering new things and an ability to do something useful were recognised as being important reasons for studying science and choosing an occupation in science. Consequently, it seems that students were unaware of the breadth of career opportunities opened up by studying science and technology, neither are they familiar with the characteristics of the careers in the field (OECD Global Science Forum, 2005). Moreover, there seems to be a strong link between interest in science and interest in careers in science and technology.

Our analysis on motivational aspects in the framework of an industry site visit is based on the theory of Deci and Ryan (2000). A person's way of thinking has an important role in the process of motivation. Motivated behaviour may be self-determined or controlled and they involve different reasons for behaving. Self-determined or autonomous behaviour is behaviour which arises freely from one's self. Controlled behaviour, in contrast, means that the behaviour is "controlled by some interpersonal or intra-psychic force, e.g., a curriculum. The motivation styles in their self-determinant theory (SDT) are extrinsic and intrinsic motivation. Intrinsic motivation has positive effects on the quality of learning. Intrinsically motivated behaviour originates from the need to feel competent and self-determined (Deci & Ryan, 2000). Extrinsically motivated behaviour is instrumental in nature. Such action is performed for the sake of some expected outcome or extrinsic reward or in order to comply with a demand.

Central to SDT is the concept of basic psychological needs that are assumed to be innate and universal. These needs are the need for autonomy, competence, and relatedness (the need to belong to a group). In addition, self-determined learning could occur when the content or context of the activity is considered as interesting, enjoyable or personally valuable by a learner. According to Schiefele (1999, p. 259), students could engage in a specific learning activity if the activity itself is interesting, enjoyable, or otherwise satisfactory. Schiefele suggests that interest is a content-specific concept and a facilitator of learning, and that it consists of two kinds of valences: feeling-related and value-related valences. Feeling-related valences are feelings that are associated with a topic or an object, for instance, feelings of enjoyment and involvement. Value-related valences refer to the attribution of personal significance to an object.

Lavonen, Juuti, Byman, Uitto and Meisalo (2006) examined lower secondary school students' interests in the characteristics of occupations or careers as a part of the theoretical problem analysis of designing an industry site visit. Students, in general, appreciated an occupation allowing them time for their interests, creative activities, and social interaction. They concluded that student interest could be influenced during the site visit through different role models.

Student motivation and interest could be influenced by learning activities in the site visit. Before the visit, students could orientate to the site and form questions. During the visit, the feeling of autonomy, social relatedness, and competence could also be emphasised by guiding students to work as a journalist and to prepare a report of interviews in a small group. The students could search for related information over the Internet, read relevant texts online, and write their reports on word processors with process writing techniques.

In summary, the following four features of a learning activity could motivate students in learning:

1 autonomy-supporting activities or support for choices the student-centred learning methods or use of ICT; co-planning of the learning activities with students having choices of how to study;
2 support for students' feeling of competency choice of tasks, which are possible for the student to solve; choice of tasks, which guide students to recognise their previous knowledge and build up new knowledge; choice and use of constructive evaluation methods;
3 support for students' social relatedness choice of activities and ICT use which help students to feel close to peers; feeling that the students can trust each other;
4 support to interest awakening feeling related components of interest (enjoyment) awakening value related components of interest (usefulness) utilising new material science content and context.

A site visit could guide a teacher to emphasise these four features of learning. Moreover, the reading and writing activities above could emphasise these features. What should be noted is that these features of a learning activity should be emphasised and taken into account while planning an activity.

During our DBR project we collected quantitative and qualitative data about student motivation (Loukomies *et al.*, 2009). Students with different motivation styles mentioned in the interview that the site visit had a positive influence on their interest and motivation to learn science. However, students' opinions showed that they had few possibilities to plan their own studying during the site visit. Therefore, the students should engage more in the planning, even though they were satisfied with everything being arranged for them. Based on the interview data, the most important features for the students' motivation and interest were as follows:

- **Support for social relatedness.** It was important to the students to work in groups and spend time together with their classmates.
- **Interesting context for studying.** It was important to the students to get outside the school and see physics and chemistry applied in real life.
- **Interesting contents.** The students remembered well the device they were presented.
- **Possibility for hands-on-learning,** which supports both feelings of autonomy and competence.

CHALLENGES AND FUTURE RESEARCH DIRECTIONS

We have presented here an industry site visit as a framework for enhancing scientific literacy and motivating students in science learning. Based on our preliminary studies and evaluation cycles we have clarified a versatile approach toward student learning and motivation. For further ascertaining our findings, we suggest further studies employ more empirical data e.g., video recordings made during visits. In DBR, testing of the prototypes reveals more focused problems to answer by further testing. In ongoing studies, the following questions need to be answered concerning the site visit model:

- What indications of motivation and interest can be identified during site visits?
- How can feelings of autonomy, competence and social relatedness, as well as interest and motivation be supported during site visits?
- How does learning about careers affect motivation and interest?
- How can students be guided to learn through reading and writing activities?
- What is the role of ICT in this process? How much relevant information can pupils find on the Internet and from other sources?
- What and how do they learn in reading and writing activities?
- How can students be guided to learn about science careers?
- How can students be guided to learn about materials, their properties and behaviour as well as their structure?
- What is the role of inquiry activities in learning about materials?
- How can inquiry activities be combined with the site visit?

We have already designed and evaluated a few inquiry activities by the second design and evaluation cycle. Our preliminary analysis suggests that students learn science, when they meet materials in different situations like in everyday contexts, at an industry site, and within inquiry activities. Likewise, the site visit provided students with possibilities to recognise properties and behaviour of materials in different situations.

When considering motivation, there are two main challenges: 1) The first problem is taking into account the feeling of autonomy. Even though teachers should make some preliminary preparations, it is challenging for them to discuss with students the preliminary goals of the visit. It also seems to be difficult for teachers to use more time for planning of the tasks and there are further difficulties related to reporting (structure of the report, ICT use in reporting, the evaluation of the visit and the report). 2) The second problem concerns social relatedness. Even though all the interviewed students emphasised the significance of working in a group together with friends, the significance of working together in the context of science learning did not emerge from the students' answers.

CONCLUSION

In this chapter, we have first discussed a functional and contextualised definition of scientific literacy as an element of multiple-literacy required for modern societies. Then we have introduced a model of the industry site visit with some related activities, and analysed how it might contribute to students' scientific literacy as conceptualised by us. We have also presented some empirical results of both cognitive and affective outcomes, based on the first design and evaluation cycle of our DBR project.

As a form of out-of-school education, the site visit model appears to be well suited for enhancing several elements of scientific literacy: knowledge *about* science, interconnections of science to society (and especially to industrial life), and student's interests and attitudes to science. Our empirical results indicate that during the visit students learn mostly about the professions and practical applications, and less about the "pure" science. This finding is in accordance with the objectives of the model; our aims were concentrated in facilitating students' entry to working life instead of learning content knowledge. Students considered the visit as relevant to their interests, which was also essential in our approach to scientific literacy. The site visit had a positive influence on students' motivation to learn science.

Furthermore, our model included activities of learning through reading and writing that contributed to *the fundamental sense of scientific literacy* (Norris & Phillips, 2003). The articles, written by the students, demonstrate that students searched for materials through the eyes of the staff: science and technology had become more understandable by being related to real life, therefore becoming more humanistic to students.

Finally, several challenges were identified for developing the site visit further in the perspective of motivation and interest. It is clear that the teacher has the overall responsibility for organising the visit, contacting parents, finding necessary resources, and considering all safety precautions. However, she should support students' feeling of autonomy, social relatedness and competence. Therefore, her role in the practical phase of an industry site visit should be as an adviser and a consultant, and students should be involved at all stages of organising the visit. The activities will begin with careful co-planning, including all aspects of choosing an interesting site.

ACKNOWLEDGMENT

The work described in this paper has been supported by the Materials Science project (SAS6-CT-2006-042942-Material Science)

REFERENCES

Baker, L. (1991). Metacognition, Reading and Science Education. In C.M. Santa & D.E. Alvermann (Eds.), *Science Learning: Processes and Applications*. Newark: International Reading Association, IRA.

Bell, P., Hoadley, C. M., & Linn, M. C. (2004). Design-based research in education. In M. C. Linn, E. A. Davis, & P. Bell (Eds.), *Internet environments for science education* (pp. 73-85). Mahway, NJ: Lawrence Erlbaum Associates.

Bentley, D., & Watts, M. (1989). *Learning and Teaching in School Science: Practical Alternatives*. Milton Keynes, UK: Open University Press.

Braund, M., & Reiss, M. (2006). Towards a More Authentic Science Curriculum: The contribution of out-of-school learning. *International Journal of Science Education*, *28*(12), 1373–1388. doi:10.1080/09500690500498419

Braund, M., Reiss, M., Tunnicliffe, S. D., & Moussouri, T. (2008). *Getting the most from 'Out-of-School' learning in science: What should teachers know?* Retrieved July 15, 2008, from http://www.ase.org.uk/htm/conferences/atsi/pdf/F9.pdf

Braund, M., & Reiss, M. J. (2004). *Learning science outside the classroom*. London: Routledge.

Bruce, S. P., & Bruce, B. C. (2000). Constructing images of science: people, technologies, and practices. *Computers in Human Behavior*, *16*(3), 241–256. doi:10.1016/S0747-5632(00)00004-2

Brunton, M., & Coll, R. K. (2005). Enhancing Technology Education by Forming Links with Industry: A New Zealand Case Study. *International Journal of Science and Mathematics Education*, *3*, 141–166. doi:10.1007/s10763-004-1516-1

Deci, E. L., & Ryan, R. M. (2000). The "what" and "why" of goal pursuits: Human needs and self-determination of behavior. *Psychological Inquiry*, *11*, 227–268. doi:10.1207/S15327965PLI1104_01

Design-Based Research Collective. (2003). Design-Based Research: An Emerging Paradigm for Educational Inquiry. *Educational Researcher*, *32*(1), 5–8. doi:10.3102/0013189X032001005

Durant, J. R. (1993). What is scientific literacy? In J. R. Durant & J. Gregory (Eds.), *Science and culture in Europe* (pp. 129-137). London: Science Museum.

Edelson, D. C. (2002). Design research: What we learn when we engage in design. *Journal of the Learning Sciences*, *11*, 105–121. doi:10.1207/S15327809JLS1101_4

Erkkila, K. (1996). *Enterprise Education in the Case of Finland*. Paper presented at the World Congress of Comparative Education Societies, Sydney, Australia.

EU. (2004). *Europe needs more scientists!* Brussels: European Commission, Directorate-General for Research, High Level Group on Human Resources for Science and Technology in Europe. Retrieved November 7, 2007, from http://www.tekes.fi/EU/fin/6po/tutkijaliikkuvuus/gago_report_final_en.pdfwww.tekes.fi/EU/fin/6po/tutkijaliikkuvuus/gago_report_final_en.pdf

EU. (2005). *Europeans, science and technology. Eurobarometer 224*. Brussels: The Directorate General Press and Communication of the European Commission. Retrieved November 7, 2007, from http://ec.europa.eu/public_opinion/archives/ebs/ebs_224_report_en.pdf

Falk, J., & Storksdieck, M. (2005). Using the Contextual Model of Learning to Understand Visitor Learning from a Science Center Exhibition. *Science Education*, *89*, 744–778. doi:10.1002/sce.20078

FCCS. (1994). *Framework curriculum for the comprehensive school (in Finland)*. Helsinki: State Printing Press and National Board of Education.

Global, O. E. C. D. Science Forum. (2005). *Papers presented in the conference Declining enrolment in science and technology in November 2005*. Retrieved November 7, 2007, from http://www.oecd.org/document/52/0,3343,en_2649_34319_35398132_1_1_1_1,00.html

Griffin, J. (2004). Research on Students and Museums: Looking More Closely at the Students in School Groups. *Science education*, *88*, 59–70. doi:10.1002/sce.20018

ITEA. (2000). Standards for technological literacy. Content for the study of technology. *Technology Teacher*, *58*(5), 8–13.

Jenkins, E. W. (1994). Scientific literacy. In T. Husen & T. N. Postlethwaite (Eds.), *The International Encyclopedia of Education* (Vol. 9, pp. 5345-5350). Oxford, UK: Pergamon Press

Jenkins, E. W. (1997). Scientific and technological literacy: Meanings and rationales. In E. W. Jenkins (Ed.), *Innovations in science and technology education* (Vol. 6, pp. 11-42). Paris: UNESCO.

Jones, G. M., Howe, A., & Rua, J. (2000). Gender differences in students' experiences, interests, and attitudes toward science and scientists. *Science Education*, *84*(2), 180–192. doi:10.1002/(SICI)1098-237X(200003)84:2<180::AID-SCE3>3.0.CO;2-X

Kuitunen, H., & Meisalo, V. (1988). Science and technology education and industry. In C. Layton (Ed.), *Innovations in science and technology education 2*. Paris: UNESCO, 141-154.

Langsford, S. (2002). Museums as resources for science teaching. *Australian Primary & Junior Science Journal*, *18*(1), 17–19.

Lankshear, C., & Knobel, M. (2003). *New Literacies: Changing Knowledge and Classroom Learning*. Buckingham, UK: Open University Press.

Laugksch, R. C. (2000). Scientific literacy: A conceptual overview. *Science Education, 84*(1), 71–94. doi:10.1002/(SICI)1098-237X(200001)84:1<71::AID-SCE6>3.0.CO;2-C

Lavonen, J., Juuti, K., Byman, R., Uitto, A., & Meisalo, V. (2006). Job characteristics found important for their future career choice by ninth grade students. In S. Yoong, M. Ismail, A. Nurulazam, F. Salleh, F. S. Fook, L. C. Sam, & M. N. L. Yan (Eds.), *Proceedings of the XII IOSTE SYMPOSIUM: Science and Technology Education in the Service of Humankind,* School of Educational Studies, Universiti Sains Malaysia, Penang, Malaysia.

Loukomies, A., Lavonen, J., & Juuti, K. (2009). *The Motivational Features of an Industry Site Visit*. Paper submitted to ESERA 2009 Conference. Istanbul.

Martin, L. (2004). An Emerging Research Framework for Studying Informal Learning and Schools. *Science Education, 88*, 71–82. doi:10.1002/sce.20020

McEneaney, E. H. (2003). The Worldwide Cachet of Scientific Literacy. *Comparative Education Review, 47*(2), 217–219. doi:10.1086/376539

NCCBE. (2004). *National Core Curriculum for Basic Education 2004* [Electronic Version]. Helsinki: National Board of Education. Retrieved from http://www.oph.fi/english/page.asp?path=447,27598,37840,72101,72106

Norris, S. P., & Phillips, L. M. (2003). How literacy in its fundamental sense is central to scientific literacy. *Science Education, 87*, 224–240. doi:10.1002/sce.10066

OECD. (2007). *Assessing Scientific, Reading and Mathematical Literacy: A Framework for PISA 2006.* Paris: OECD Publications.

Osborne, J., & Dillon, J. (2008). *Science Education in Europe: Critical Reflections. A Report to the Nuffield Foundation*. London: The Nuffield Foundation.

Osborne, J., Simon, S., & Collins, S. (2003). Attitude towards science: a review of the literature and its implications. *International Journal of Science Education, 25*(9), 1049–1079. doi:10.1080/0950069032000032199

Parvin, J., & Stephenson, M. (2004). Learning Science at Industrial site. In M. Braund & M. J. Reiss (Eds.), *Learning science outside the classroom* (pp. 129-137). London: Routledge.

Rivard, L. P. (1994). A Review of Writing to Learn in Science: Implications for Practice and Research. *Journal of Research in Science Teaching, 31*, 969–983. doi:10.1002/tea.3660310910

Roberts, D. A. (2007). Scientific literacy/Science literacy. In S. K. Abell & N. G. Lederman (Eds.), *Handbook of research on science education* (pp. 729-780). Mahwah, NJ: Lawrence Erlbaum Associates.

Ryder, J. (2001). Identifying science understanding for functional scientific literacy. *Studies in Science Education, 36*, 1–44. doi:10.1080/03057260108560166

Schiefele, U. (1999). Interest and learning from text. *Scientific Studies of Reading, 3*, 257–279. doi:10.1207/s1532799xssr0303_4

Schreiner, C., & Sjøberg, S. (2005). Et meningsfullt naturfag for dagens ungdom? [A meaningful school science for today's youth?]. *Nordina: Nordic Studies in Science Education, 2*, 18–35.

Shamos, M. H. (1995). *The myth of scientific literacy.* New Brunswick, NJ: Rutgers University Press.

Sjøberg, S. (1997). Scientific literacy and school science – Arguments and second thoughts. In S. Sjøberg & E. Kallerud (Eds.), *Science, technology and citizenship* (pp. 9-28). Oslo, Norway: NIFU Rapport 10/97.

Sjøberg, S. (2000). *Science And Scientists: The SAS-study. Cross-cultural evidence and perspectives on pupils interests, experiences and perceptions. - Background, Development and Selected Results*. Department of Teacher Education and School Development, Acta Didactica 1/2000, University of Oslo. Retrieved November 7, 2007, from http://folk.uio.no/sveinsj/SASweb.htm

Tynjälä, P. (1999). Towards Expert Knowledge? A Comparison Between a Constructivist and a Traditional Learning Environment in University. *International Journal of Educational Research, 31*(5), 357–442. doi:10.1016/S0883-0355(99)00012-9

Woolnough, B. E. (1996). Changing Pupils' Attitudes to Careers in Science. *Physics Education, 31*(5), 301–308. doi:10.1088/0031-9120/31/5/020

Zeidler, D. L., Sadler, T. D., & Simmons, M. L. (2005). Beyond STS: A Research-based Framework for Socioscientific Issues Education. *Science Education, 89*(3), 357–377. doi:10.1002/sce.20048

Chapter 14
The Networked Pupil and the Vanishing Paper Trail: A Re-Culturing of the Learning Process in the 21st Century

Andrew Marks
Independent Researcher, UK

ABSTRACT

This chapter offers a theoretical discussion of the impact upon the processes of teaching and learning by the increasing predominance of ICT in the school scenario, and posits other future scenarios where the social realm of the classroom could disappear altogether. Questions are raised as to what the consequences of this could be, both for teachers and learners (as agents), but also for the notion of school science as an entity (and indeed the socio-structuralism of the school itself).

INTRODUCTION

I am 'not a teacher', but have grown to understand the teaching scenario based upon having taught in the university sector, teaching sociology. I might also add that neither am I a scientist, I merely operate as a social scientist with an ever increasing observing interest in the world of science teaching and learning. My principal position is as a theoretical researcher who has developed an increasing interest in the science teaching and learning scenario, and the questions that arise from it: how does the ever increasing role of ICT change the relationship between the learner agent and the teaching-agent within the structure that is the school?

DOI: 10.4018/978-1-60566-690-2.ch014

In years gone by, the education process was a balancing of the literal and the practical. The 20th century pupil would quite probably have had a learning day, combining the reading of books, the creation and or manipulation of physical materials during Metalwork or Woodwork classes, and possibly also a sports class of some sort. However, times are changing. The ever-greater dependence upon technology in all aspects of the practical and social world in the current century appears to be changing the process of education. Is education becoming something of a 'Practical'/'Literal'/'T echnical' triad – having previously merely been a 'Practical'/'Literal' axis - with skills being imparted by the education process now being conveyed from all points of this triad and received as necessary from it?

Was it, perhaps, always this way, and it is merely that with the ever-increasing dependence upon ICT, all forms of learning are becoming technical (and quite possibly, 'virtual')? Could, for example, a study of literature be partaken without ever having recourse to read a book – such as by downloading each and every chapter as (and when) required?

As such, the technical/ICT related aspects of the learning processes could be becoming dominant within the classroom – and the triangle alluded to above would certainly not be equilateral in proportion.

- What would or could the consequences of this be?
- Are the processes of learning to be improved?
- What effects, moreover, could it have on the learning environment for the pupils themselves?
- Would any cultural changes related to expansion and dominance of ICT as a teaching and learning tool necessarily work for the good of the pupils in their wider social networks?
- How will the ever-greater dominance of ICT in the learning sphere alter the relationship between pupils and teachers?

Could the ICT based aspect of the learning process indeed be (eventually) becoming the centrality of the learning culture – and indeed the centrality of childhood development as the 'classroom' becomes less and less of a norm? Moreover, will the alteration of the physicality of learning also change the power relationship between teacher and pupil – and to what end? Finally, it must be asked, what will the effect of all this be for the teaching and learning of science?

ICT AND THE CULTURE OF LEARNING

Every society sets up a certain ideal of man, of what he should be as much from the intellectual point of view as from the physical and moral. This ideal is, in some degree, the same for all members of society, but it becomes differentiated beyond a certain point, *according to specific groupings that every society contains* in its structure. It is this ideal, which *both integral and diverse*, that is the focus of education. (Durkheim, 1922, p.203, *emphases added*)

'Computers don't have to be motivated to do what you tell them to. But *children do*.' (Sax: 2007, p.34, *emphasis added*.)

Law (1991, p.165) argues that no social relations are purely 'social' in character; rather they are 'heterogeneous', being embodied in a series of corporeal, textual, natural and technical materials. Thus, the social relationship of the classroom experience becomes dependent upon not only the verbal and physical interaction of teacher and pupil (and pupil with pupil) and teacher and pupil with books, material etc., but also (above all for the case of this discussion) with *technology*. Will the ever-increasing predominance of ICT in the learning environment always necessarily have a positive effect for the participants[1]?

ICT AND SOCIAL CLASS

There is evidence to suggest that ICT can have a downward effect on learning outcomes of pupils when in combination with social class, especially when factored against computer ownership (and corresponding familial wealth). Wainer *et al* (2008) found that ...

Frequency of computer use had, in general, a negative effect on test result, and the negative effect increased for younger and poorer students. Computer ownership had, in general, a small but positive effect on the test results for older students, and no effect for 4th graders. Internet access has a negative effect for younger and poorer students and a positive effect for 11th graders. Finally, whether the teacher used computers or Internet as pedagogical tools had no effect on the student's test results for all social economic classes and grades. (Wainer et alWainer et alWainer et alWainer et al: 2008, p.1417, emphases added)

This last point seems especially interesting, since it seems to be suggesting that the teachers' impartment of ICT based learning has neither relevance to nor reflection upon the learning of pupils. Is this further indication of an ever-widening generation gap in technological prowess in society? (This is the 'Digital Inequality' alluded to by Iske *et al*, 2008 p. 131).

Is, for example, the 40-year-old teacher less technologically gifted than his/her 8-year-old son or daughter? Moreover, the claims that computer *use* has a generally negative effect upon the young and the poor students whilst ownership produces improved test results seem to be self-asserting and somewhat naive statements: if you are poor and young you are less likely to get access to nor practice upon the necessary ICT, whilst if you're older and from a richer background you're much more likely to have the machinery in your home. However, Angeli and Valanides (2009, p.154) have suggested that ICT based Technological-Pedagogical-Content-Knowledge (ICT-TPCK) can be utilised in making difficult-to-understand topics more easy to present and to comprehend (thus easier both for the teacher and the pupil). Thus this particular argument appears to cut both ways, having evidence to support both charges.

ICT AND GENDER

Furthermore, there could also be gender – and gender/class - related issues here, since male students are more likely to want to pursue the 'practical' – especially if that course can immediately be seen as saleable in the labour market – then practical courses are far more likely to have a masculine bias among their uptake.

Not only is access to the world of work assumed to lie through education but compulsory schooling is an ideological practice which engineers society's consent for the operation of a ruthless, dehumanised labour market in which access to privileged employment is governed by the possession of cultural skills and qualities … *which cannot be gained in school.* (Branson, 1991: 92, emphasis added.)

That working class males specifically regard education as having little or no utility value is well documented and has a long historical, and international, strand to it (See Hoggart, 1957, 1995: Keddie, 1971: Willis, 1977: Hurt, 1979: Payne, 1983: Heath, 1987: Murray, 1990, 1991: Haythorne, 1991: Sherman, 1991: Muncie, 1999: Marks, 2000a, b, 2003: Riddell *et al*, 2001. As Chapman (2000, p 8) reasonably asks, should education appeal to 'male' attitudes by offering only work related courses? Or should we try to break the linkage? Is there more to education than merely training for the world of work? (This latter is an ironic question.)

What differences would or could ICT have for the gender issues in the modern classroom? Sax (2007, p 58) found that computer games have a far more masculine uptake among those children who play them[2]; this stemming (he suggests) from the illusions of 'power' that such games can give these young males. Would this perhaps offer the possibility of a greater masculine uptake of ICT usage in schools? And if so, what would/could this mean for the school system?

THE LEARNING PROCESS AND THE DECLINE OF THE 'PHYSICAL'

It had been alluded-to elsewhere (Marks, p.2005) that university environments are becoming ever more 'virtual' and less 'physical' in their interactive processes, with 'on-line' learning forming an ever greater part of the learning 'culture' of the higher education sector. Could this also be becoming true of the ever-more-virtualising *school* context?

Cultural Changes

An old science fiction cliché exists that nothing can date an era like its vision of the future. Are we offering dateable socio-scientific clichés of early 21st century analytic method for future observation? In our ever faster movement towards virtualisation of everyday lifestyles with ever more work being facilitated from home via the home computer and 'home office' scenario, can we foresee a scenario where the physical domain of the school will no longer exist to any great extent (no longer being necessary)? Will we reach a point in time when children will be able to download all of their necessary lessons and read and write their own coursework (on computer) to e-mail to the teacher (whom they may never have recourse to meet)? Were this to be the case, what 'culture' would the learning world form? Friere (1993, p 54) offered the following as examples of the 'culture' of the learning/ education processes and the interactions involved.

- The teacher teaches and the students (pupils) are taught;
- The teacher *knows everything* and the students *know nothing*;
- The teacher thinks and the students are thought about;
- The teachers talk and the students listen – *meekly*;
- The teacher disciplines and the students are disciplined;
- The teacher chooses and enforces his/her choice and *students comply*;
- The teacher acts and the students have the illusion of acting through the action of the teacher;
- The teacher chooses the programmes content, and the students (who were not consulted) adapt to it;
- The teacher confuses the authority of knowledge with his or her own professional authority which s/he sets *in opposition* to the freedom of the students;
- The teacher is the Subject of the learning process, while the pupils are mere objects (*emphases added*).

At each point Friere appears to be offering an image of a culture of conflict in the school scenario. This is clearly a dated scenario and becoming ever less relevant as ICT comes to further predominate in the learning process. If Iske *et al's* (2008: ibid) 'digital inequality' is (as they seem to be suggesting) a largely age-based phenomenon? As such, is the teacher in a learning process behind the knowledge and skills already acquired behind their pupils? And thus could they well have less useful knowledge for the manipulation of the ever-expanding ICT culture than their pupils already do? Would/could the pupils, in these circumstances, find the teacher somewhat redundant? Is, therefore, the classroom becoming ever more relegated to the twentieth century and earlier?

Learner/Teacher Power Relations in 21st Century (and Beyond)

Clearly, if one wishes in this way to explore the construction and distribution of powers to act, or indeed to exercise discretion, then an important part of the analytic trick is to avoid reductionism. To assume that the boss *has* power is only helpful if we ask how that power is constituted relationally. (Law: 1991, 183)

If Law's argument relating to the role of technology in the social realm (see earlier in this piece) is to be accepted as a given for this discussion, then one must take on board the notion that the ever increasing embracement of ICT in the learning process will change the balances of power within pedagogical relationships and with pupil's social networking generally.

Can a scenario be envisioned where the school as we currently visualise it no longer exists? Will, as was posited earlier in this piece, a social realm exist whereby the learning process is conducted entirely 'virtually', with the need for pupil/teacher interaction being minimal (if it is needed *at all*)? Could the 'classroom' be a physicality experienced at home as the learner employs their home-hardware to download their classes as necessary and possibly at a speed that suits them? Then returning completed work to an abstracted 'teacher' persona on the other end of the 'virtual' network. *And never the twain shall meet?*

Could we be moving, as was suggested earlier, from a 'teacher' based class sociality, as envisioned by Friere (ibid.), to an ever-greater 'technically' biased heterogeneity (as Law offers)? Furthermore, from this position, could the learning sociality be becoming ever less human in its construction as its very human interaction diminishes in comparison with technical interfacing and usage? Could we have downloadable, non-physically-specific, non-comparative and non-physically-interactive learning processes to come?

What could the consequences of this be for science teaching and learning? Will this, ultimately *be* scientific?

POWER RELATIONS OF ICT AND SCIENCE TEACHING

It must surely be taken as a given that 'science' is a process of networking via experiments, comparisons of data, acceptance or rejection of findings and the reading and comparison of other experiments with one's own work. How would a scenario with ever-greater detachment of pupils from the socially structured classroom setting, with a teacher (at least nominally) controlling the work if for no other reason than the safety of all concerned, and with social interaction between pupils *within* that setting, remain scientific?

If pupils reach a point whereby all that they learn is merely picked up from a computer file, and never actually practiced, then how can this process be experimental, and thus how can it be scientific? Surely the consequences of this for school science and its teaching and learning would be phenomenal?

Could we be foreseeing a time whereby the learning of science becomes merely the adoption of pre-governed and dated data, with no recourse to experimentation? Will Science, in such a scenario, be seen becoming as purely 'literal' in construction as (for example) Classical Studies or English Literature? Whilst eventually the paper trail vanishes forever into the computer (for better or worse) taking the 'literal' with it, so might it also take the 'practical', as there remains no need for anything to actually be practiced – merely looked at.

DISCUSSION

An interesting discussion piece was offered by Waller (2007), where he brought into consideration five myths regarding the status of technology as a social entity:

Firstly, that technology is neutral; secondly, that placing technology in the school leads to 'learning gains'; thirdly, that greater access for teachers to educational technology makes them 'more proficient and efficient'; fourthly, the greater equipping of schools with ICT 'leads to school improvement', and finally that students need to have greater 'technological literacy in order to be employable' (p2).

The first four of these myths are particularly relevant to this discussion. Let us examine them in turn.

The Neutrality of Technology

Note that to ask about the distinction between people and machines is, in part, an inquiry into the character of agency: what it is or what it takes, in part, to be a human being. (Law, 1991)

Waller himself (2007) argues against the very neutrality of technology; pointing out, firstly, that the majority of the technology currently offered to schools is made by large corporations localised in the USA (p7) and that this could have the prospect of a 'privatising' of the education sector. This does, however, neglect to a large extent to consider the concept of agency in the learning process: how far are pupils (and indeed their teachers) in control of the learning destiny of the learning agent? It is probably far too early to gain a reasoned answer to such a speculative question, yet it needs to be held in consideration nonetheless.

Is the learner-agent more likely to get an education embracing freedom from structurally dominant issues (for example class, race or gender)? Will rising levels of ICT and virtuality in the learning process bring about greater levels of equality in the learning processes – or could entirely the reverse scenario occur? Certainly the geographical dominance of particular schools could vanish entirely, as a pupil from (for example) Manchester would have no recourse not to attend a school in (for example) London as the physical specificity of the location of the school *will have no* necessity for attendance in the first place.

Indeed, much discussion has been made over the years of the Australian 'School of The Air'. After 1961, television increased in importance for teaching science and in rural areas and the historical educational trend with regard to rural education in Australia suggests that mass media is favoured. (Higgins, 1981) Higgins' account suggests that for the geographically isolated (which Australia can offer a significant and disparate proportion) the 'virtual' learning experience (though this is not a term he uses) is the one to be favoured (See also McAtee and Zani, (1975), Tomlinson et al (1985) and Rogers (1999). – though he goes on to argue that finances would be better spent on enhancement of radio and telephone communication as the primary means of distance education going on to suggest that the country's educational finances ought to be used on improved transportation links (as well as 'increased regionalisation and centralisation' as offered by Kitt (1983) though in some ways this might go towards defeating the purpose intended.) and citizens-band radio systems – though this being a relatively old piece, and published before the rise of the Web, such an argument might now be redundant.

A more recent discussion of the 'School of the Air' concept, offered by Sanderson (1998) describes the concept as combining:

... correspondence study, radio lessons, home visits, and mini-school meetings, (p31)

It could be argued that in the latter two examples this seems to be missing the point somewhat, since distance is what is being aimed for (and removal of geography as a constraint of power). However, access to Internet facilities might not be found everywhere in remote areas, and as such this may not yet be an option for all. However, the great advantage of Web based learning over telephone and television/radio learning is the lack of temporal confinement to it. The Web based learner can learn as and when s/he likes, rather than at the cue of the teacher. This can provide all kinds of social advantages.

Technology and Learning Gains

Again, Waller (2007, p7) offers the claim that:

Used well digital technologies have the potential to improve achievement in our schools and col-

leges, to boost the prospects of British industry and commerce, to offer opportunities to all learners, particularly those who would otherwise be excluded, and to significantly enhance our quality of life.

However, he goes on to argue (by way of self-contradiction) that there are no automatic gains to be made from incorporating ICT into the classroom – noting that technology will regard significant pedagogical change, which will not occur by itself, but needs to be created. Will this occur? If so, how? (And by *whom*?)

Greater Access to Educational Technology for Teachers Makes them More Proficient and Efficient

Technology affects education at all levels. Children are naturals at using technology. They use technology in all aspects of their lives; for that reason it is imperative that *teachers not only keep up with the latest technology,* but also find ways to integrate technology, including the media, into classroom instruction. (Lacina, 2005, p 118 *emphasis added*)

Waller (2007) is suspicious of data offered by the DfES regarding the 'improved professional development status' and the increased capacity and opportunity provision for career development and status. Suggesting that the data offered in this instance is self reported by school heads (interested in gaining improved reputations) and not indicative of possible/probable increased workloads and lowered work standards in the teaching profession. This, however, may be open to change as the technology itself becomes increasingly predominant within the profession and the learning culture.

Indeed, what is this likely to make of the debate between digital-literacy and multiple-literacy? The former currently surely implies the latter, as one requires levels of linguistic (and thus book competent) literacy to become digitally literate. However, is there an argument to be made for suggesting that as time passes, and as paper based learning becomes increasingly redundant and outmoded – with essays etc. being written on a computer and teacher commentaries upon them being mailed back the same way – technological learning will become increasing singularised? Only time will tell.

Increased ICT in Schools will Lead to School Improvement

This point assumes that the teachers themselves (as in the above case offered by Waller) are improved to the extent that they are far more digitally literate than the pupils themselves. It also assumes greater neutrality in the status of the technology itself. Waller, as shown here was depressed regarding the control of most school technology by American corporations, and Law (ibid.) has posited that no technology is neutral, it remains a heterogeneous (and powerful) agent in the social world (Law, 1991, p165), and its increasing 'virtualisation' may only serve to complicate this relationship further.

Only in scenario where schools are to be given socially pure technological tools, to be used by technologically proficient educators where the pre-existent social power divisions have been removed can the ICT as educator become a pure force for good.

REFERENCES

Angelis, C., & Valanides, N. (2009). Epistemological and Methodological Issues for the Conceptualisation, Development and Assessment of ICT-TPCK: Advances in Technological Pedagogical Content Knowledge (TPCK). *Computers & Education, 52*(1), 154–168. doi:10.1016/j.compedu.2008.07.006

Branson, J. (1991). Gender, Education and Work. In D. Corson (Ed.), *Education for Work: Background to Policy and Curriculum*. Clevedon, UK: Open University Press/Multilingual Matters Ltd.

Chapman, J. H. (2000). *The Final Report of the Research into Male Participation: Male Participation and the Lifelong Learning Agenda*. Northern College, Barnsley.

Department for Education and Science. (2003). *Fulfilling the Potential: Transforming Teaching and Learning Through ICT in Schools*. London: DfES.

Durkheim, E. (1922). Education et Sociologie. In A. Giddens (Ed.), *Emile Durkheim Selected Writings*. Cambridge, UK: Cambridge University Press.

Friere, P. (1970, 1993). *Pedagogy of The Oppressed* (rev. ed.) (M. B. Ramos, Trans.). New York: Penguin Books.

Haythorne, E. (1991). *On Earth To Make The Numbers Up*. Leeds, UK: Yorkshire Arts Circus.

Heath, A. (1987). Class in the Classroom. *New Society*, *81*, 13–15.

Higgins, A. H. (1981). Distance Education and Pupils: From Horseback to Satellite. *Education in Rural Australia*, *10*(2), 31–36.

Hoggart, R. (1957). *The Uses of Literacy*. London: Penguin.

Hoggart, R. (1995). *The Way We Live Now*. London: Pimlico.

Hurt, J. S. (1979). *Elementary Schooling and The Working Class*. London: Routledge and Kegan Paul.

Iske, S., Klein, A., Kutscher, N., & Otto, H.-W. (2008). Young People's Internet Use and Its Significance for Informal Education and Social Participation. *Technology, Pedagogy and Education*, *17*(2), 131–141. doi:10.1080/14759390802116672

Keddie, N. (1971). Classroom Knowledge. In M. F. D. Young (Ed.), *Knowledge and Control*. London: Collier-MacMillan.

Kitt, J., Davies, N. G., & Gillam, J. A. (1983). *School of the Air by Satellite. A Study of the Improvement of Distance Education in North-west Queensland Using the Australian Communications Satellite System*. (Evaluative Report) Queensland Dept. of Education, Brisbane (Australia). Abstract ED247883 from ERIC database.

Lacina, J. (2005). Media Literacy and Learning. *Childhood Education*, *82*(2), 118–120.

Law, J. (1991). Power, Discretion and Strategy. J. Law (Ed.), *A Sociology of Monsters: Essays on Power, Technology and Domination* (Sociological Review Monograph 38). London: Routledge.

Marks, A. (2000a). Lifelong Learning and The Breadwinner Ideology. *Educational Studies*, *26*(3), 304–320. doi:10.1080/03055690050137123

Marks, A. (2000b). Men's Health in the Era of "Crisis" Masculinities - A Health Policy Role for UK Lifelong Learning? *Adults Learning*, 11–13.

Marks, A. (2003). Welcome to the new Ambivalence: Reflections on the current and historical antagonism between the working class male and higher education. *British Journal of Sociology of Education, (Carfax), 24*(1), 84-94.

Marks, A. (2005). A Post- Gesellschaft-Gemeinshaft: A Sociological account of the 'PIPS' Project. In S. Rodrigues (Ed.), *International Perspectives of Teacher Professional Development* (pp. 149-163). New York: Nova Science Publications.

McAtee, W., & Zani, T. L. (1975). *The Education of Isolated Children in Western Australia* (Research Report). Education Department of Western Australia, West Perth.

Muncie, J. (1999). *Youth and Crime: A Critical Introduction*. London: Sage.

Murray, C. (1990). *The Emerging British Underclass.* London: Institute of Economic Affairs.

Murray, C. (1991). *Underclass: The Crisis Deepens*. London: Institute of Economic Affairs.

Payne, I. (1983). A Working Class Girl in a Grammar School. In D. Spender & E. Sarah (Eds.), *Learning to Lose: Sexism and Education.* London: the Women's Press.

Riddell, S., Baron, S., & Wilson, A. (2001). Gender and Post-School Experience. In B. Francis & C. Skelton (Eds.), *Investigating Gender: Contemporary Perspectives in Education*. Buckingham, UK: Open University Press.

Rogers, G. (1999). Reflections on Teaching Remote and Isolated Children. *Education in Rural Australia, 9*(2), 65–68.

Sanderson, F. (1998). Teaching Experience in a School of the Air. *Education in Rural Australia, 8*(2), 31–33.

Sax, L. (2007). *Boys Adrift: The Five Factors Driving the Growing Epidemic of Unmotivated Boys and Underachieving Young Men*. New York: Basic Books.

Sherman, R. R. (1991). Vocational Education and Democracy. In D. Corson (Ed.), *Education for Work: Background to Policy and Curriculum.* Clevedon, UK: Open University Press.

Tomlinson, D., Coulter, F., & Peacock, J. (1985). *Teaching and Learning at Home: Distance Education and the Isolated Child* (Research Series No. 4). Queensland Dept of Education, Brisbane (Australia). Abstract ED262927 from ERIC database.

Wainer, J., Dwyer, T., Dutra, R. S., Covic, A., Magalhaes, V. B., & Ferreria, L. R. R. (2008). Too Much Computer and Internet Use is Bad for Your Grades, Especially if You are Young and Poor: Results from the 2001 Brazillian SAEB. *Computers & Education, 51*(4), 1417–1429. doi:10.1016/j.compedu.2007.12.007

Waller, T. (2007). ICT and Social Justice: Educational technology, global capital and digital divides. *Journal for Critical Educational Policy Studies, 5*(1). Retrieved March 1, 2009, from http://www.jceps.com

Willis, P. (1977). *Learning to Labour: How Working Class Kids Get Working Class Jobs*. London: Gower Press.

EDITOR NOTE

It is with sadness that we note that Andrew Marks died suddenly at his home in October 2009.

ENDNOTES

1. In this case pupils *or* teachers.
2. Unfortunately he does not offer statistics.

Chapter 15
Learning, Old Age and Email

John S. Murnane
The University of Melbourne, Australia

ABSTRACT

This chapter describes a small research project to place Internet-linked computers in a retirement complex in Melbourne, Australia (Murnane 2007, 2008). The aim was to research the existing computer skills of the residents, provide lessons in the use of email and general Internet and computer use, investigate the most appropriate type of lessons, document the problems encountered by very senior people encountering computing and the Internet for the first time, collect research on how they used the computers, the attitudes of financial management, nursing and occupational staff towards the activity, and the involvement of peers, family and friends.

LEARNING, OLD AGE AND EMAIL

The Internet is widely regarded as a young person's playground, but it can be even more important for the aged, supplying several elements scarce or completely missing from their lives. The onset of age-related problems can be delayed by activity and mental stimulation. The Web, and particularly email, can provide by far the best link to family, friends and the World. The Web is a powerful and interesting source of activity which can be indulged in without the need for organised group activity, and it could support public engagement with science. Experience with a small research project at the Old Colonists' Association of Victoria shows over 80's in supported Hostel accommodation with no experience of computers can learn to use email and the Web independently. However, even the latest products of user-friendly technology do not negate the mental problems associated with aging and attention must be paid to appropriate social and technological environments, and particularly to teaching methodology.

DOI: 10.4018/978-1-60566-690-2.ch015

INTRODUCTION

There is an increasing volume of material in the literature involving the use of computers and the Internet concerned with clinical aspects, diagnostics and monitoring of the aged (Nugent 2007), but comparatively little on those in this age group learning to use computers themselves. Shapira, Barak and Gal (2007) report on factors relating to the psychological impact of learning to use the Internet, as does White, McConnell, Clipp, Branch, Sloane, Pieper and Box (2002), but these and other researchers say little about the actual learning process itself, its effect on individuals, and virtually nothing on its social context. The research described here, although limited, suggests that difficulties people over 80 have in learning to use computers have been considerably glossed over. Although the published findings are quite strong, the papers provide few details on the actual teaching methodology or the practice employed, and provide little or no guidance to anyone starting an Internet program in a low-care retirement complex. With the exception of Fokkema and Knipscheer (2007), who provide no detail other than that instruction was given at home by volunteers, most studies (Seals, Clanton, Agarwal, Doswell & Thomas 2008; Tse, Choi & Leung 2008; White, *et al.* 2002) are confined to groups taught in semi-formal classes, whereas our experience shows the need for older seniors is for individual lessons and one-to-one help, with extensive support from family and peers.

Beginners over the age of 80 face several problems in learning to use computers. The most significant of these is an age-related deterioration in mental condition. This affects everyone over 80, though very unequally, and causes problems in understanding a modern graphical interface and in remembering the set of manipulations required to read and send email. Poor hand-control causes trouble with both keyboard and mouse, and there is a noticeable feeling among them that computers are a young person's domain, a perception that can sap confidence.

On the other hand, these retirees have three significant factors on their side. The first is their burning need for a fast, versatile connection to family and friends that does not require them to physically travel, and which is not dependent on anyone but themselves. The second is a community spirit of collective help and support. The last is ample time to experiment.

In future years most people entering a retirement home in Australia can be expected to be familiar with computer and Internet technology, but there is clearly a five to ten year gap, perhaps more, in which residents will have to be taught, and retirement home Management (those responsible for budget and the diversional staff responsible for activities) will need to be convinced of the need to provide Internet access. Given the reported benefits the Internet has for the aged, it is not good enough to leave the problem to solve itself: people now in retirement deserve to experience them. The problem is not in establishing these, but in teaching people not acquainted with computers to send and receive email. It is not the advantages of Internet access that now require research, it is the practicality and the practice. Hence this project concentrated on individual residents as they learned to use computers. As such it is also useful in terms of exploring the nature of technical literacy we often overlook.

BACKGROUND

Why weren't we offered this ten years ago?

Comment from a resident at our first meeting. (Field Notes)

Despite the fashion in some quarters to disparage the ability of anyone not born into 'the computer age' to take up, or even see the utility of the computer and its inherent communication possibilities,

(Prensky 2005) the Internet, in its World Wide Web form, might have been especially designed for the elderly or disabled. Potentially liberating in ways even surpassing its importance for the young and mobile it can provide a dimension otherwise totally lacking in the life of someone limited in their mobility or confined by circumstances to a single building. Not all older retirees will be able to learn to use computers by any means, but those who still can, should be enabled to.

Seniors health is closely tied to their mental stimulation (Leonard 1993, Rönnberg 1998, Tse, *et al.* 2008), and an Internet connection to the World, along with the other applications routinely provided by a computer, can make a very positive contribution to this. Rönnberg (1998, p 393) asserts that 'only rarely do nursing homes meet all the psychological needs of their residents. The environment has been described as one of limited privacy and isolation from the outside world.' Regular contact with relatives and friends outside the retirement home is another area often connected with the promotion of senior's health (Downs, Ariss, Grant, Keady, Turner, Bryans, Wilcock, Levin, O'Carroll & Iliffe 2006). Yet in Australia, retirement complexes rarely provide Internet access for their residents. Sum, Mathews, Pourghasem and Hughes (2008, p 209), in a study of 222 Australian Internet users over the age of 55, found 'nearly all (91.4%) of the participants most often used a computer at home, 5.9% at work and 1.5% at other places like school and library.' They report no significant level of responses from those in a retirement home.

The profusion of 'seniors' websites shows that older people *can* use computers and the Internet, yet commentators and 'experts' frequently downplay the possibility. Prensky (2005, p 29) refers disparagingly to anyone too old to be in school in 2005 as a 'digital immigrant,' implying an immediate disadvantage. Butler (2005, p 7) considers everyone born later than 1982 has 'an intimate' relationship with the Internet, implying those born before this date do not. TV and radio talk-back hosts sometimes express the opinion that older people cannot use communication technology. As late as March 2006, Mark Colvin, host of 'PM,' the Australian Broadcasting Corporation's flagship current affairs program, remarked that he 'still can't program [his] VCR' (Colvin 2006). This was apparently meant as a light, 'throw away' remark. However, it and similar utterances do nothing to encourage the non-computer literate to try and use the technology. In fact they help build a culture of fear amongst older people that they would *not* be able to cope, a feeling expressed by some participants in this research. To overcome this perception, examples of successful Internet use in retirement homes are required.

This research has therefore mainly concentrated on teaching-related aspects of Seniors and the Internet. Not only are these almost totally disregarded in the literature, there is an underlying element of doubt in some quarters that non-computer literate Seniors can learn to use this technology. This must simultaneously be dispelled if any real progress is to be made. There is a concomitant and just as important question of 'training' for relatives. If internet communication is established they must understand the need to actively and regularly participate and respond.

SETTING

One association in Victoria agreed to participate in the project and two computers were installed in their Park (a pseudonym) aged care complex in Melbourne. The Park aged care complex consists of an extensive area with 118 independent living cottages, a low-care Hostel with 52 residents including a 12 room dementia wing, and a high-care Nursing Home. The *ICT in Education and Research Group* from the University of Melbourne arranged to supply two computers linked to the Internet on a wireless network, arrange and finance an Internet Service Provider (ISP) and provide training for the residents for 12 months. At the end of this time the equipment would become the property of the Park aged care complex. In return some staff and some residents agreed to contribute to data collection. This involved short questionnaires, but came predominantly from participant observation with residents learning and using the machines.

Residents were offered instruction in computing and communication skills, initially the use of email. Due to the small research budget, instruction was limited to residents of the Hostel plus a few from the independent units, and averaged three hours a week. All residents, whether they join the 'research' or not, were free to use the computers and take advantage of the instruction provided. Data was gathered in two waysParticipant research within a teaching program organised for residents in the use of computers and the Internet, preceded by a questionnaire, and Staff interviews and questionnaire.

The project involved very informal and lightly-structured educational program. Field notes were taken while the lesson was in progress and overall observations at the end of each session recorded. It should be noted that the author was both teacher and researcher, and the challenge to think, teach, talk and record was considerable. The research did tend to suffer in consequence, but the quality of the 'research experience' was correspondingly high. All following data is from field notes unless otherwise documented.

RESEARCH METHODOLOGY

The research methodology was strictly that of Participant Action Research (Chien, Cook and Harding 1948), something dictated by three considerations:

- This was a low-budget project, the researcher being the teacher. Only three hours of instruction could be scheduled in most weeks. This severely limited participant numbers. Consequently, the data supports only general observations relating to the group, or specific observations relating to an individual.
- The nature of the project dictated that desires of the Hostel Residents took precedence over any research considerations.
- The researcher was the teacher and no separation of events and observation was possible.

Participant Action Research provides a framework here, albeit one under challenge from time to time (Costello 2003, Hodgkinson 1988). Not only a research paradigm where *all* participants are collaborators in the research, it is specifically a strategy whereby research drives change directly, establishing a cycle of planning, doing, analysing and changing (Sanford 1970). Further, the researcher is required to actually 'control the course of social action,' at least to some extent (Chien, *et al.* 1948, p 60). It also requires that all participants are aware that a research-change cycle is taking place and actively participate. This in fact happened, with the Hostel residents themselves determining the form of the 'lessons.'

RESULTS

The research was based in the low-care Hostel. The first meeting with twelve residents took place in April 2007. The project was explained to the residents by the facility Diversional Therapist. Seven immediately expressed interest in email, though only one of the twelve had any previous contact with computers. Several did indeed express doubt at their ability to learn to use a computer, with three doubtful of their ability to even use a keyboard.

Lessons began in August 2007 and the programme is now well into its second year. A member of the diversional (activities) Staff arranged the roster. We began with three residents attending at any one time for forty-five minute sessions. The idea was that one could sit at a computer and the other two could observe, subsequently taking their own turn. This did not work. The two computers were installed in adjoining corrals set up for individual computer access (not as part of this project). It proved impossible to provide instruction on one screen while another person worked at the second machine. In addition, there just was not enough space. Furthermore, only the person actually using the machine could really see what was going on, even when the computers were moved to a more appropriate location. One-to-one instruction was continued for newcomers and we are now convinced that for people of this age and mental characteristics, one-to-one instruction is the only good teaching method. Even if the environment did permit easy observation of the screen, with a video projector say, we would still advocate taking one student at a time. This has poor implications for retirement home staff helping residents to learn: one-to-one sessions are very expensive, though it is likely that some volunteer help will be available. A culture of mutual support and training amongst the residents is by far the best hope for the long term.

The 'Students'

Most retirees enter the Hostel over the age of 80. Those in our teaching program have numbered fifteen: thirteen female, two male, ages 84 to 97. Seven of these had never used a computer before including four who had never used a keyboard. The others had a range of skills, from minimal to advanced with specific applications. It is in the nature of this environment that new residents will come, and others will move on. Two of the original six are no longer resident in the Hostel.

Ergonomic Issues

The original position of the computers was cramped, being in a Unit built to take two machines in a small library. When the project began this contained one old computer not linked to the Internet. Staff information suggests this was used only by one resident to play games. This Unit was very badly designed. The screens were set too high, there was no space for the mouse next to the keyboard and the 'pull-out' drawers that held them were prone to move in and out when stability was required, and consequently needed the drawer to be propped to prevent it moving. Fortunately, a requirement to return the library to a bedroom forced a move of the computers to a pleasantly sunny and very wide corridor. We now have two plain tables which allow good placement of the equipment and room to move. Importantly, this position encourages passers-by to stop and chat. Interest and involvement of staff and other residents is vital to long term success. The move to a new area bought the ability to usefully utilise both machines at once and we also went to one-hour sessions. For people in this age group all computer work is slow, but hourly bookings are long enough to allow for most activities. Longer times would be too great a strain even if they could stay longer, and almost all are ready to quit after 50 to 60 minutes.

Specialised computer desks are not only unnecessary but, in this environment, have been found to be a positive hindrance. A large flat desk with adequate space and suitable seating is required. Seating height should be adjustable but standard commercial 'work station' chairs may not suit the elderly.

Without any prompting, the students all brought notebooks with them in which they wrote notes on operating procedures, email addresses, and the important details of their accounts, and this needs some space too.

Teaching

From the beginning the students themselves determined the form of the lessons by quietly and firmly indicating their desire to sit at the computer and operate it independently. They all made it very plain from their first session that they wanted direct control, a determination most probably linked to a desire to learn to use email independently. Other instructional possibilities had been considered and indeed planned, but were never implemented in the face of the residents determination to use the keyboard and mouse themselves, one at a time. Operationally, problems with short-term memory would appear to dictate as much immediate hands-on experience as possible.

Of the Hostel residents in the original group, seven had no computing experience and four had never used a keyboard. In practice there was very little difference here, even in understanding the purpose of Shift and Enter. A group of three from the Independent Living Units all had their own computers but still experienced some problems with the research machines, either because of a different version of Windows and/or a different arrangement of icons, though this was fairly quickly overcome.

The range of interests between the Hostel residents in the first six months was very small indeed. With one exception they really only wanted access to email, but their interest in it was intense. Despite never having used email they all knew about it, particularly its ability to carry photographs. None initially showed any real understanding or interest in the Web or a Search Engine, although in the last six months this has altered. Interest in searching for historical and genealogical information in particular is growing quickly, and two who recently joined are using the computers to watch older-versions of televised program.

None of the Hostel Residents expressed any confidence that they would learn to use email independently, and three expressed outright doubt. From their conversation the idea that you can be too old to learn about computers definitely seemed to be playing a significant part here. Although they were all very sensitive to the problems advancing age brings to learning new things, and they did have a general conviction, often expressed, that young people can and do learn to use the technology with ease.

General mental ability is remarkably constant across the active period of a life-span. A long term study by Deary, Lawrence, Whalley, Crawford and Starrc (2000) found considerable stability between the ages of 11 and 77. There does seem to be a relationship between healthy old age and decline in working (short-term) memory function (Small 2001; Small, Stern, Tang & Mayeux 1999) and cognitive speed (Brosseau, Potvin & Rouleau 2007; Christensen 2001) although the extent and causes are still open to question (Foster 1999; Johnson, Reeder, Raye & Mitchell 2002; Larrabee & Crook 1994; Rinn 1998). It certainly appeared in this research that here was the outstanding difference between the retirees and younger people learning to use email: things a younger person will generally find easy to retain required re-teaching week after week.

The Hostel group ranged in age from 84 to 97, well above the age of 77 at which Deary *et el* (2000) consider deterioration may be expected to set in. Indeed, all Hostel residents except one displayed the age-related mental problems above, but with no particular pattern, these being too variant and individual to form one, in so small a group. To date, age considered, the outstanding 'student' has been the 97 year old. However, now even she is starting to find the challenge of operating a computer something of a strain.

This is really the only serious block to the aged learning to use of the Internet independently, but it is not a problem with the technology as such. Differences in ability to remember new things between the Hostel residents and those from the Independent Units is striking: the 'Independents,' ages 66 to 87, retain lesson content far more easily and show no obvious signs of stress.

It is easy for an experienced user to forget the sheer volume of data displayed on a computer screen. Most of it totally ignored most of the time, but someone new to Windows must first learn what to pay attention to. Besides multiple

Windows icons, the number of sensitive areas in a typical browser page is enormous. There is the information hidden in pull-down menus, and things like scroll-bars and their tiny controller icons are not only hard to see and hit with the pointer, but difficult for a beginner to control, requiring either the mouse to be repeatedly clicked, held down for a period pre-determined by experience, or dragged-and-dropped, Different sensitive areas of the screen have different ways of indicating the mouse is over them: changing colour, being underlined, flashing an I-beam, popping-up a window, or not doing anything at all. Links are single-clicked, but icons double-clicked. File names differ from URLs and email addresses. Passwords have their own rules. All this even a moderately experienced user knows.

Setting up a Hotmail account consistently took an hour or even more. It was actually a mistake to select Hotmail. Hotmail requires a complete Username, such as aperson@hotmail.com to be entered. A system that only needed the Username itself, for instance Gmail, would have eliminated one source of error—particularly the need to type the '@' sign.

Spaces were another problem. Inexperienced users instinctively revert to standard spelling and grammar by placing spaces between word-units. The worst problem is in the entry of passwords. Because these appear as asterisks any error other than the wrong number of characters cannot be diagnosed. Passwords should therefore be as simple as possible, the need for security notwithstanding. The residents all brought note-pads to their sessions and wrote down their Username and Password without needing to be prompted. Considering the progressive mental difficulties they all experience, being able to remember these details has not proved to be as difficult as was initially anticipated and most no longer need to consult their note-pads, but problems with keystrokes for log-ins continue.

It is not always at all obvious why an attempted log-in fails. Even carefully watching the keystrokes does not always give any clues. Failed log-ins are not only frustrating, they almost certainly work against a willingness to try logging in outside 'class' hours. At times, bafflement was so acute that I have taken over and typed in the password myself. Since this has always been successful, the reason for failure of the log-in remained obscure. However most are probably caused by touching two keys at once, or a misinterpretation of written characters. Modern keyboards with very thin keys (as on a Laptop) would probably help a lot here since they reduce the possibility to touching two keys simultaneously. Two of the six original Hostel residents in the programme were observed to log in to their accounts completely unaided after six lessons over 10 weeks.

Photographs attached to an email were at once the most valued part of the communication and the most problematic. Photos that were embedded in the email itself, appearing automatically on the screen, only caused problems with the need to scroll a long way, and the presence of a photograph further down was not always recognised. Opening a photo attached to the email was a problem which no Hostel resident had as yet really mastered. All had trouble identifying the attachment with its 'paper-clip' icon as the place to 'double-click,' and the file must then be opened by another application. This second step routinely caused bafflement, no matter how many times it had been encountered and the steps recorded in a notebook. (It would help if all photos were sent as JPGs as Windows treats as TIFFs differently.)

All residents taking lessons showed excitement and enthusiasm, every week. With the exception of three who had just joined the teaching group, all participating Hostel residents had sent emails and received them from either family or friends. The volume of email received varies widely, both in quantity and length. Undoubtedly the most satisfying part of this research is to witness the receipt of the first email from family, particularly from grandchildren or a niece. (Very little, to date, from nephews!!) No attempt was made to quantify this enthusiasm, but it was by far the most outstanding feature of the project. Even though the group was small, the roster for places each week usually cannot accommodate everyone who would like to be included. Essentially, everyone with an email account would like a booking every week.

The benefits to residents can also be seen in their desire and determination to access email independently. Sometimes they were successful, but from Staff reports, failure due to log-in problems were still significant. On one occasion at the six-month mark, two residents were reported as spending an hour and a half unsuccessfully trying to log-in to their accounts. While disappointing and frustrating to all concerned, this level of persistence does illustrate the benefit residents perceive they are deriving.

Email composed by residents varied considerably in length, running from a sentence or two to three or four paragraphs: perhaps 20 or 30 lines of text. Length depended noticeably on the subject matter, not necessarily the size of the message received. Interesting information prompted a longer answer, particularly if accompanied by photographs. 'Conversations' almost exclusively concerned family matters. Celebrations such as births and weddings naturally received much attention, as did Christmas and Easter and messages concerned with jobs and travel were always appreciated.

Generally there was not a lot for the residents to write about themselves, unless they had been on an excursion. Family outings in the days immediately preceding the email often produced the longest text. Life in the Hostel was fairly predictable and did not lend itself to much description. Consequently, the subject matter more often concerned things related to messages received.

Spam was a small problem. Not being familiar with the nature of the Internet, residents were baffled and confused by its appearance. Two residents regularly received multiple emails containing attachments of items, most often PowerPoints found on the Web, from younger friends confined to their home. To one resident they are mostly just a nuisance but the other shows intense interest in a proportion of them.

For three weeks in January 2008 the computers were off-line due to problems with the telephone system. This caused an observable amount of distress within the Hostel group, emphasising the very high value they placed on the facility. As well as the temporary loss of contact itself, staff reported they expressed concern they would forget how to use email if the connection was not quickly reestablished.

Initially, only one resident showed an interest in the World-Wide-Web. His interest was in family history and genealogy. He only asked to set up an email account after three lessons and he only sent a few to a friend, receiving one in return. He was the only resident to enquire about the structure and meaning of a URL. Three others began asking about these same Web areas after four months and interest is now growing quite strongly among the rest of the group. Another developing area is Google Earth, to view places the resident has lived, and where younger relatives are now living or vacationing. Sometimes this follows mention of a place in an email, but so far, none have been observed using information from Goggle in one of their messages.

DATA ENTRY

The Keyboard

The easiest place to observe the need for considerable retraining from week to week in this age-group was the keyboard. The four who had never used one were still struggling to some extent with the Shift, Tab, Enter and arrow keys after twelve months. Even those who had used a keyboard were still experiencing the occasional problem. However, it was noticeable that things like using the arrow keys to move a few characters or lines to correct mistakes in typing were requiring less and less reteaching with each lesson.

Computer keyboards still carry their typewriter inheritance and not just in their QWERTY keyboard. The Shift key caused puzzlement, the word 'shift' being no longer physically related to a mechanical typewriter mechanism. The Space Bar, not being labeled, was initially a mystery, as was the Enter key. For some residents the use of 'Enter' when a new line is required still needs an occasional prompt.

The keyboard Repeat function was turned off completely after week three, or keys held down while attention transferred itself elsewhere made progress almost impossible. This now makes the use of the arrow keys awkward since they have to be repeatedly touched to move the cursor more than one letter or line, though it makes scrolling an email line-by-line easy. For the six original students, it could now profitably be turned back on, but has been left off because of new students joining the group.

Pointing Devices

An item highlighted in the initial Staff survey as likely to cause problems was 'hand control' and this has proven to be the case. Brosseau *et el.* (2007, p 600) note that 'few studies have examined the influence of aging on motor skill learning.' Their study on 'tracking' tasks found a slowing in learning ability with age.

The mouse represents one of the two main operational problems to older people, particularly the standard two-button-and-scroll-wheel PC type. All the students appeared to quickly grasp the pointing function, if indeed they did not understand it before lessons began. Its purpose and operation was something else. A mouse is more than a controller of a moving point on the screen.

Teaching games which used the mouse, such as Solitaire, as a training exercise, did not prove attractive: the Hostel residents wanted to get straight to email. However, subsequent games of Solitaire when the computers went off-line for three weeks did lead to an observable improvement in confidence.

Comprehension of what the mouse did was soon learned and the only problem, which continues, is whether a single or double-click was required. However where to point it did, and continues to be a problem. Pointing to the (often) very small targets is not at all easy for old and possibly arthritic figures. Compounding the problem was a tendency to look at the mouse itself when a click was required. This often resulted in the pointer moving off-target between the 'point' and the 'click.' Drag-and-drop maneuvers in particular presented problems. Three different types of pointing device have been trialed, all with their own advantages and disadvantages.

The ultimate way around the 'mouse' problem is to dispense with it altogether. Czaja (1998) reports a successful project which by-passes the problem completely by providing an all-text based email system. The system was reported as being 'easy to use' (Czaja 1998, p 148), but the need to provide a customised system with its inability to provide a wide range of applications, notably Web pages, instead of a standard graphical interface, mitigates heavily against this approach. It would also magnify problems with the keyboard.

The PC Mouse

Worst of the three pointing devices trialed was the standard PC mouse and this happened to be the first one used. All the students except one were right-handed and, as do most right-handers, they instinctively grasped the mouse with the last two fingers and thumb. This poised the index and middle fingers over the top of the mouse. However, due to their poor hand control, it either caused them to come down on both buttons at once, bringing up the Edit Menu, or locking down the scroll-wheel. Either of these events stopped the lesson until rectified. For the right-handed, clicking the left button requires a certain level of dexterity not required of those who use a mouse left-handed. Grasping a mouse with the left hand naturally positions the index finger over the left-hand button. (The left-hander also uses the mouse right-handed, even though the original set-up placed the mouse centrally, in front of the keyboard, and she continues to use it that way.)

Mouse movement just prior to 'a click' was a real problem. Perhaps because they grew up in an era of mechanical devices which required a forceful activation, and were not acquainted with the micro-switch and its acceptance, and indeed need, of a delicate touch, button clicking tended to be on the 'firm' side.

Encountering problems with double-clicking for the first time, one student commented with considerable conviction, that 'I know what the trouble is. I'm not pushing it hard enough am I?' Her original 'clicks' were already much too vigorous. Because the problem is essentially physical/mechanical, just telling the student to keep the mouse still between clicks is of limited value. The problem can be partly overcome by the teacher holding the cord just in front of the mouse body to steady it, but this does little to promote independent use. Teaching a technique of using the right hand to point the mouse and 'clicking' with a finger on the left was partly successful, and it did help reduce mouse movement between the point and the click *somewhat*. The time-interval at which Windows will accept a double-click has been lengthened but this too only partly helped, possibly because it also lengthened the interval in which the mouse could move off-target. Sometimes after several unsuccessful attempts to activate an icon the only way to keep the lesson going was for me to surreptitiously double-click the mouse myself—progress being preferable to continued puzzlement and disappointment. Progress for these students depends just as much on confidence as it does on dexterity. The need to occasionally pick the mouse up and reposition it on the mouse pad also caused problems. The 'orientation' of the mouse caused problems, students found difficulty associating the direction in which the pointer moved with the orientation of the mouse on the pad. As well as problems with 'hand control,' there also appears to be a cognition problem here: people of this age may have difficulty correlating physical hand movement and the orientation of the mouse with what happens on the screen, something for further investigation.

The Apple Mouse

A partial solution to some of these problems was the (original) one-button, USB, laser, Apple mouse. This worked happily with Windows and the absence of a Menu via the right button is an advantage, since if invoked, it tends to confuse and has to be removed, needing another problematical mouse-click, and there is no scroll-wheel to get in the way. At the time it was trialed there were six Hostel residents in the program, and three, on first acquaintance, volunteered the opinion that it was easier to use and that they preferred it. Two others, when asked, agreed. Only the student who had previous computer experience via games said he preferred the "black," i.e. PC mouse, but later he changed his mind.

The Apple mouse does have disadvantages, the most obvious being the loss of the pop-up menu via the right-hand button. The Apple Operating System invokes this function if the mouse is 'held down' for a second and Windows does not, but since the idea was to *prevent* the Menu's appearance, this loss was not significant. The students found difficulty understanding that the surfaces of the PC mouse were 'buttons.' (The Apple mouse is worse in this respect.)

The Track-Ball

A track-ball was trialed last and proved to be by far the best alternative, something suggested in previous studies (White, *et al.* 2002), though the advantages were not actually analysed. The Infogrip BIGtrack was selected on the basis that its large ball was postulated to be easier for older hands and brains to cope with, and the buttons are placed quite separately. It has seven advantages, largely stemming from the movement function being *physically* distanced and distinguished from the buttons.

It is easier to understand its functionality because each actuator is physically separate and very visible. The BIGTrack ball is bright yellow and the buttons bright blue, so it is easy to distinguish different functions by colour as well as by their respective movements. When a click is required, it is easy to say 'Hit the blue button.' The ball, with its comparatively high moment of inertia, was easy to manipulate. Conveniently, it can be moved with the palm or the fingers. This provided better control and removed the need to occasionally pick a mouse up and move it back to the centre of the mouse-pad.

'Move' is a separate operation to 'click,' and here track-ball has great advantage. All students instinctively used the palm or figures of the right hand to move the ball and the index finger of the same hand to push the button. This usually moved the hand completely off the ball prior to the click, completely obviating unwanted pointer movement, *particularly between double-clicks.* The incidence of 'missed' clicks where the pointer was not quite in the required place decreased immediately to a quite tolerable level, although they did not disappear entirely.

Drag-and-drop required a different strategy. Our students were taught to move the ball with the right hand and hold the button down with a finger on the left, something they took to very naturally. The use of drag-and-drop for email was non-existent as none were as yet up to selecting or moving blocks of text, but it was found to work very well in Solitaire, which required it. The very obviousness of the second (right) button stimulated interest in its function. While the right button an a PC mouse only brought up distracters, and consequent avoidance, the second bright blue button on the track-ball has produced interest, in particular its use to suggest correct spelling.

One of the students, Molly, is quite dependent on lip reading. It is not possible to watch the keyboard, the screen and lip read at the same time. When she was learning to use a mouse we slowly developed a system of hand signs. All students easily and naturally responded to the teacher pointing to a spot on the screen, but Molly, not being able to hear what to do there, required more informative hand movements. The track-ball, with its physically distinguished control movements of palm and finger, rapidly lead to a much more sophisticated and useful sign language. Pointer movement was conveyed by a stroking motion of the hand held flat in front of the screen in the desired direction of travel; a single or double button click by a up-and-down jabbing finger; and 'drag-and-drop' by jabbing down with a finger on the left hand and a simultaneous stroking motion with the palm of the right, corresponding with the required actions on the track-ball itself.

Molly was introduced to the track-ball in the third week the Internet was disconnected, so she played Solitaire, which required drag-and-drop. Learning to drag-and-drop had been a real problem to her, not just in using the mouse, but in remembering how to do it from week to week, though the time she needed for revision progressively shortened. As it happened Molly was the first student to use the track-ball and it was 15 minutes before she was able to perform an unassisted drag-and-drop. For the rest of the lesson she happily manipulated her cards unaided, though she often 'dropped' a card prematurely and had to re-acquire it, and she continued to occasionally multiple-click the 'deck,' thereby often missing a directly playable card such as a King or Ace (particularly frustrating for the observer as she never managed to win a game).

This particular track-ball has disadvantages. Point and click with the same hand makes for slow screen manipulation, two-handed operation is slightly clumsy, and it does have a rather course operating 'feel' compared with a contemporary mouse. None of these matter particularly in this environment, and in subsequent weeks, the Internet restored, all but one resident continued to prefer it, even though there was a mouse-equipped computer set up next to it. Convinced of its popularity and utility, Hostel Management have now purchased another track-ball for the second computer.

DISCUSSION AND CONCLUSION

Despite the long list of problems enumerated above, it would be a mistake to dismiss the research as in any way frustrating, or even difficult. None of those involved: residents, staff or myself have any doubt of the benefits of email and the Web. With one exception, residents have shown nothing but enthusiasm and an absolutely unshakable desire to continue learning. The exception was, sadly, Molly. After seventeen lessons over five months she decided that the strain of using the computer was too great and, after sending one last message, dropped out.

All participants would like more frequent lessons. Statistically the sample is tiny and essentially self-selected, but the conclusions are unmistakable. Not only are residents writing longer letters, they are now correcting mistakes in spelling and grammar, and mostly use capital letters correctly. They are all good readers and show an eagerness to share thoughts and other information about the email they receive and the relatives and friends who send them with me. Photos attached to emails are particularly welcome and are explained and discussed at length. Early in the program the Hostel purchased a printer so copies could be made. Occasionally I bring in a camera so they can attach photographs of themselves at the computer to include in theirs, something they also appreciate. There are indications that the strain of reading from a computer screen are adding to the stress problems of operating a computer. Two residents have gone to the trouble and expense to obtain glasses suited to reading a screen. Given the cost, this is significant.

Some residents have gained sufficient confidence to log-on by themselves, and often pair-up to do so. This is probably to help each other solve problems, but it could also be connected with a desire to share messages and experiences. The objective here is to have experienced residents begin to induct non-users into email rather than have an external teacher. To date none are confident enough to do this, but the development of a continuing 'school' for current and new arrivals, run by residents, is the most important medium-term aim of the project.

Relatives and friends have just as vital a part to play. If there is excitement when a new email arrives, there is corresponding disappointment when no new messages appear, especially if the resident has sent an email to a relative in the past week. Families must be prepared to accept responsibility here and follow up messages consistently. Even a brief reply elicits a high level of excitement and satisfaction.

Visitors have been invited, via the Park Newsletter, to go to the computers during visits, log-in and enjoy the emails with the resident. Staff report this is now happening more frequently, but *both* these supports are expected to be important to long-term success.

Other staff in the facility are taking an increasing interest, stopping to chat to residents using the computers and assisting them with any problem they happen to be encountering. Apart from the help, the opportunity to discuss family doings is always highly valued (Leonard 1993; Rönnberg 1998; Tse, *et al.* 2008). Staff time is a factor here and they are often too busy to stop, particularly in the mornings, but if a self-help culture does develop, residents doing their own thing on a computer are quite likely to be more mentally active and involved than when participating in group activities.

This project and the other research cited suggests Internet access should be a standard feature of all retirement facilities, particularly those which cater for residents limited in their ability to travel outside the premises. Evidence of the benefits have been established to the point that they should be seen to outweigh costs. In the next five to ten years new residents will begin enter with an *expectation* of Internet access, indeed the three residents in this study now in independent units expect to be able to use the Internet when they have to move to the Hostel. The ultimate goal is to sensitise, not only retirement home operators, but the State and Federal authorities responsible for them, to the need for Internet availability as a standard requirement and expectation of Seniors.

REFERENCES

Brosseau, J., Potvin, M. J., & Rouleau, I. (2007). Aging affects motor skill learning when the task requires inhibitory control. *Developmental Neuropsychology*, *32*(1), 597–613.

Butler, G. (2005). Who's in our classrooms? *Teacher*, *156*, 6–9.

Chien, I., Cook, W., & Harding, J. (1948). The field of action research. *The American Psychologist*, *3*, 43–50. doi:10.1037/h0053515

Christensen, H. (2001). What cognitive changes can be expected with normal ageing? *The Australian and New Zealand Journal of Psychiatry*, *35*(6), 768–775. doi:10.1046/j.1440-1614.2001.00966.x

Colvin, M. (2006). *Govt plans digital TV shake-up*. Australian Broadcasting Commission. Retrieved March 16, 2006, from http://www.abc.net.au/pm/content/2006/s1591580.htm

Costello, P. J. M. (2003). *Action research*. London: Continuum.

Czaja, S. J. (1998). Gerontology Case Study: designing a computer-based communication system for older adults. In V. J. Berg Rice (Ed.), *Ergonomics in Health Care and Rehabilitation*. Boston: Butterworth-Heinemann.

Deary, I. J., Lawrence, J., Whalley, H. L., Crawford, J. R., & Starrc, J. M. (2000). The Stability of Individual Differences in Mental Ability from Childhood to Old Age: Follow-up of the 1932 Scottish Mental Survey. *Intelligence*, *28*(1), 49–55. doi:10.1016/S0160-2896(99)00031-8

Downs, M., Ariss, S. M. B., Grant, E., Keady, J., Turner, S., & Bryans, M. (2006). Family carers' accounts of general practice contacts for their relatives with early signs of dementia. *Dementia*, *5*(3), 353–374. doi:10.1177/1471301206067111

Fokkema, T., & Knipscheer, K. (2007). Escape loneliness by going digital: A quantitative and qualitative evaluation of a Dutch experiment in using ECT to overcome loneliness among older adults. *Aging & Mental Health*, *11*(5), 496–504. doi:10.1080/13607860701366129

Foster, T. C. (1999). Involvement of hippocampal synaptic plasticity in age-related memory decline. *Brain Research. Brain Research Reviews*, *30*(3), 236–249. doi:10.1016/S0165-0173(99)00017-X

Hodgkinson, H. L. (1988). Action Research—a critique. In S. Kemmis & R. McTagggert (Eds.), *The Action Research Reader* (3rd ed.). Geelong, Australia: Deakin University Production Unit.

Johnson, M. K., Reeder, J. A., Raye, C. L., & Mitchell, K. J. (2002). Second thoughts versus second looks: An age-related deficit in selectively refreshing just-active information. *Psychological Science*, *13*(1), 64–67. doi:10.1111/1467-9280.00411

Larrabee, G. J., & Crook, T. H. (1994). Estimated prevalence of age-associated memory impairment derived from standardized tests of memory function. *International Psychogeriatrics*, *6*, 95–104. doi:10.1017/S1041610294001663

Leonard, J. (1993). *Cutting to the root of ageism: The effect of mental stimulation on the elderly*. OH: The Union Institute.

Murnane, J. S. (2007). Retirement, Mobility and the Internet. In A. Tatnall, J. B. Thompson, & H. Edwards (Eds.), *Education, Training and Lifelong Learning, IFIP WG 3.6 and 3.4 Joint Working Conferences*, Prague. Heidelberg, Australia: Heidelberg Press.

Murnane, J. S. (2008). Age, mobility and eMail. *Journal of Assistive Technologies*, *2*(4), 16–25.

Nugent, C. D. (2007). ICT in the elderly and dementia. *Aging & Mental Health*, *11*(5), 473–476. doi:10.1080/13607860701643071

Prensky, M. (2005). Digital natives, digital immigrants. *Gifted*, *135*, 29–31.

Rinn, W. E. (1998). Mental decline in normal aging: a review. *Journal of Geriatric Psychiatry and Neurology*, *1*(3), 144–158. doi:10.1177/089198878800100304

Rönnberg, L. (1998). Quality of life in nursing-home residents: an intervention study of the effect of mental stimulation through an audiovisual programme. *Age and Ageing*, *27*, 393–397. doi:10.1093/ageing/27.3.393

Sanford, N. (1970). Whatever happened to action research? *The Journal of Social Issues*, *26*(4), 3–23.

Seals, C. D., Clanton, K., Agarwal, R., Doswell, F., & Thomas, C. M. (2008). Lifelong learning: Becoming computer savvy at a later age. *Educational Gerontology*, *34*(12), 1055–1069. doi:10.1080/03601270802290185

Shapira, N., Barak, A., & Gal, I. (2007). Promoting older adults' well-being through Internet training and use. *Aging & Mental Health*, *11*(5), 477–484. doi:10.1080/13607860601086546

Small, S. A. (2001). Age-related memory decline. Current concepts and future directions. *Archives of Neurology*, *56*, 360–364. doi:10.1001/archneur.58.3.360

Small, S. A., Stern, Y., Tang, M., & Mayeux, R. (1999). Selective decline in memory function among healthy elderly. *Neurology*, *52*, 1392–1399.

Sum, S., Mathews, M. R., Pourghasem, M., & Hughes, I. (2008). Internet technology and social capital: How the Internet affects seniors' social capital and wellbeing. *Journal of Computer-Mediated Communication*, *14*(1), 202–220. doi:10.1111/j.1083-6101.2008.01437.x

Tse, M. M. Y., Choi, K. C. Y., & Leung, R. S. W. (2008). E-health for older people: The use of technology in health promotion. *Cyberpsychology & Behavior*, *11*(4), 475–479. doi:10.1089/cpb.2007.0151

White, H., McConnell, E., Clipp, E., Branch, L. G., Sloane, R., Pieper, C., & Box, T. L. (2002). A randomized controlled trial of the psychosocial impact of providing internet training and access to older adults. *Aging & Mental Health*, *6*(3), 213–221. doi:10.1080/13607860220142422

Chapter 16
Virtual Worlds for Science Learning

Mick Grimley
University of Canterbury, New Zealand

Trond Nilsen
University of Canterbury, New Zealand

Roslyn Kerr
University of Canterbury, New Zealand

Richard Green
University of Canterbury, New Zealand

David Thompson
University of Canterbury, New Zealand

ABSTRACT

This chapter proposes that the use of virtual worlds for science education is warranted and fits well with contemporary learning theory in the context of constructivist instructional approaches being desirable and that learners learn best when they are engaged in active mental processing. Over recent years, games have become increasingly social, supporting massively multi-player online game experiences and then evolving into virtual worlds, such as Second Life, which show significant promise for educational uses. This chapter introduces the field of virtual worlds, and then discusses relevant theory and research. The authors describe the potential of virtual worlds for education by emphasizing how they can be leveraged as an effective tool for constructivist teaching techniques. In addition, the authors present some of the literature that supports their use for science education. This chapter concludes with practical concerns and some possible solutions in the context of future research directions.

INTRODUCTION

When personal computer games began to appear in the mid 1980s, both teachers and game developers began to struggle to find ways to apply them to education. Despite isolated successes, however, games have not significantly affected the educational landscape, particularly in the sciences. Various reasons may be cited for this:

DOI: 10.4018/978-1-60566-690-2.ch016

Firstly, games have chiefly been applied as drill exercises for rote learning designed to complement in-class activities. As such, they are a marginal activity, and their greatest advantage is ignored - that of creating experience.

Secondly, science oriented games typically utilize models of real systems. These exist in a tension between fun and game balance, and realism, accuracy, and rigor. Experiences tend to be inaccurate but meaningful, or accurate but meaningless.

Finally, games are generally perceived to isolate learners from one another.

Over recent years, games have become increasingly social, supporting first multi-player, then massively multi-player online game (MMOG) experiences. Along similar lines, general purpose platforms now exist that share many of the attributes of games, but without the focus on simple fun. These platforms are known as virtual worlds, and though new, show significant promise for education in both the sciences and other fields. This chapter introduces the field of virtual worlds, then discusses relevant theory and research, and finally presents practical concerns and some possible solutions in the context of future research directions.

BACKGROUND

The term 'virtual world' is very broad; colloquially, it can cover anything from shared text based adventure games, to large scale 3D computer games such as *World of Warcraft* (Blizzard Entertainment, 2004) and to user constructed worlds such as Second Life. In this chapter, we are chiefly concerned with worlds of the latter sort. For clarity's sake, however, we first address the breadth of virtual worlds in order to arrive at a working definition by briefly examining three example virtual worlds, then identifying the key attributes they share. Finally, we contrast these examples with one another to see the different ways in which they can be applied.

The first games with any claim to the name 'virtual worlds' were text based multi-user dungeon games, or MUDs, for short. The first was MUD1, written by Richard Bartle (Bartle 1983) in 1979. Throughout the 1980s and early 1990s, MUDs developed further, finding a home on university networks, dial-in bulletin board systems and, eventually, the internet. In them, multiple players could explore a fantasy world together, killing monsters (known as mobiles, or mobs for short), collecting treasure, and so forth. In this way, they were largely inspired by table top role-playing games such as Dungeons and Dragons.

Unlike modern games, however, MUDs were entirely text based – players saw the world through written descriptions instead of graphics and, to interact with it, they would enter short text commands, much as in regular text adventure games. For example, the game might tell a player that they are in a room with green curtains, a sword on the floor, a goblin cowering in the corner, and Antonius (a character belonging to Tony, another player) standing by the door. From here, they might type 'leave room', 'pick up sword', 'attack goblin', or 'wave at Antonius'. As commands are received the MUD would react to them and then propagate messages describing any changes in the world to all players. So, in our example, if the original player chose the latter action, Tony might read on his screen 'Albion waves at you'.

Though text based MUDS lack immersive content such as graphics and sound, players can find them highly engaging. They incorporate social interaction through text, shared narrative, goal based game play and generally a level-based progression - these are effectively the same core features that make modern massively multi-player online role-playing games (MMORPGs) so addictive. Unfortunately, however, text based interfaces require effort to learn, and are unappealing or inaccessible to many players. Consequently, MUDs were popular primarily with people with a strong computing background and a passion for games.

Beginning in the late 1990s, MUDs began to acquire graphical interfaces through the use of specialized client software. Furthermore, server technology and the internet allowed games with many more players. MUDs had been primarily non-commercial, but the addition of graphics, sound, and a simpler user interface made them accessible to a much wider market, and they became viable as commercially published games. The modern MMORPG was born, as games such as Everquest, Asheron's Call, and Ultima Online became extremely popular.

In addition to allowing them to attract much larger numbers of players, the switch to graphical user interfaces allowed games to become much more complex and diverse. Various forms of visualization allowed players to interpret more game information at once, and mouse-based input proved much faster for complicated interactions that previously would have required players to remember long, complex commands. Furthermore, graphics allowed games to convey much larger areas of continuous space; whereas in text, worlds had been limited to small numbers of discrete places such as rooms linked together in a network. Graphics allowed the game to present vast terrains, which increased both realism and splendour, making it much easier for players to feel that they really inhabited a virtual space.

Another development that supported the development of MMORPGs was the introduction of personal avatars. In text games, the presence of other players' characters was conveyed by single statements. Other than the player's name and possibly a short text description on examination, there was little opportunity for customization and self expression. Avatars, on the other hand, had been long present in computer games as the generally anthropomorphic graphical representations of a player in a game. In early games they were simple animated graphics called 'sprites' such as, the well-known characters *Mario* (Nintendo, 1994) the Plumber and *Sonic the Hedgehog* (Sega, 1994). Later, they became rendered 3D models complete with rich textures and procedurally modifiable shapes; that is, their pose could be modified or extended on the fly as they moved, they could be fitted with items of costuming and equipment, and their textures could be customized to meet the specific desires of each player. This meant that given only a small number of base avatars (perhaps humans, elves and dwarves, each with a male and female variant), a vast range of specific avatars could be constructed by players, often as a means of self-expression.

Though players were able to customize their own representation in the world, in most MMORPGs, players were unable to directly customize the world itself. Some games allowed the construction of 'guild-halls' for groups of players, but these were typically based on simple standardized designs or with the assistance of the technical staff running the game world. This all changed in worlds such as *Second Life* (Linden, 2003), where players, now more appropriately termed users (or Residents, in official *Second Life* terminology), became able to own land, create buildings, sculpt terrain, and design objects.

Within most MMORPGs, object design and creation requires the use of external 3D modelling software that produces models in a standardized format that can then be imported into the game, though often only by the game developers, not by players. Most 3D modelling tools have a very steep learning curve, and are beyond the skills and time available to most players. Second Life, however, incorporates a simplified 3D modelling interface built directly into the world's standard graphical user interface. It is optimized for ease of use by allowing users to design objects as collections of linked 'primitives', each being some variant of a sphere, cube, torus, or cylinder. Each primitive may be stretched, resized and moved by clicking and dragging in the game interface, then bound together with several other primitives to create a complex object which interacts with other objects in the world according to the rules of a reasonably realistic physics engine. Advanced design still requires substantial patience and effort, particularly if one wishes to include scripted behaviour in an object design, but there is no longer any major initial hurdle that must be overcome, meaning that new users may more easily experiment and start to design objects of their own.

One concern that arises in such worlds is that of creative ownership. Since Second Life incorporates powerful design tools, users may create designs of their own that could have substantial value. Consequently, there is a desire for users to have ownership over these designs and, by incorporating a sophisticated series of ownership and rights management controls, Second Life does just this. If a user creates designs, they may trade them with other players using the world's in-built currency, or they might offer them for free. They may also prevent unauthorized distribution or modification to their designs and, importantly, they may exchange the in-game money gained from their design work with real money via a floating currency exchange to US dollars. Consequently, users in Second Life are able to set up and operate design businesses catering to other players within the world, even to the extent of making substantial profits.

As well as affording users the ability to design and construct virtual objects, Second Life allows users to sculpt the virtual land itself and build structures upon it. In most MMORPGs, the land within the game world is not owned to any meaningful extent by the players, but is shared amongst all players or perhaps assigned as territory to some group of players by the game designers. In Second Life, however, virtual land may be bought and sold by players, and those owning land have significant control over that land. The owner of a virtual plot can, for example, limit access to the land to some small subset of users, or lock the design and placement of all objects in that land (including buildings) so as to create a permanent environment for some purpose.

By allowing users to modify and design the world to such a substantial level, Second Life has become more than simply a virtual world offered by some provider to users – it has become a genuinely user-designed general purpose virtual world with no fixed theme and no fixed purpose, much as space is in the real world. However, unlike the real world, virtual worlds are infinitely more flexible, freer from onerous restrictions, and unlimited by physics. It is for these, among other reasons, that many researchers see great potential in the use of virtual worlds for learning.

TOWARDS A DEFINITION

So far, we have discussed three classes of virtual world: text based multi-user dungeons, massively multi-player role playing games, and general purpose virtual worlds such as Second Life. Though these span a wide range of technology and design, they all share five key common attributes. Taken together, we consider these the essential features of online virtual worlds. For the purposes of this chapter, we define virtual worlds as:

- **Simulated:** Virtual worlds are complex systems modelled by a computer and displayed to the user in some way using a combination of text, graphics, and audio. The effect of actions is determined according to fixed rules within the simulation. These rules may correspond to the way the real world works, but this is not necessarily the case – their simulated nature affords virtual worlds great flexibility of theme and expression, as well as strong control over both the intended and side effects of user actions.
- **Spatial:** Virtual worlds incorporate and represent some space. This space may be small, discrete or limited such as in a MUDs representation of a building as a set of discrete rooms, or vast and continuous,

such as in space based MMORPGs such as EVE Online. Almost all virtual objects are represented spatially within the world, and user avatars (see below) are situated within it and able to move according to some pre-defined rules of the simulation.

- **Shared:** A virtual world is shared simultaneously by multiple users. These users may interact within the world's space, and the effects of one user's actions on the world will be experienced by others, depending on the simulation rules and their spatial location. Typically, users are also able to communicate with one another through the world.
- **Persistent:** Virtual worlds model places both real and fantastic over time. Objects, avatars, and structures persist across time, perhaps subject to rules of the simulation such as physics or garbage collection, and changes made to these objects also persist over time.
- **Embodied:** Users are represented somehow within the world. In early virtual worlds, they are represented as text or simple graphics; whereas in more modern worlds they are represented using animated 3D models. These representations are called avatars and, in more recent worlds, are often highly customizable and offer an avenue for self-expression.

Throughout this chapter, we are chiefly concerned with general purpose virtual worlds. These have the following additional features that distinguish them from virtual worlds primarily intended as games:

- **User modification:** Users may create and modify most types of objects within the world. As well as facilitating self-expression and creativity, this allows the construction of specialized environments to satisfy the particular needs of any user or group or users. This is particularly important for educational uses, as educators can tailor the world to meet the needs of the lesson rather than tailoring the lesson to meet the needs of the technology as is often required when using computer game engines.
- **Economic system:** Users may trade their creations with others for in-world currency which can then be exchanged for real world currency. This facilitates the creation of a design economy which in turn promotes the availability of a wide range of re-usable objects and designs. For educators, this means that they have access to a much wider range of objects and structures for use in educational environments without having to construct them themselves. Furthermore, this affords the development of a market for specifically educational objects, designs, and environments.
- **No predefined usage:** Unlike games, general purpose virtual worlds do not pressure users into a particular mode of usage. In particular, unlike games, they do not contain any pre-defined narrative or goal structure that may interfere with a lesson. However, though pre-imposed narrative and goals can be an impediment, they may become powerful motivational tools when tailored specifically to a lesson and can lead to increased engagement and improved learning (Malone, 1981; Malone & Lepper, 1987).

VIRTUAL WORLDS AND EDUCATION

This chapter aims to convince the reader that the use of virtual worlds for science education is warranted and their use fits well with contemporary learning theory. In this chapter we take a similar view to many contemporary learning theorists: arguing that constructivist instructional approaches are desirable and that learners learn best when they are engaged in active mental processing (see for example Bransford, Brown & Cocking, 1999; Mayer, 2004). We describe the potential of virtual worlds for education by emphasising how, if used to their full potential, they can be leveraged as an effective tool for constructivist teaching techniques. In addition, we present some of the literature that supports their use for science education.

There are several arguments for the use of virtual worlds in education. They have been shown to encourage higher levels of motivation and be conducive to constructivist learning. In the context of science education and scientific literacy, virtual words have been found to be enormously useful owing to their potential to extend the already crucial reliance on modelling and experimentation that exists in the sciences. There are numerous studies confirming the usefulness of virtual environments to facilitate learning in a variety of different ways (e.g. Gazit, Yair & Chen, 2005; Limniou, Roberts & Papadopoulos, 2008; Zacharia, 2007; Zumbach, Schmitt, Reimann & Starkloff, 2006).

Constructivism and Virtual Worlds

No matter what type of constructivism you adhere to, it is widely recognised that humans construct their own understanding of the world around them by assimilating and accommodating information gained from their personal experiences. It is therefore reasonable to conclude that in order for humans to learn effectively they should engage in activities that are mentally active and appropriate for the material that they are trying to learn and understand. This also implies the use of instructional approaches that encourage mental activity and sense-making. Jonassen (1994) gives a compelling list of learning environments that would support the constructivist notion:

- provide multiple representations of reality, thereby:
- avoiding oversimplification of instruction by representing the natural complexity of the real world;
- focus on knowledge construction, not reproduction;
- present authentic tasks (contextualizing rather than abstracting instruction);
- provide real-world, case based learning environments, rather than pre-determined instructional sequences;
- foster reflective practice;
- enable context- and content-dependent knowledge construction; and
- supporting collaborative construction of knowledge through social negotiation, not competition among learners for recognition. (p.35)

Virtual worlds are naturally oriented toward active learning. For example, when users move around in-world they must make sense of the changing graphics; therefore they are perceptually active. In addition, processing is multi-sensory affording better memory and understanding (Mayer, 2001; Barraclough & Guymer, 1998). Greater engagement with virtual worlds compels higher order thinking skills such as reasoned decision making and problem-solving. Clearly this occurs in any learning environment, but many learning environments are much more abstract and fail to allow concrete authentic and meaningful interactions with the material being learned. Classrooms tend to be detached from reality, whereas virtual worlds reflect a multidimensional space that can be crafted to simulate and depict almost anything. Learning in a virtual world is much more experiential than traditional learning.

Experiential and Situated Learning

Experiential learning is defined as "the sense-making process of active engagement between the inner world of the person and the outer world of the environment" (Beard & Wilson, 2006). Theorists such as Dewey, Kolb, Rogers and Lewin all subscribe to the potential of experiential learning. Experiential learning is proposed as an effective form of learning because it makes learning meaningful and allows the learner to establish connections between the learned material and the real world. Proponents of experiential learning techniques propose that the learner should be immersed in an authentic environment and be encouraged to reflect upon their experiences. In addition, situated learning is an approach derived from situated cognition and shares many of the ideas put forward by experiential theorists. Situated learning supporters recommend that learning should take place in a situation that is authentic and representative of the way that knowledge is used in everyday life (Griffin, 1995). For advocates of situated learning, it is impossible to separate learning and doing. In the case of scientific literacy, it is desirable that students spend time around scientists and doing scientific work. Through this, they acquire understanding of not only the content, but how science takes place and the accompanying jargon (Zumbach et. al. 2006). In a virtual world, it could potentially be easy for students to spend time in a virtual laboratory in order for situated learning to occur (Boellstorff, 2008).

The idea of experiential and situated learning is very appealing. For those of us who have taught in schools, however, true experiential and situated learning seems unsustainable and impractical for obvious reasons. On the contrary, virtual worlds allow students to experience real world phenomena without leaving the classroom. Although not perfect reproductions of the real world, they can potentially be much more useful. It is possible to experience things in a virtual world that are either impossible or impractical to experience in the real world. Firstly, in a virtual world students can experience dangerous situations that are too unsafe in the real world. The idea of working in a lab with dangerous chemicals or equipment suddenly becomes much more feasible, and significantly more affordable. Secondly, students can experience normally impossible locations such as outer space, or locations in another time period (Ramasundaram, Grunwald, Mangeot, Comerford & Bliss, 2005). For example, Second Life currently includes a thematic recreation of the city of ancient Rome, allowing participants the opportunity to experience many aspects of the ancient city and its culture. Thirdly, students can experience different perspectives that are normally difficult to achieve or to experience, such as viewing closed systems from the inside, flying over terrains or experiencing many different representations of the same thing. For example, Antoniette and Cantoia (2000) demonstrate how a virtual world allows students to experience a painting from the inside by exploring and interacting with it.

Students can find it very helpful to identify with one part of a phenomenon and then view the entire phenomenon from that perspective (Gee 2008). For example, Chittaro and Ranon (2007) describe:

> *...the Virtual Big Beef Creek project, where a real estuary has been reconstructed to allow users to navigate, get data and information to learn about ocean science. Users can explore the environment using different avatars (i.e., embodiments of the user in the VE) whose viewpoints and navigation constraints are different. For example, if the user chooses to be a scientist, she can move as a human being and acquire data such as water temperature, while if she chooses to be a fish, she can swim deeply in the ocean and cannot surface. (p. 10)*

However, Gazit, Yair and Chen's (2005) study suggests that it can be difficult for students to comprehend and fully understand phenomena from multiple viewpoints. They suggest that while virtual environments have enormous potential for students to construct their own learning, in the scientific disciplines misconceptions can easily result unless accompanied by well thought out guidance and mentoring.

Finally, transformations of distance, size or speed can be manipulated in relation to an avatar. For example it may be possible to personally view and interact with a substance at the atomic level in the virtual world, something not possible in the real world except in the abstract (Chittaro & Ranon, 2007). Another example might be that instructors are free to build a scale model of a human that can be explored from the inside, or allow students to interact with different organs of the body to give a greater understanding of each organ. In a virtual world, students are free to construct their own learning through interpreting their personal perceptual experiences (Limniou, Roberts & Papadopoulos, 2008). However, it is the challenge of the instructor to build content that allows meaningful and effective experiences that challenge the students' thinking and support long lasting and effective learning.

Discovery Learning and Virtual Worlds

Discovery learning, sometimes called inquiry learning, is an instructional technique developed to serve constructivist ideas and their application to instruction (Mayer, 2004). Discovery learning allows learners to discover information and concepts for themselves. However, if learners are left to discover things for themselves they can sometimes foster misconception and spend inordinate amounts of time for little reward. As Mayer (2004) contends, guided discovery rather than pure discovery is a much better prospect, still allowing active mental processing but supported to avoid misconceptions and wasted time.

Inquiry based learning or discovery learning as described above is an approach used frequently in schools, especially within science education. The approach is however limited by many factors including time, money, space, equipment, consumables, realistic settings, ethics and practicality. A virtual world allows for many of these obstacles to be overcome. Students are free to engage in discovery learning with few limitations. Time factors are reduced as worlds are perpetual, even the most expensive equipment can be simulated relatively cheaply, you never run out of consumables, settings can be realistic or imaginary, studies can be performed on simulated animals without fear of ethical consequences; anything is possible. It is also feasible to utilize an agent within a virtual world to guide learning or for the instructor to be (virtually) present to provide guidance, or both. In addition learning can be collaborative, encouraging groups of students to work together on a project through sophisticated voice or textual communication. One advantage here is that students do not have to be in the same room, school, city or even country to work together.

Nelson and Ketelhut (2007) argue that virtual environments are immensely useful for teaching scientific inquiry. Nelson and Ketelhut (2007) describe how despite scientific inquiry being part of US curricula for many years, it remains a controversial and difficult area to teach, particularly in classrooms that rely heavily on textbook based learning. By contrast, they argue that virtual environments are ideally designed for students to learn scientific inquiry. For example, they describe how students can observe avatars with a particular disease, communicate with these avatars and others to find out more about the disease, discuss the disease amongst themselves and use virtual tools like microscopes.

Interaction and Collaboration

The constructivist notion includes the idea that humans construct knowledge through personal experiences and that an important part of these experiences are those experiences gained through communication with other people (Jonassen, 2002). In addition, there are many other learning theories that emphasize the importance of social interactions for meaningful learning (Schunk, 2008, p 253). It is therefore important to discuss how virtual worlds can accommodate learning through interaction and collaboration. This type of interaction is also of interest when considering experiential and situated learning because most real world scenarios involve some form of human–human interaction.

It is common for computer oriented learning to be thought of as generally not collaborative or interactive; however this is not necessarily true. Many computer applications such as virtual worlds now actively encourage collaboration. Students can communicate with teachers, other students, experts and people from other countries. Within a virtual world these interactions are no longer performed by exchanging voice or text messages alone; they also allow visualisation, usually in the form of a person's chosen bodily representation (their avatar). Virtual worlds allow the possibility of a much more sophisticated form of communication and interaction by allowing collaborators to interact physically with a shared environment (through their personal avatars).

Chittaro and Ranon (2007) argue that learning through interaction with others can be far more stimulating and inspiring than learning through reading or listening. They argue that within a virtual world, other humans can take on numerous roles to facilitate learning through adopting different avatar representations. Role play activities are far more realistic when played out in a semi-authentic environment and conversations, discussions and debates are far more realistic when you can see the other characters involved. In addition, research reveals that learning experiences are enhanced when the learner is guided by a virtual character whether controlled by another human or as an automaton (Lester, Converse, Kahler, Barlow, Stone & Bhoga, 1997; Johnson, Rickel & Lester, 2000). These characters within a virtual world are an improvement on traditional text based e-learning environments as they introduce lifelike settings and social elements. These virtual tutors can also demonstrate aspects of the environment visually (Chittaro & Ranon, 2007). Therefore virtual worlds that incorporate these lifelike agents to assist learning are generally going to be more effective than text based e-learning scenarios.

Another unique aspect to virtual worlds is the fact that they are generally user created. In Second Life for example, users may own their own virtual land and build whatever they wish on this land. In this respect, the world grows according to the wishes of its residents. This is an important aspect of user created worlds, such as Second Life, in that they can be positively harnessed for collaborative learning. Kirner, Kirner, Kawamoto, Cantao, Pinto and Wazlawick (2001) report how Brazilian school children developed their own educational material within a virtual world. The children were creating a virtual world based around given educational themes. This type collaborative construction of content by users is a prime example of cooperative learning in an authentic setting.

Finally, Second Life facilitates "on the spot" learning between residents. Potentially, this kind of learning can be more effective as residents can acquire information immediately, "on demand" at precisely the time they require it, rather than attending a lecture or tutorial at a specified time (Ondrejka, 2008).

Motivation and Virtual Worlds

It is essential that educators engage students and instill intrinsic motivation to learn. For without such motivation, most learning environments are unproductive. Without the motivation to learn and engage with the material in ways described above the learner is no longer a learner but merely a bystander (Lepper & Chabay, 1985; Malone & Lepper, 1987). Ormrod (2008) describes the following advantages gained by intrinsically motivated learners compared to extrinsically motivated learners:

- learners take initiative
- increased cognitive engagement

- willingness to undertake more challenging tasks
- a will to understand the learning content
- ability to undergo conceptual change
- creativity
- persistence
- experience pleasure in the learning tasks
- self regulation
- motivation to continue the task
- high achievement (p.454)

Anyone who has teenage sons or daughters knows the power of digital games or virtual worlds to instill immense amounts of intrinsic motivation. The new generation of school children were raised in a computer age. By the time these current students graduate almost all would have played computer games at some time with the vast majority being regular digital game players or inhabitants of some virtual world (Oblinger, 2004; Jones, 2003). Massively multiplayer online role-playing games are attracting millions of players across the world and new virtual worlds are being created daily. Players invest massive amounts of time, money and energy into playing these games. These games are particularly advanced socially, interactive and highly motivational (Gee, 2003; Prensky, 2002). Computer games and virtual worlds are motivationally intrinsic and it would be remiss of educators to fail to try to capitalize on such a concept. Learning can be fun and should be fun. Researchers have proposed a design structure for intrinsically motivated instruction, based on observations of computer games, as one that incorporates challenge, fantasy, control, curiosity, cooperation, competition, recognition (Malone, 1981; Malone & Lepper, 1987).

Csikszentmihalyi (Hektner, Schmidt & Csikszentmihalyi, 2007) introduced the concept of 'flow' where 'flow' is the highly motivational and pleasant state experienced when you are engrossed in an activity. This state has been related to the feeling that is experienced by users of computer games and has also been related to effective learning environments (Kiili, 2005). Although virtual worlds do not instil intrinsic motivation per se, they do have the potential for including all of the elements of an engaging computer game. As such, they present potential for embodying intrinsic motivational learning environments and should be of interest to science educators wishing to engage their students in the subject matter.

USE OF VIRTUAL WORLDS FOR SCIENCE EDUCATION

There are a number of studies that confirm the usefulness of virtual worlds specifically for teaching science. The subject of science has been identified as having several distinctive features that make virtual worlds particularly useful, as the studies described below indicate.

Chittaro and Ranon (2007) argue that virtual worlds such as Second Life are uniquely suited to providing educational experiences that are difficult to emulate in the real world. As Zacharia (2007) observes, many science educators advocate the use of virtual environments as a surrogate for the 'real' thing, such as during any of the following conditions:

1. a 'real' laboratory is unavailable, too expensive or too intricate;
2. the experiment to be conducted is dangerous;
3. the techniques that are involved are too complex for the students; or
4. there are severe time constraints. (Zacharia 2007: 122).

In Ramasundaram et al.'s (2005) study, a virtual field laboratory of a Florida ecosystem was created using real world data and tested on fourth year graduate students. Students were able to negotiate the virtual laboratory through animations, focus questions, navigation through hyperlinks, exploration of 3D models and adaptive selective simulations. It was found that the virtual laboratory mimicked real world learning experiences in addition to providing simulations that are not possible in the real world. Similarly, Gazit, Yair and Chen (2005) used a virtual environment in order to further student understanding of the workings of the solar system, an environment that is impossible for students to experience in the real world. They used a non-immersive virtual environment which showed the planetary objects as they revolve in their orbits against the constant background of the Milky Way and the stars where students could move around and change their viewpoint using the mouse. Poland, La Velle and Nichol (2003) created a virtual model of a Greek island using specialized 3D landscaping software in order for English A' Level biology students to experience the ecology and conservation of sea turtles. A 3D environment was created that allowed students to take on the role of sea turtle conservators without the need for students to travel to the Mediterranean. Jelfs and Whitelock (2000) describe how the United Kingdom Open University designed several virtual environments to allow science students to undertake virtual field trips to places they would not normally be able to go.

Zacharia (2007) tested the usefulness of virtual worlds through a comparative study. Forty five undergraduate students enrolled in introductory physics were broken into a control group, who used only a real world environment, and an experimental group, who used both virtual and real world environments. The virtual environment used the Virtual Laboratories Electricity software where students were provided with an area to build and buttons to select circuit parts. It was found that the experimental group not only had a greater conceptual understanding than the control group, but that certain parts of the curriculum were taught with much greater effect using the virtual environment instead of the real environment. Zacharia (2007) argues that these results suggest that the importance of physicality in learning science is overrated, with virtual manipulation proving more effective for teaching some aspects of science.

The effectiveness of virtual manipulation was confirmed by Suh and Moyer (2007). In contrast to Zacharia (2007), Suh and Moyer (2007) worked with far younger, third grade students, testing their understanding of algebra through using real and virtual manipulatives. Like Zacharia (2007), Suh and Moyer (2007) found several unique features of the virtual that were not matched by the physical, including explicit linking of the visual and symbolic and immediate self-feedback and checking.

Zumbach et al. (2006) used a virtual biology environment "lifelab" for high school students to perform virtual experiments with plant vaccines. The environment was a 3D world which included guidance in the form of a comic figure called "Dr Drop". Zumbach et al. (2006) found that not only were the virtual experiments highly effective for teaching these concepts, but that it was effective regardless of prior learning level or level of interest in the subject.

Limniou, Roberts and Papadopoulos (2008) also performed a comparative study which compared student understanding between a 2D desktop model and a 3D virtual environment. The 3D virtual environment allowed students to gain a full experience of molecular structure and changes in a chemical reaction through full immersion. They found increased understanding from the 3D environment due to the limitations of human vision being overcome, and furthermore that students were more motivated about their learning as a result of the 3D immersive experience.

Shim et al. (2003) compared understanding and knowledge of the biological structure and function of the eye between two groups of gifted year 10 students. One group used a virtual reality system to learn about the eye, consisting of a 3D world where various components and viewpoints could be manipulated, while the control group used standard 2D pictures and diagrams as used in traditional classroom teaching. In a test following the experiment, taken by all students on the information taught, the experimental group scored significantly higher. It was further observed that the students using the virtual reality system were more motivated, stimulated and engaged in the topic than those in the control group.

FUTURE RESEARCH DIRECTIONS

In the preceding section, we presented evidence suggesting that virtual worlds can enhance teaching and improve learning. Unfortunately, like any new technology or approach, there are many practical hurdles. In this section, we discuss a range of these and offer some suggestions for how they might be alleviated, either now or in the future.

As an illustration of some of these practical difficulties, consider the Teen Second Life Pilot project (Schome, 2007): a large scale study of learning in Second Life conducted from 2006 to 2007 with approximately 100 gifted high school students from across the UK. It examined students' willingness to engage with a virtual world, evaluated both the skills required to use Second Life and students' ability to learn them, and evaluated the extent to which they were able to retain knowledge. Despite buy-in and direct support from Linden Labs (the makers of Second Life), there were numerous practical difficulties including high setup costs, high technical requirements for access to Second Life, problems with commercial content usage, support problems and technical issues arising during project creation.

We first address issues of cost and equipment, followed by issues related to content construction and availability. Next, we discuss the integration of virtual worlds with a traditional classroom, focusing particularly on additional burdens placed on the teacher. Finally, we discuss political issues, such as acceptance by schools and governments, and issues of child safety.

Cost

As pointed out by Chittaro and Ranon (2007), it can often be challenging to acquire the necessary funding for students to participate in virtual worlds. There are several potential sources of cost, all of which must be considered by anyone intending to use virtual worlds in a classroom environment.

Computers: Ideally, each student should have their own computer, but this is not always practical. Depending on the activity, students may be able to work together in groups of two or three; more than this and the benefits of virtual worlds are likely to be limited. Furthermore, since many virtual worlds are graphics intensive, they often require computers more powerful than those available in the classroom. Since it is not normally possible to purchase new machines, the best approach is usually to carefully evaluate and choose a virtual world with requirements matching those provided by the computers that are available.

Server: As discussed below, it may sometimes be preferable to host the virtual world within the school or school district. In this case, server cost must also be addressed. The requirements of virtual world server software vary from platform to platform, but typically a regular computer is sufficiently powerful to serve up to about 30 students sharing the same virtual world. Furthermore, this server need not be dedicated – the server software could be started and stopped as is needed, leaving the computer free for other tasks when the world is not in use.

Network: If the virtual world is situated outside the school on the public internet, internet bandwidth may become an issue. Alone, the bandwidth required to connect to a virtual world is comparatively low but if twenty students connect simultaneously, the user experience may be diminished to the point of the world being unusable, and the cost of the traffic may become prohibitive. The scope of this problem varies depending on the nature of a school's connection and so, to prevent unwelcome surprise bills, it is important to test the amount of traffic used in advance.

Another important consideration is that schools and school service providers may employ secured networks or firewalls to filter their public internet access for security, safety and privacy reasons. Connecting to an external virtual world may require special configuration (and possibly administrative dispensation) to address the internet security policies of the school or service provider.

Software: For some virtual worlds, there is no cost for software – both server and client software is available for free. Unfortunately, this may often imply that there is no support available, that content is either sparse or expensive, or additional setup effort is required.

Technical Support: As with any new technical approach, when setting up and using virtual worlds, there may be many technical hurdles to be overcome. In some cases, a motivated teacher may have or be able to develop the skills necessary to address these, but in other cases some technical support is required. Some schools may be lucky enough to have full time technical support staff available; if this is the case, they will still need to buy into the idea of using virtual worlds lest they view them as an unwarranted burden. If internal technical support is not available, teachers may be able to find technical support online, but some technical skill is normally required nonetheless. All of these factors are costly in either time or resources. Furthermore, these costs may not be fixed or even predictable.

World resources: Some public virtual worlds (notably Second Life), allow users to purchase and control regions within the world. Normally, there are costs associated with this, and educators may find it necessary to share space in order to keep costs down.

Virtual worlds are a novelty in education, and consequently there is a lack of infrastructural support that exaggerates these costs. As time passes and they become more prevalent, costs will most likely drop. For now, however, the key strategy for dealing with cost is to plan ahead and consider costs upfront.

Content

Most virtual worlds contain a variety of content, consisting of virtual objects, images, sounds, and environments. Unfortunately, they are not typically designed with educational use in mind, and some level of customization and content construction may be necessary for them to be useful for learning. To some extent, individual teachers may be able to take pre-existing content and customize it themselves, based around their own lesson plans, but for virtual worlds to gain widespread acceptance, educational content within them needs to be standardized, much as textbooks are today.

In today's environment, then, a teacher wanting to use a virtual world to teach will need to design and build the virtual environment themselves. Thankfully, many virtual objects and imagery can be re-purposed and used, so this doesn't necessarily imply that teachers must start again from scratch. However, the effort required to construct an environment is non-trivial, and there are many questions concerning how this should be done that are still unanswered. Questions include what teaching strategies are best adopted, what level of realism is required, and how an environment might simultaneously support multiple learners without them interfering with one another (Zumbach . 2006).

As a result, virtual worlds are for now mostly appropriate for early adopters, researchers, or teachers wanting to experiment with them in the classroom. As their educational use becomes more mature, however, it can be expected that educationally suitable content will become available for application in the classroom, greatly reducing the amount of effort required of the teacher.

Classroom Integration

Another significant question is how lessons using virtual worlds can be integrated into the classroom. Though it is not clear how this will be best done, it will most likely entail a significant change in style for many teachers, and may even eventually change the fundamental ideal of the classroom (Chittaro & Ranon 2007).

Though this reordering may not be a good thing in every way, it seems clear that the prominence of digital media is already pressuring teachers to change classroom practices accordingly. Unfortunately however, this is happening slowly if at all (Dodge, Barab, Stuckey, Warren, Heiselt & Stein, 2008). Despite studies suggesting children are less engaged in learning and less motivated, schools continue to operate within a cultural logic that focuses on the teacher and print based materials as expert and does not engage with new technologies (Dodge et al. 2008).

Some researchers have argued that teaching within virtual worlds is time consuming and that teachers will struggle to find time to accommodate this teaching style. For example, Nelson and Ketelhut (2007) discuss the usefulness of virtual worlds for teaching scientific inquiry, describing how it can be time consuming for students to visit the virtual environment and build up knowledge gradually. However, in another study (Steen, Brooks & Lyon, 2006), where virtual manipulatives were used to teach basic mathematical and geometric concepts to 1st and 2nd grade students, anecdotal evidence was presented that suggest that virtual worlds can be less time consuming: "Students did not have to clean-up rubber bands from geoboards, they did not have to put away pattern blocks, they did not have pass out manipulatives, and re-doing an activity was not an ordeal." (Steen, Brooks & Lyon 2006: 287)

Safety

Both games and virtual worlds have been the subjects of concern and criticism regarding child safety, and the appropriateness of content. This includes concerns over predatory adults, offensive, obscene, or otherwise inappropriate content, right down to the general requirement of privacy. Obviously, these concerns must be convincingly addressed if they are to be employed in the classroom.

Problems primarily arise in public virtual worlds – ideally, virtual worlds used for teaching will be maintained in such a way that they are under the control of the teacher or school that is employing them. Unfortunately, these require significant levels of support, and publicly operated virtual worlds are currently more practical and cost effective. Before these can be used, teachers should carefully consider whether the following minimum standards can be met.

- Students must be protected from predatory behaviour by other world users.
- Students must not be exposed to inappropriate content.
- Learning environments must be private and free from intrusions by other parties not participating in the lesson.
- Care should be taken that students cannot use the world to access copyrighted or otherwise illegal material.
- It may be desirable that students are prevented from leaving the learning environment to visit other parts of the world.

Second Life has addressed these concerns by explicitly dividing into two separate worlds. One is for general use by adults, while the other is restricted to teenagers (ages 13-17), and authorized educators. Unfortunately, Second Life makes no provision for users younger than 13 years old. A further protection is that users of Second Life can own regions of the world and explicitly control who may access them. Other virtual worlds employ similar restrictions and measures, but teachers should nonetheless carefully address concerns for student safety on a case by case basis.

CONCLUSION

While there are a number of practical issues related to the use of virtual worlds for enhancing science education, there are clear advantages over and above those described in this chapter and related to pedagogy. Firstly, students tend to be of the Digital Native Era and thus intimately familiar with technology to the extent of expecting its use in education (Prensky, 2001). Secondly, the use of virtual worlds allows enormous flexibility in the learning environment allowing students to work in almost any environment imaginable without risk and with little inconvenience. Thirdly, working with virtual worlds offers a much more student focussed learning environment conducive to constructivist learning. However, it should be stressed that it is also possible to employ virtual worlds in Science Education for the "same old, same old" material that will neither inspire nor enhance student learning. It is possible to create a lecture theatre within a virtual world and to have students sit around listening to a lecturer spout words of wisdom, but how is this using the potential of a virtual world? It is much better to allow students to explore and interact with the material they are being taught.

REFERENCES

Antoniette, A., & Cantoia, M. (2000). To see a painting versus to walk in a painting: An experiment on sense making through virtual reality. *Computers & Education*, *34*(3-4), 213–223. doi:10.1016/S0360-1315(99)00046-9

Barraclough, A., & Guymer, I. (1998). Virtual reality-A role in environmental Engineering education? *Water Science and Technology*, *38*(11), 303–310. doi:10.1016/S0273-1223(98)00668-4

Bartle, R. (1983). A Voice from the Dungeon. *Practical Computing, December*, 126-130.

Beard, C., & Wilson, J. P. (2006). *Experiential learning: A best practice handbook for educators and trainers*. London: Kogan Page.

Blizzard Entertainment. (2004). *World of Warcraft* [computer game]. Designers: Pardo, Kaplan & Chilton.

Boellstorff, T. (2008). *Coming of Age in Second Life: An Anthropologist Explores the Virtually Human*. Princeton, NJ: Princeton University Press.

Bransford, J. D., Brown, A. L., & Cocking, R. R. (1999). *How people learn: Brain, mind, experience, and school*. Washington, DC: National Academy Press.

Cheal, C. (2007). Second Life: hype or hyperlearning? *Horizon*, *15*(4), 204–210. doi:10.1108/10748120710836228

Chittaro, L., & Ranon, R. (2007). Web3D Technologies in Learning, Education and Training: Motivations, Issues, Opportunities. *Computers & Education*, *49*(1), 3–18. doi:10.1016/j.compedu.2005.06.002

Dodge, T., Barab, S., Stuckey, B., Warren, S., Heiselt, C., & Stein, R. (2008). Children's Sense of Self: Learning and Meaning in the Digital Age. *Journal of Interactive Learning Research*, *19*(2), 225–249.

Gazit, E., Yair, Y., & Chen, D. (2005). Emerging Conceptual Understanding of Complex Astronomical Phenomena by Using a Virtual Solar System. *Journal of Science Education and Technology*, *14*(5-6), 459–470. doi:10.1007/s10956-005-0221-3

Gee, J. P. (2003). High score education [Electronic Version]. *Wired, 11*. Retrieved June 3 2004, from http//www.wired.com/wired/archive/11.05/view.html?pg=1

Gee, J. P. (2008). Learning and games. In K. Salen (Ed.), *The ecology of games: connecting youth, games, and learning* (pp. 21-40). Cambridge, MA: MIT Press.

Griffin, M. M. (1995). You can't get there from here: Situated learning, transfer and map skills. *Contemporary Educational Psychology, 20*, 67–87. doi:10.1006/ceps.1995.1004

Hektner, J. M., Schmidt, J. A., & Csicksentmihalyi, M. (2007). *Experience sampling method: Measuring the quality of everyday life*. London: Sage.

Jelfs, A., & Whitelock, D. (2000). The Notion of presence in virtual learning environments: What makes the environment "real". *British Journal of Educational Technology, 31*(2), 135–152. doi:10.1111/1467-8535.00145

Johnson, W. L., Rickel, J. W., & Lester, J. C. (2000). Animated pedagogical agents: Face-to face interaction in interactive learning environments. *International Journal of Artificial Intelligence in Education, 11*, 47–78.

Jonassen, D. H. (1994). Thinking technology. *Educational Technology, 33*(4), 34–37.

Jonassen, D. H. (2002). Learning as activity. *Educational Technology, 42*(2), 45–51.

Jones, S. (2003). *Let the games begin: Gaming technology and entertainment among college students*. Retrieved June 3, 2004, from http//www.pewinternet.org/

Kiili, K. (2005). Digital game-based learning: Towards an experiential gaming model. *The Internet and Higher Education, 8*, 13–24. doi:10.1016/j.iheduc.2004.12.001

Kirner, G., Kirner, C., & Kawamoto, A. L. S. Canta˜o, J., Pinto, A., & Wazlawick, R. S. (2001). Development of a collaborative virtual environment for educational applications. In S. Diehl & M. V. Capps (Eds.), *Proceedings of the 6th international conference on 3D web technology* (pp. 61-68). New York: ACM Press.

Lepper, M. R., & Chabay, R. W. (1985). Intrinsic motivation and instruction: Conflicting views on the role of motivational processes in computer-based education. *Educational Psychologist, 20*(4), 217–230. doi:10.1207/s15326985ep2004_6

Lester, J. C., Converse, S. A., Kahler, S. E., Barlow, S. T., Stone, B. A., & Bhoga, R. S. (1997). The persona effect: Affective impact of animated pedagogical agents. In S. Pemberton (Ed.), SIGCHI conference on human factors in computing systems (pp. 21-27). New York: ACM Press.

Limniou, M., Roberts, D., & Papadopoulos, N. (2008). Full immersive virtual environment CAVETM in chemistry education. *Computers & Education, 51*, 584–593. doi:10.1016/j.compedu.2007.06.014

Linden Labs. (2003). *Second Life*. Retrieved from http://lindenlab.com/

Malone, T. W. (1981). Toward a theory of intrinsically motivating instruction. *Cognitive Science, 4*, 333–369.

Malone, T. W., & Lepper, M. R. (1987). Making learning fun: A taxonomy of Intrinsic motivations for learning. In R. E. Snow & M. J. Farr (Eds.), *Apptitude, Learning, and Instruction. Volume 3: Conative and Affective Process Analyses* (pp. 223-250). Hillsdale, NJ: Lawrence Erlbaum Associates Inc.

Mayer, R. E. (2001). *Multimedia learning*. New York: Cambridge University Press.

Mayer, R. E. (2004). Should there be a three-strikes rule against pure Discovery learning? The case for guided methods of instruction. *The American Psychologist*, *59*(1), 14–19. doi:10.1037/0003-066X.59.1.14

Nelson, B. C., & Ketelhut, D. J. (2007). Scientific Inquiry in Educational Multi-user Virtual Environments. *Educational Psychology Review*, *19*, 265–283. doi:10.1007/s10648-007-9048-1

Nintendo of America, Inc. (1994). *Donkey Kong Country* [Computer Game]. Developer: Rare Ltd.

Oblinger, D. (2004). The next generation of educational engagement. *Journal of Interactive Media in Education*, *8*, 1–18.

Ondrejka, C. (2008). Education unleashed: participatory culture, education, and innovation in second life. In K. Salen (Ed.), *The ecology of games: connecting youth, games, and learning* (pp. 229- 251). Cambridge, MA: MIT Press.

Ormrod, J. E. (2008). *Human Learning* (5th ed.). Upper Saddle River, NJ: Pearson.

Poland, R., La Velle, L. B., & Nichol, J. (2003). The virtual field station (VFS): Using a virtual reality environment for ecological fieldwork in A'-level biological studies--Case Study 3. *British Journal of Educational Technology*, *34*(2), 215–231. doi:10.1111/1467-8535.00321

Prensky, M. (2001). *Digital Game-Based Learning*. New York: McGraw-Hill.

Prensky, M. (2002). Not only the lonely: Implications of 'social' online activities for higher education. *Horizon*, *10*(4), 1–10.

Ramasundaram, V., Grunwald, S., Mangeot, A., Comerford, N. B., & Bliss, C. M. (2005). Development of an environmental virtual field laboratory. *Computers & Education*, *45*, 21–34. doi:10.1016/j.compedu.2004.03.002

Schunk, D. H. (2008). *Learning theories: An educational perspective* (5th ed.). Upper Saddle River, NJ: Pearson.

Sega of America, Inc. (1994). *Sonic the Hedghog* [Computer Game]. Developer: Sonic Team.

Shim, K. C., Park, J. S., Kim, H. S., Kim, J. H., Park, Y. C., & Ryu, H. I. (2003). Application of virtual reality technology in Biology education. *Journal of Biological Education*, *37*(2), 71–74.

Steen, K., Brooks, D., & Lyon, T. (2006). The Impact of Virtual Manipulatives on First Grade Geometry Instruction and Learning. *Journal of Computers in Mathematics and Science Teaching*, *25*(4), 373–391.

Suh, J., & Moyer, P. S. (2007). Developing Students' Representational Fluency Using Virtual and Physical Algebra Balances. *Journal of Computers in Mathematics and Science Teaching*, *26*(2), 155–173.

The Schome Community. (2007). *The schome-NAGTY Teen Second Life Pilot Final Report: a summary of key findings and lessons learnt*. Milton Keynes, UK: The Open University. Retrieved July 20, 2007, from http://kn.open.ac.uk/public/document.cfm?docid=9851

Zacharia, Z. C. (2007). Comparing and combining real and virtual experimentation: an effort to enhance students' conceptual understanding of electric circuits. *Journal of Computer Assisted Learning*, *23*, 120–132. doi:10.1111/j.1365-2729.2006.00215.x

Zumbach, J., Schmitt, S., Reimann, P., & Starkloff, P. (2006). Learning Life Sciences: Design and Development of a Virtual Molecular Biology Learning Lab. *Journal of Computers in Mathematics and Science Teaching*, *25*(3), 281–300.

Chapter 17
Representational Inquiry Competences in Science Games

Rikke Magnussen
Aarhus University, Denmark

ABSTRACT

This chapter considers the enactment of competences in a particular science learning game Homicide, which is played in lower secondary schools. Homicide is a forensic investigation game in which pupils take on the role of police experts solving criminal cases in the space of one week. The Homicide game-based learning environment was designed with the aim of supporting scientific inquiry through a simulation of elements of a professional forensic practice situation. The game is thus designed to support work with genuine scientific inquiry and to meet the seventh- to tenth grade curriculum objectives for science and Danish education in Danish schools. This chapter presents the results of a long-term empirical study conducted with four school classes who played the game. The focus of the study has been to understand what competences are enacted when a professional inquiry is played out in schools. The chapter considers how students constructed visual representations of the cases they investigated and how they used these representations to establish hypotheses and evidence. The term 'representational inquiry competences' is developed; it refers to the students' ability to construct, productively use, transform and criticize visual representations as an integrated part of conducting an inquiry in the science game.

INTRODUCTION

In the past decade, science education has been the focus of studies and development of a new generation of theory-based learning games (Squire, 2002; Barnett *et al.*, 2004; Shaffer *et al.*, 2005; Magnussen & Jessen, 2006). It has been argued that active and critical learning about rich semiotic systems and complex problem-solving that well-designed games are thought to involve, are similar to science learning when understood as an active process of inquiry involving real-life science (Gee, 2004, 2003). In spite of the boom in game learning research, relatively little is known about how learning occurs

DOI: 10.4018/978-1-60566-690-2.ch017

through game-play or what interaction occurs when complex game-based learning environments are brought into a school culture. The objective of the research in this chapter is to provide a detailed understanding of practices and competences that are seen and developed in the science game environments in school. In this chapter I present results from a three year long study of the science game Homicide (Magnussen, 2009).

BACKGROUND: GAME SIMULATION OF SCIENCE PROFESSIONS

The empirical focus is the science game *Homicide*, which can be grouped as part of a new generation of theory-based learning games that simulate professional environments. One common trait that these games share is that they simulate elements of the objectives and environments in a specific profession by using and making available the technology, tools and/or methods of that profession.

Some examples of simulations are environmental engineers trying to find a polluted site (Squire & Klopfer, 2007), urban planners redesigning the central pedestrian street in a town (Shaffer, 2006), or criminal investigators investigating a murder using forensic techniques (Magnussen, 2007). An objective for creating these types of games is to apply the game media to designing complex settings based on the learning environments of real-life professionals, thus allowing students to engage in the complex, creative, and innovative problem solving and learning processes of these professionals.

The motivation for developing these types of games stems from a critique of the teaching of standardized skills to children in today's school system. The skills acquired in this system do not prepare them for a future that involves a constantly changing, complex work life (Shaffer & Gee, 2005). Critics believe that under the current system, students do not learn to deal with problems that do not have ready-made answers and that they do not solve problems using creative, innovative thinking or collaboration. An objective of designing this type of game is to use the game media to create environments with simulations of complex real-life situations where students have to think like professionals and solve problems in innovative ways as professionals do (Shaffer, 2007).

Simulating professions is not new. Commercial games such as Counter Strike or the game version of CSI simulate the professional practices of counter-terrorists and forensic detectives. The difference between the commercial games and this new generation of learning games is that the designs are based on learning theories and/or detailed studies of the learning processes and the tools of real-life professionals. Some of these games are also designed to meet the goals of the formal school science curriculum (Magnussen & Jessen, 2006), while others, "create the epistemic frame of a socially valued community by re-creating the process by which individuals develop the skills, knowledge, identities, values and epistemology of that community" (Shaffer, 2007, p164). The latter class of games is defined as 'epistemic games'.

The empirical field work in this chapter consists of observations of the game Homicide being played in four schools. The game is an IT-supported game where players role-play forensic experts. It is designed for cross-disciplinary science education in lower secondary schools in Denmark and was developed by a game development group (including the author of this paper) at Learning Lab Denmark at The Danish School of Education and was published by the Danish school book publisher Malling Beck. The game takes five days to play (if you start at eight in the morning and finish around one or two pm in the afternoon). It is organized as a combination of work in investigative groups (each working on their individual case) and meetings where groups share information about their cases and are encouraged by the chief of police - the teacher - to set new goals in their investigation.

After each meeting, the teacher makes new parts of the game accessible on the computer, which gives the investigators new information to work with. The game ends with the groups presenting their theories to the other teams and writing an indictment based on the evidence and testimonies of suspects and witnesses. The interaction in the game is primarily between the students in the classroom and not a computer-student interaction as in most computer games. Instead the computer is used as an extended police database.

The game's interface provides the players with access to videotaped interviews with 'suspects', reports from the local police, maps and pictures of and information about evidence found at the crime scene. With this information at hand the players have to plan their investigation process. The investigators analyse the evidence through laboratory work and analytical processes using technological and scientific, theoretical and practical methods that are available in the Forensic Handbook which is part of the data base. Some of this work includes pen and paper tasks and some practical analysis in the school laboratories. The laboratory work includes chemical analyses of samples from a suspect's hands to determine whether that person has gunshot residue on his or her hands, which would indicate that the suspect has fired a gun lately; and measurements of shooting angles to determine the height of the shooter. The students have to handle different types of data and use different skills, including critical thinking when they analyze interviews with the suspects and empirical competencies in handling the data from the technical investigations.

The educational goals of the game are closely integrated with the fictional investigation process. The overall goal for cross-disciplinary science learning in the game is that the game should support working with, and learning of, the process of inquiry as the basis of scientific investigation. The process contains different steps: problem definition, establishing hypotheses, conducting investigations, making observations, collecting data and explaining results. The methods used to solve the murders, e.g. fingerprint technique are as close to the real forensic professional methods used by forensic experts as was possible. The students have, of course, seen movies and read about police work, and getting a chance to use the professional tools are meant to further their motivation.

Figure 1. The policeman's desk with different investigation resources (character gallery, videotaped interrogations etc.) is the initial interface students meet when they enter the game Homicide. © 2003. Danish School of Education, Aarhus University. Used with permission.

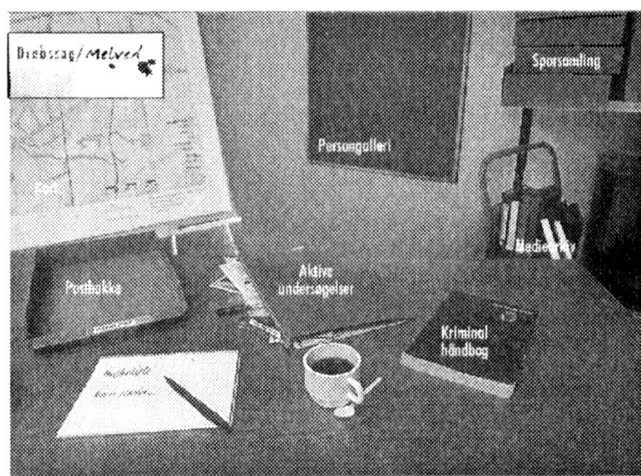

The empirical field work was carried out as a series of design interventions in the classrooms settings in which the game is played. The methodological approach in the studies undertaken is design-based research characterized by iterative cycles of problem definition, design, interventions, and analysis (See: Brown, 1992; DBRC, 2003).

Design-based research (DBR) was thus used to develop and study the game Homicide in iterative cycles of problem definition, design, intervention and analysis in the different classes (Magnussen, 2009). In the course of these designs and interventions an observational focus was established on the representations constructed by participating students and on how they were used in the investigation processes. In this chapter I present analysis of situations from the game play to illustrate the representational inquiry practices taking place in the game.

REPRESENTATIONAL INQUIRY IN CLASS B

The study presented here is based on a larger study of four different school classes who have played Homicide (Magnussen, 2009). It is based on four school weeks of video observations; in total 100 hours of observations conducted from 2004 – 2006. The students in the four classes were between 13 and 15 years old. Apart from the video observations, group interviews were done in all the classes with students and teachers as well as registration of the gradual development of the visual representations the students produced during the game. In this chapter I present data from one of the four classes, Class B, showing how the students used the representations in their investigation of the criminal case in the forensic investigation game. Examples and results from the studies in the different classes are used for discussing what competences are enacted in this type of game-based spaces in school.

REPRESENTING INQUIRY

Class B was an eighth grade class in a public school 30 km outside Copenhagen. The majority of the group of teachers running the game in this class had run the game before in another class. In Class B, I mainly focused my observations on Group B1, which investigated a case concerning a boy named Moussa who had been found dead in the local swimming club. The reason for focusing on this group was that its members introduced the use of representations in the class and used these (as other groups also did) actively in their investigations throughout the entire game. By focusing on one group's work, I was able to document and trace the development of their work in detail. The game ran for a full school week as intended in the game material and was set up in two classrooms. The teachers had set up separate workspaces for each group with tables next to display boards. Each group was located next to display boards, but was not encouraged by the teacher to use them before the first status meeting. The reason for arranging with the teachers not to encourage using the boards was that I was interested in knowing whether students would initiate the construction and use of representations under these conditions. The cordless network in the class was not working properly so each group was given access to one computer in the computer room away from the classroom. They were encouraged to print out game documents and use them at their workspaces in the classroom. The status meetings in Class B were also held around the presenting group's workstation. All laboratory work was done in the school laboratory away from the classrooms.

I would like to start by providing an overview of the first phase of Group B1's development of their representation and then present a snapshot exemplifying how the group used the representation in their group discussions. As described, Group B1 was the first group to start using the display boards behind their table in their workspace. As was the case in Class A, Group B1 used their first representation in the presentation at the first status meeting after 15–20 minutes of playing the main case.

At the first status meeting Tobias presented the victim's personal data and the data on when, how and in what condition the victim was found. This information is available at the game interface in the autopsy report and the report from the local police. The first stage of the representation, which only contained a minor part of the available information (see Figure 2 above), included the Pictures of the victim from the autopsy report (bottom of the board), a description of the victim, Moussa Alaleh (center), the victim's brother Raat Alaleh (right) and the caretaker, Teddy 'Hal' (top), at the swimming club where Moussa was found dead. Tobias presented the characters by pointing at them. At the meeting, the teacher encouraged the other groups to use boards in the same way as Group B1. Other groups said that they had been on the verge of doing the same before the meeting.

After the meeting, most of the groups started to put files on the boards. In Group B1 Tobias left to go to the computer room to take notes from the videotaped interrogations of witnesses and suspects while the remaining members of the group worked on the reorganization of the data on the board (see Figure 3).

At this stage, the group organized the data on the board in a system that generally remained the same until the final phase of the game. The descriptions of the suspects were aligned over one another on the left side of the board with the data or new information from the technical analysis about that character next to the descriptions. Data on clues found at the crime scene that had not yet been analyzed were placed in the middle of the board. The right side of the board seemed (in the first phase of the game) to host tools for the investigation (such as the timeline) or analyzed data which could not be or had not yet been connected to a specific character. An example of this was the lifted fingerprint the group had glued on a red piece of paper. It had been lifted from the personal belongings in the victim's locker at the swimming club and a suspect with a matching print had not been identified at this stage in the game.

After the four group members had finished organizing the documents on the board, Tobias returned from the computer room to give the group the information obtained from viewing the videotaped interrogations. In these recordings, the characters are interviewed by a police officer about where they were at what time on the night of the crime as well as their relationship and that of other characters to the victim. The group was assembled in front of the board while Tobias systematically went through his notes concerning the different characters, and either he or Mark pointed at their Picture on the board while the different group members commented on the information.

When Tobias gave the information concerning time of the different events and at what time the character Teddy Hal was at the swimming club, Katja suggested that they should make one timeline to plot the time of the different events that had occurred and the whereabouts of all the different characters. This started a long dispute between Tobias and Katja in which Tobias argued they should make separate timelines for each character and that the 'report dude' should do that. At this point, he pointed at Dorthe and told her that she was the 'report dude'.

Katja again argued that they should make one large timeline containing all the data concerning time. She was supported by Mark, who also wanted to make one large timeline. Tobias argued they had a lot of data on time and Katja replied that this was why having the data collected in one place was a good idea. Tobias then contended that making all the time data fit on one paper would be difficult. This ended in a dispute about who should do what, at which point Tobias told Dorthe to make a list of everything that had to do with time. The group did not seem to make much use of this timeline during the game though. Their representation kept this overall structure during most of the game.

Character profiles are aligned one over the other on the left side of the board. Data or new information from the technical analyses is placed next to the descriptions. Unanalyzed data on clues found at the crime scene were placed in the middle of the board. The right side of the board hosted tools for the investigation (such as the timeline) or analyzed data that could not or had not yet been connected to a specific character (e.g. the red piece of paper with the fingerprint lifted from the belongings in the victim's locker at the swimming club).

The objective of this chapter is to understand what competences students demonstrate while

Figure 2. Tobias, a B1 group member, presented the first overview of the Moussa case at the first status meeting. © 2003. Danish School of Education, Aarhus University. Used with permission

Figure 3. Katja (black jacket) reorganized the data on the board with the help of the other group members Dorthe (grey shirt), Mark (green shirt) and Jannik (red shirt). © 2003. Danish School of Education, Aarhus University. Used with permission

Figure 4. Reading from his notes, Tobias pointed at the character Teddy 'Hal', who was the caretaker at the swimming club where the victim was found. © 2003. Danish School of Education, Aarhus University. Used with permission

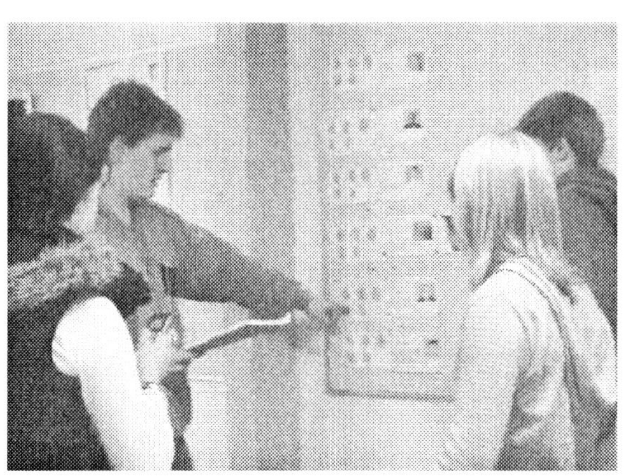

Figure 5. Katja pointed to the spot on the board where she thought the timeline should be. © 2003. Danish School of Education, Aarhus University. Used with permission

using the game, and what role representations play in this process. Before entering in to the analysis of this I present a snapshot from the group's investigation process which involves a group discussion concerning what happened at the crime scene in the hours leading up to the death of the victim, Moussa. At this point the group had had access to several different sources of information. One example is that they knew (from the interrogations of the victim's girlfriend) that the victim had an argument with a boy he regularly swam with named Peter. The night he was killed the victim had made fun of Peter, which had made Peter angry. The group had also

Figure 6. The B1 representation, early day 2. © 2003. Danish School of Education, Aarhus University. Used with permission

Figure 7. Reconstruction of the B1 representation in the first phase of the game, early on day two in the game week. © 2003. Danish School of Education, Aarhus University. Used with permission

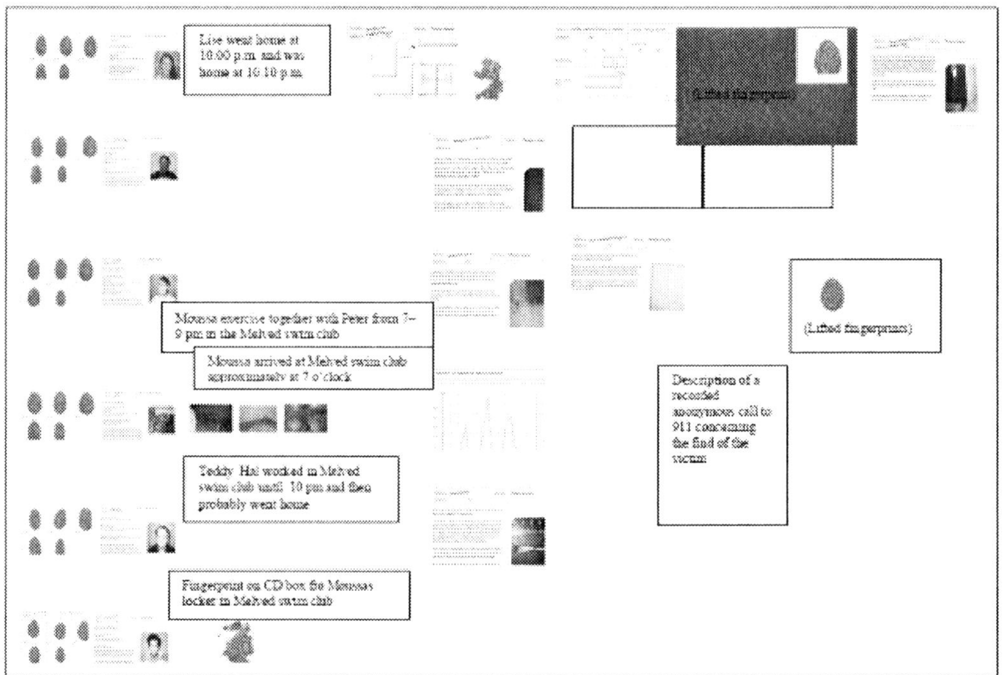

had access to interrogations of Peter in which he told the police that he and the victim also had an argument in the dressing room before Peter went home just before ten. The suspect, who proves to be the murderer in the end, had not been introduced at this point. At the start of this situation, in an attempt to reconstruct the exact time of the crime, the group members named Dorthe, Katja, Tobias and Mark sat around the table in their workspace and discussed what time the victim, Moussa, went into the dressing room.

Tobias: But he (the victim) didn't go into the dressing room until about ten, because he talked to his girlfriend Lise first.

Mark: Oh, yes, right. Was it at exactly (points to Tobias) nine or ten?

Dorthe: 9 – 9:48.

Tobias: Yes.

Mark: In that case, I think it's early 10 (says something inaudible).

Tobias: Yes, that might very well be, because...

(Everybody in the group starts talking all at once.)

Mark: No, listen up. He's wearing his boxers, so he must have taken a shower.

Tobias: Or was on his way to take one.

Mark: No, because he's didn't have his swimming trunks on when he came back

Tobias: Oh, yes, that's right. That is actually very good thinking (Tobias points at Mark)

Katja: Who says he took a shower?

Tobias: Come on, that's what you do!

Mark: You do!

Tobias (to Katja): Shhh.

(Some members of the group laugh.)

Mark: You do! You go and wash off the chlorine, any swimmer knows that.

Katja: Yes, well, I don't shower (gets up and leaves).

Dorthe: Me neither. Yuck.

Ramus: In that case, we can shorten it a little because, what did she say again – hang on.

Mark: That means he definitely died between, errr...

Tobias: Just a second.

Tobias: This means that... (gets up and walks over to the left side of the representation where the profiles and the other data are posted) What was it she said again?

Tobias: (points to the handwritten note next to Lise Bondesen's data indicating her whereabouts around the time of the crime (Figure 7) She had left for home at about ten.

Mark: She had already left a little before ten.

Tobias: Yes, shortly before ten, which means that this is also when Moussa went into the - where do we keep the... (he looks at the board and points to the drawing of the swimming club (Figure 8). Yes, in the dressing room, and that's where he met Peter, and they had an argument. That would probably have been a little after ten.

Mark: (points to the board in the direction of Peter's profile (Figure 9) Peter left – he says that he left a little before ten.

Tobias: Yes, I know.

Dorthe: Well, obviously he didn't.

(Several people speak at once.)

Figure 8. Discussion © 2003. Danish School of Education, Aarhus University. Used with permission

Figure 9. Discussion continues © 2003. Danish School of Education, Aarhus University. Used with permission

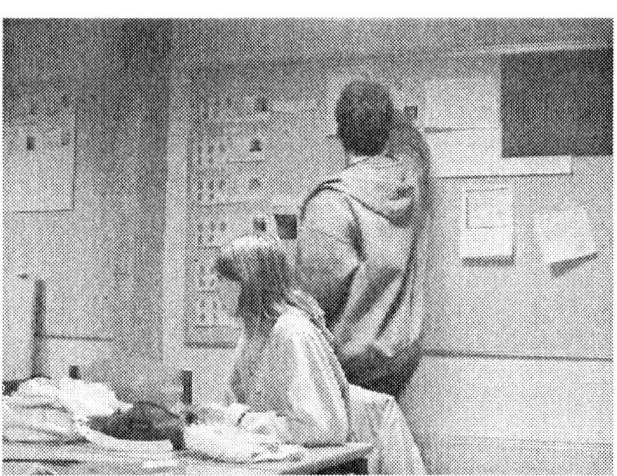

Tobias: He left at ten, so he was home at a quarter past ten.

Mark (interrupting Tobias): Can I show all of you something (gets up)? If we could go down and print that one out with the surroundings, we'd see that there is... actually, his house is 500 meters from the road – or from the swimming club - 500 to 700 meters.

Dorthe: That's not very far.

Mark: That's not far enough for it to take him 15 minutes to walk home. Unless he took a long detour.

Video observations, Class B, day 2

The situation shows the process the students followed to determine the victim's whereabouts before his death. How did the group go about the investigation in the situations described above, and what role did the representation play? In the search for answers to these questions, I leave the data for a moment to introduce a model for understanding the inquiry processes in the student's representational practices.

Kuhn suggests a model for understanding inquiry skills – or what Kuhn also defines as "scientific thinking" (Kuhn, 2005; Kuhn, 2002). Kuhn's definition of inquiry skills centers on deciphering the skills engaged in theory-evidence coordination (Kuhn, 2002). Gaining control of the process of theory (as "what makes sense to me") and evidence (as sources of knowing) coordination is, according to Kuhn, central in developing inquiry skills.

In her book, Education for Thinking, Kuhn shows how many people are unlikely to take into account multiple factors that potentially affect an outcome (Kuhn, 2005). Children and many adults often reason by making "inferences of causality about particular factors, while ignoring the potential effects of others" (Kuhn, 2005, p65). The factors which are judged causal just need to occur in the presence of the outcome on some occasion. Another often used strategy is "to attribute a phenomenon to a single causal factor, among several present, and dismiss the others" (Kuhn, 2005, p65). This phenomenon is known as discounting. Kuhn argues that these forms of faulty reasoning provide an avenue for the formation of faulty causal models – which may be resistant to change.

Figure 10. Engaged students © 2003. Danish School of Education, Aarhus University. Used with permission

Kuhn describes how inquiry learning requires students to seek new information, interpret this information, and (if warranted) requires them to have the ability to revise their existing beliefs accordingly. She argues, however, that it is entirely reasonable that the new information can be rejected because it is believed to be flawed or irrelevant or because of the strength of other data which have supported a different conclusion. What is essential, argues Kuhn, is the process leading to the decision of whether beliefs should be revised or not (Kuhn, 2005; Kuhn, 2002). Kuhn proposes three phases which inquiry activities undergo: inquiry, analysis, and inferences (Kuhn, 2005). These phases are briefly summarized below.

In the initial inquiry phase, students either do or do not identify the goal of the activity, and what they should do to achieve the goal. The task objectives at this stage are to access data, recognize its relevance to theory and formulate what questions to ask of the data. In this process theory-evidence coordination becomes explicit and intentional (Kuhn, 2005). In the second phase, the analysis phase, the task objectives are to represent evidence distinct from theory, make comparisons and seek and detect patterns. Kuhn describes how it is central, when a discrepancy between theory and evidence appears, that the evidence is represented in its own right, and the implications for the theory are examined. Once the relevance of evidence is established, Kuhn describes how students "must come to see analysis of it" (Kuhn, 2005, p. 78). Kuhn defines this as the identification of patterns and relationships and the coordination of what is observed with existing understandings. The final phase in inquiry activities is the inference phase, which Kuhn describes as the culmination of the inquiry where the investigators must come to terms with what they claim, and how they know this to be true (Kuhn, 2005). The task objectives here are to draw justified claims, reject unjustified claims, and acknowledge indeterminate claims. The inquiry, analysis, and inference strategies are described as separately here but Kuhn emphasizes that multiple strategies coexist and that the inquiry process is most often a mixture of more or less effective strategies.

Let us return to the situations in Class B and look at the different phases of the student investigation processes, and what role representations play at the different stages. In the initial phase of the game or what, with reference to Kuhn, I will call the inquiry phase, Group B1 used the boards for presenting data on the victim, and how and in what condition he was found, as well as data on the related characters. By focusing on the victim, and what had happened to him, the group had, as described by Kuhn, identified the purpose and goal of the activity and engaged in finding out what happened in the case. To achieve this goal they printed out the first data available and constructed an initial representation. The creation of the representation can thus be seen as a first strategy for doing inquiry into the case.

After the first presentation, Tobias went to access the data from the interrogations while the rest of the group collectively started to systematize the data available in the case by reconstructing the representation. When Tobias returned he began reading the notes he took on the interrogations about the different characters' relationships to the victim, and what they did at what time on the night of the crime. At this point, Katja initiated focusing explicitly on data concerning time. She suggested that the group construct one timeline with all the different data covering what time the characters did what on the night of the crime. She argued that having one place to collect all the data on time would be a good idea. Tobias disagreed, arguing that there were too many data to be able to fit them all on one paper. Instead, he argued that they should make separate timelines for each character.

This discussion can be viewed as a discussion of what strategies to use for comparing evidence concerning time and how the representation should be constructed to support this strategy. Katja suggested a tool for the graphic integration of all the data on time to be able to gain an overview of what characters did at exact times. Tobias suggested a tool to connect data to the character, but also to graphically align them with other characters' data. The construction of the representation frames the discussion. Both Tobias and Katja expressed their viewpoints in terms of how to structure the representation. These viewpoints are not expressed explicitly as means for supporting the comparison of evidence, and this strategic point is thus implicit in the discussion and might not be consciously controlled reflections on inquiry strategies as defined by Kuhn (Kuhn, 2005).

In the first inquiry phase, the group focuses on collecting and systematizing most of the data available in the case at this stage. They print out every character profile as well as unanalyzed clues from the crime scene. They focus on accessing the information from the interrogations first and not, for instance, the technical analysis of clues. They thus engage in accessing different types of data but do not yet engage in building theories about the case. Data on time are the first focal point, and the strategy for comparing what characters did what at what time is recognized. The structure of the representation is used for discussing, for example, what can be defined as representational strategies for searching for answers to the overall chapter in the game, i.e. what happened on the night of the crime and who did it?

The question of time is also central in the situation where the group analyzes at what point in time the victim died. Further into the investigation, the group is now at what Kuhn defines as the analysis phase. At the beginning of the situation, the students sat around a table where they tried to determine the exact time of the crime. As explained earlier, they had access to information from witness interrogations and knew that the victim and the boy he was training with, Peter, had had an argument both around the pool and later on while alone in the dressing room before Peter said he went home. Peter was thus under suspicion when this situation took place, and finding out whether Peter caused the victim's death is central, which Peter being mentioned midway and at the end of the situation indicate. Nevertheless, note how no theories about Peter's guilt are expressed until late in the investigation.

The first discussion concerned what time the victim, Moussa, went into the dressing room. From the notes on the interrogation with the victims' girlfriend Lise, they knew that she said that Moussa went into the dressing room a little before ten o'clock. Mark then argued that the incident that killed Moussa must have occurred earlier than ten. He concluded this based on the observation that Moussa must have taken a shower as he would otherwise have been found in his swim trunks and not in his boxers. Katja pointed out the uninvestigated link in this claim with the question, "Who says he took a shower?" Katja's question forced Mark to rethink his implicit conclusion that all people shower after going to the pool. He then defended his line of argument with the idea that swimmers have the experience or expertise to know that they have to wash off the chlorine. In the first part of the situation, the students interpreted the evidence from the interrogation of Peter and Lise and connected them to evidence about what Moussa was wearing when he was found, and theories about swimmers showering to wash off chlorine. Next, they tried to construct a hypothesis on what time Moussa encountered his murderer. They did not at any time in the discussion relate the construction of this hypothesis to theories about who the murderer could have been.

After having established the hypothesis on the sequence of events in the dressing room, the students moved on to determining the exact time that Moussa encountered his murderer. Mark was the first one to estimate the time of death but foundered. Tobias kept his focus on determining what time the victim met his murderer by using a different strategy. He tried to calculate the interval between what time the victim's girlfriend said he went into the dressing room, and the estimated time the victim must have finished showering. Seated at the table, he tried to remember the details of the evidence he wanted to relate the hypothesis to but could not reach a conclusion. He got up and turned to the representation, where he was able to check the exact information regarding when Lise said she went home and Moussa went into the dressing room. Next to Lise's information was the map of the dressing room. Tobias then began connecting the data on time with the data on the layout of the dressing room. Using established hypotheses on how long it would take Moussa to shower, Tobias concluded that Moussa would have finished showering a little after ten o'clock. In this part of the investigation, Tobias sought to relate this supposition to existing evidence about time and geographical features of the events. This is a complex action in the case with many types of exact data. At this stage, Tobias used the representation to help him connect the established hypothesis of the sequence of events in the dressing room with evidence on time and then geographical evidence.

Next, Tobias tried to coordinate the hypothesis they had constructed with the theory that Peter was involved in the murder to make inferences of what they could claim based on their investigation. In this instance, he made what Kuhn would define as "inferences of causality about particular factors, while ignoring the potential effects of others" (Kuhn, 2005, p. 65). He stated that Moussa went into the dressing room, and because evidence shows that he and Peter had an argument, Tobias concluded that this incidence took place at the time the group had estimated the victim ran into his murderer. He ignored the evidence that Peter's interrogation states he went home at ten. This statement could be untrue, but Tobias failed to construct a relation between theory and evidence by encoding and representing the evidence as an entity distinct from theory, as Kuhn defines it (Kuhn, 2005).

Mark pointed out the discrepancy by physically pointing to the evidence on the representation stating that Peter had said he went home just before ten o'clock. Tobias acknowledged this, but Dorthe tried to maintain the validity of the faulty inference. Mark then tried to construct a relation between theory and evidence by drawing attention to evidence not graphically present on the representation. He claimed that if they could calculate the distance from the swimming club to Peter's house, they would find that the distance was too short for it to take Peter 15 minutes to walk. They had thus identified another step in the investigation. Implicit in this conclusion was that Peter might not be telling the truth and might have left later than he told the police he did, which means he still might be involved in the murder. The conclusions at this stage were not related to theories or expressed as proof of a theory. Instead, by discussing the distance to the suspect's house, the students did another test to provide more evidence. The idea was expressed as evidence possibly proving that it could not have taken him 15 minutes to walk the distance. In the following section, I sum up the conclusions of this chapter and discuss what competences students engage in these investigative practices in the game.

DISCUSSION: DEFINING REPRESENTATIONAL INQUIRY COMPETENCES

This chapter has focused on understanding what competences and practices were enacted in the classes playing the game. I showed how representations help students relate evidence and coordinate it with constructed hypotheses and theory. Now I wish to discuss the results of the studies presented in relation to observations conducted in other classes playing the game homicide with the objective to define the competences enacted in this game space. Here I use the open understanding of competences as the ability to successfully meet complex demands in a particular context (Rychen & Salganik 2003).

What becomes evident after studying the inquiry processes in Class B is how aspects of constructing and using representations change at the different stages of the inquiry. Let us first revisit the different inquiry stages with the objective of getting closer to the definition of this mixed representational practice. The analysis of the various situations presented in this chapter show that the focus of the investigation is on constructing evidence-based theories of what time what events took place on the night of the crime, and what characters potentially were involved in the murder. The students' identification of the goal of the activity and of how to achieve this goal is thus in keeping with the overall objective of the game: That players should conduct inquiries by developing hypotheses and theories based on analyzed clues and data. This identification of the game objectives is the first stage of what Kuhn defines as inquiry activities (Kuhn, 2005). Accessing data and understanding its relevance to theory in order to transform implicit theory revision to an intentional inquiry is the other central aspect of the inquiry phase (Kuhn, 2005). The situation where the group collectively constructed representations shows how they access the many types of data by building a representational structure. An explicit example of this is the discussion between Tobias and Katja about timeline formats. By initiating the discussion of time in response to Tobias reading from his notes from the interrogations, Katja initially drew attention to the idea that they should ask questions concerning the time of the events and compare the characters whereabouts at different times. As argued earlier, the discussion also develops into an implicit discussion of what strategies to use to do this. This is explicitly argued in terms of what structures to add to the representation for it to support the strategies. The representation thus becomes a tool for defining, discussing and visualizing implicit inquiry strategies.

The students engage a dual competence in the process when they transform their activity to an intentional inquiry while constructing representational structures for supporting the inquiry. I see these two elements as interwoven and not as isolated steps that must be taken in a specific sequence. On the one hand, constructing a representational structure makes students discuss strategies for doing inquiry, but the representational structure also makes them aware of and defines their process. This is in line with observations conducted in other classes where graphic alignments of representational elements help students define elements such as the possible motives of a potential suspect (Magnussen, 2009).

The argument presented here is thus that the construction of representation helps students visualize implicit or intuitive inquiry strategies such as comparing evidence. Apparently, however, in these instances, the students did not move to what Kuhn defines as the meta-level of operation where they are able to select strategies in relation to the goal of the activity. In a longer study I have shown that students when presented with other formats of representations made in other classes are able to define, reflect on and evaluate the criteria and strategies for producing and using various formats of inquiry representations (Magnussen, 2009).

In what Kuhn would define as the analysis phase (Kuhn, 2005), students interpreted and related different types of evidence independent of theories about the suspects in the case. In the process of defining what time the victim encountered his murderer, the representation was used in the process of relating the hypothesis on the sequence of events in the dressing room that the students had constructed while sitting around the table with evidence on the time of events and the layout of the dressing room represented on the board. Tobias also tried to coordinate the group's hypothesis on what time the victim encountered his murderer with evidence to prove their theory that Peter was the murderer. While pointing at Peter's profile on the representation, however, Mark pointed out to him that he had ignored data from Peter's interrogation showing a discrepancy between their theory and the evidence indicating where Peter was at the time of the crime. I argue that during the process Group B1 uses the representation in their inquiry, it becomes visually evident what hypotheses or theories the students are trying to construct and what is evidence as sources of knowing. The main part of the evidence is on the representation in Group B1, whereas the hypotheses and theories are not. As a result, when the students discuss hypotheses or theories of the case, they are able, as Mark did, to physically point to evidence relating to a hypothesis or, as in the situation just described, to other evidence proving a discrepancy between theory and evidence. In other classes playing the game *Homicide* it has been shown how representations are used for representing hypotheses and how students present these hypotheses as not fully supported by evidence by pointing to the graphic elements that represents them (Magnussen, 2009).

This use aligns with McGinn and Roth's definitions of the functions of representations in scientific work and how visual representations provide pictorial or graphical inscriptions of the central aspects of the phenomenon under investigation (McGinn & Roth, 1999). Visual representations such as graphs, x-ray images, maps, models, diagrams and hybrids of these are central in creating and communicating science. Scientists create initial representations that undergo transformations in the process of making sense of a phenomenon to themselves and later to others in scientific publications. They describe how representations serve at least three functions in scientific work: as inscriptions, conscription devices, and boundary objects. As inscriptions, representations constitute the phenomenon of interest. Visual representations "provide pictorial or graphical inscriptions of important aspects of the phenomenon under investigation" (1999, p. 21). The visual representation makes the object readable, presentable, moveable and combinable with other inscriptions. Visual representations also serve as conscription devices, providing focused attention and conversation. Scientists and engineers' discussions are often about and in the presence of these inscriptions. Last, visual representations also act as boundary objects that coordinate work across groups, time and space.

McGinn and Roth (1991) argue that this type of representational practice that is a central part of professional science should also be central in science education. Suggesting that classroom environments should be organized to allow students to collaborate on constructing visual representations which should serve as a means for communication in the community, McGinn and Roth write:" Within these environments, students would be expected to use visual re-presentations as integral aspects of arguments, and to adapt and transform their representations to make more convincing arguments" (McGinn & Roth, 1999, p. 21). Aspects of this use of representations for constructing convincing arguments are also evident in the situations in this chapter. Additional studies in Class B and other classes playing the game Homicide also showed that transformations or redesign of representations is an integrated part of the inquiry process in this game (Magnussen, 2009).

Andrea diSessa and Bruce Sherin present a broad definition of the competences involved in representational science practice (2000) which is specifically interesting when seeking to define what competences are enacted in Class B. Sherin and diSessa use the term meta-representational competences (MRC) to describe the full range of capabilities students have for constructing and understanding scientific representations. MRC includes the ability to "select, produce and productively use representations but also the abilities to critique and modify representations and even to design completely new representations" (diSessa & Sherin, 2000, p. 386). They state that letting students develop representations allows them to do and learn about what scientists do. They point out that science and mathematics have evolved by refining and accumulating representational forms. Letting students do MCR-based tasks might also open up science teaching to a group of students who are not otherwise very attracted to the sciences. The classic focus on right and wrong in science might fade into the background in favour of innovation, creativity, and contextual judgment (diSessa & Sherin, 2000).

This chapter show that the same innovative and creative skills go into constructing representations as the ones DiSessa and Sherin describe. Students in the classes that were observed, have independently invented and used formats for supporting their inquiry. Content-wise, the students in Class B do not produce what the authors above exemplify as scientific representations. The representations in Class B both consist of data concerning the characters as well as technical and scientific data. We thus have representations which have a cross-disciplinary content. Another difference is the nature of what is being represented. The phenomena which are being described in the MRC research cases mentioned above are well-defined objects, their meaning unlikely to change radically in the same sense as the meaning of an un-investigated case might. In the studies presented by Sherin and diSessa's it is established scientific phenomena which are being represented from different perspectives and with different means. The un-investigated case is an unknown object that is expected to change. The understanding of relations can radically change when new information is gained about what happened at the crime scene. As presented in this chapter, students use their representation constructing hypotheses of an undefined case. The inquiry process necessary for understanding the nature of the undefined phenomenon is thus tightly integrated with the representational practices.

Based on the discussion in this chapter I propose the term 'representational inquiry competence' to describe capabilities that students have for using representations in inquiry activities. This definition contains elements of both inscriptional practice (see McGinn & Roth, 1999), MRC (diSessa & Sherin, 2000) and inquiry competences (Kuhn, 2005; Kuhn, 2002):

Representational inquiry competence:

The ability to construct, productively use, transform and critique visual representations as an integrated part of doing inquiry into defining an unknown phenomenon.

Includes the ability to visually represent the phenomena subjected to inquiry and the inquiry strategies that go into investigating the phenomenon.

Includes the ability to construct and use representations for investigating, coordinating and relating evidence, hypotheses, and theory.

The definitions of representational inquiry competences enacted in the learning environment of the game Homicide contribute to the field of learning game research by supporting existing definitions of epistemic games or games simulating elements of professions as environments which support innovative thinking (Shaffer, 2007). The findings further contributes to the field by qualifying definitions and specific understandings of what competences are enacted in game-based learning environments that simulate professional science practice. These competences are defined based on the analysis of the game Homicide played in schools but, as discussed above, we see elements of representational inquiry processes integrated in different professions. McGinn and Roth describe how scientists create initial representations in the field or laboratories, and how these formats undergo a series of transformations as the scientist tries to make sense of the phenomenon (1999). They describe how these transformations "are part of the 'sanitization' process in scientific publications that move physical specimens in the field to squiggles on a page to graphs and equations that can be shared with and interpreted by other scientists" (McGinn & Roth, 1999, p. 20). In the extended study of Homicide played in different classes it became evident that transformations of representations are an integrated part of representational inquiry (Magnussen, 2009). The practice of using representations in a cross-disciplinary inquiry into defining something unknown, however, may also bear resemblance to processes in other professions. Professions involving design or innovation processes are examples of areas where elements of these competences are central. The processes, which often involve defining and developing something unknown, draw on complex systems of knowledge from many areas of life, and the creativity involved in visualizing the undefined object has to be part of the creative use of tools for supporting the object defining processes. Tools for innovation processes are numerous and include mind maps, which bear resemblance to some of the representations produced by students in other classes in the three year study (Magnussen, 2009), or sketching tools such as the ones used by industrial designers, software engineers, usability engineers, etc. (Buxton, 2007). Competences used for doing representational inquiry are thus central in the context of professional scientific practice and professions involving creative innovation processes.

It is of interest to look at how the elements of representational inquiry competences relate to official curriculum objectives. In Denmark the official curriculum goals are described in the Ministry of Education's overall guidelines titled Common Goals (Fælles mål) (Ministry of Education, 2009). An example of this is the descriptions of the final objectives for chemistry, physics and biology education which are described as guidelines and the point of orientation for science education in upper school (Hansen, 2006) Here references are made to student learning of inquiry methods as 'working methods and mindset' For biology, it states:

'The teaching should lead to students acquiring knowledge and skills which make them capable of:

Identifying and formulating relevant thesis and outlining hypotheses

Planning, completing, and assessing inquiries and experiments in the field and in the laboratory

Using information technology in connection with seeking information, data collection, analysis, and presentation…'.

(Ministry of Education, 2008)

These and similar descriptions in the final objectives for physics and chemistry correspond with the overall inquiry stages students go through in their investigations in the game Homicide. They formulated theses and hypotheses and planned, completed and assessed the inquiries and experiments done during the game. There are, however, central differences between the final objectives described in the official curriculum and practices and competences we have seen enacted in Homicide. The descriptions in the guidelines are subject-specific whereas with Homicide, students handle inquiries with many types of data from different disciplines, thus requiring multiple literacy. The inquiry in Homicide is cross-disciplinary and operates on two levels. On one level, students have to plan and carry out, for instance, laboratory tests. On another level, they have to connect and coordinate the results from these tests with their theories. The students use inquiry representations when they coordinate the information.

The revised version of the curriculum goals focuses extensively on inquiry methods, whereas the use of representations or inscriptions is not mentioned in the objectives for individual subjects. An important finding in this chapter is that when Homicide is played in schools, the inquiry practices and competences we have seen are closely integrated with the competences used for innovating representational formats. Representational inquiry competences thus involve central skills described in the science curriculum objectives. But the competences defined in this chapter are also an extension of science curriculum inquiry descriptions because they involve capabilities for constructing and using representations for doing inquiry. We need a deeper understanding of how the settings science games are played in are influential on the student enactment of competences. Extended studies of Homicide played in different classes show that the complexity of game design, resources in the classroom, teachers, fictional elements, and other students are influential on the enactment of representational inquiry competences (Magnussen, 2009). Future studies should be done into understanding the characteristics of the competences and the settings they are enacted in to understand how we design new types the game-based science learning environments which support new types of competences.

REFERENCES

Barnett, M., Squire, K., Higgenbotham, T., & Grant, J. (2004). Electromagnetism Supercharged! In *Proceedings of the 2004 International Conference of the Learning Sciences*. Los Angeles: UCLA Press.

Brown, A. (1992). Design experiments: Theoretical and methodological challenges in creating complex interventions in classroom settings. *Journal of the Learning Sciences*, *2*(2), 141–178. doi:10.1207/s15327809jls0202_2

Buxton, B. (2007). *Sketching User Experiences: Getting the Design Right and the Right Design*. Redmond, WA: Morgan Kaufmann Publishers.

Design-Based Research Collective (DBRC). (2003). Design-based research: An emerging paradigm for educational inquiry. *Educational Researcher*, *32*(1), 5–8. doi:10.3102/0013189X032001005

DiSessa, A. A. (2000). *Changing Minds: Computer, Learning, and Literacy*. Cambridge, MA: MIT Press.

DiSessa, A. A., & Sherin, B. (2000). Meta-representation: An introduction. *The Journal of Mathematical Behavior*, *19*(4), 385–398. doi:10.1016/S0732-3123(01)00051-7

Gee, J. P. (2003). *What Video Games Have to Teach Us About Learning and Literacy*. New York: Palgrave Macmillan.

Gee, J. P. (2004). *Situated Language and Learning: A Critique of Traditional Schooling*. London: Routledge.

Hansen, K. F. (2006). *Fælles Mål i skolens hverdag - hvordan? Inspirationsmateriale til kommuner og skoler om implementering af Fælles Mål i skolens hverdag* [Common goals in schools – how? Inspiration material for municipalities and schools on the implementation of common goals in schools]. Copenhagen: Ministry of Education, Department of Primary, Lower Secondary and General Adult Education (Ministry of Education's Handbook series 13).

Kuhn, D. (2002). What is scientific thinking and how does it develop? In U. Goswami (Ed.), *Blackwell Handbook of Childhood Cognitive Development* (pp. 371-393). Malden, MA: Blackwell Publishing.

Kuhn, D. (2005). *Education for Thinking*. Cambridge, MA: Harvard University Press.

Magnussen, R. (2007). Games as a Platform for Situated Science Practice. In S.de Castell & J. Jenson (Eds.), *Worlds in Play: International Perspectives on Digital Games Research*. New York: Peter Lang.

Magnussen, R. (2009). *Representational Inquiry in Science Learning Games*. Unpublished doctoral dissertation, Copenhagen: Danish School of Education, University of Aarhus. Retrieved from http://www.dpu.dk/site.aspx?p=6604&init=rma&msnr=3&lang=eng

Magnussen, R., & Jessen, C. (2006). Naturfaglig Praksis og Spil-lignende Læring [Science praxis and game-like learning]. *MONA: Matematik og naturfagsdidaktik,* (2), 7-26.

McGinn, M. K., & Roth, W. M. (1999). Preparing students for competent scientific practice: Implications of recent research in science and technology studies. *Educational Researcher*, *28*(3), 14–24.

Ministry of Education. (2008). *Fælles Mål*. [Common goals]. Retrieved June 26, 2008, from http://www.faellesmaal.uvm.dk/

Rychen, D. S., & Salganik, L. H. (2003). A holistic model of competence. In D. S. Rychen & L. H. Salganik (Eds.), *Key Competencies for a Successful Life and a Well-Functioning Society* (pp. 41-62). Cambridge, MA: Hogrefe & Huber Publishers.

Shaffer, D. W. (2006). Epistemic frames for epistemic games. *Computers & Education*, *46*(3), 223–234. doi:10.1016/j.compedu.2005.11.003

Shaffer, D. W. (2007). *How Computer Games Help Children Learn*. New York: Palgrave Macmillan.

Shaffer, D. W., & Gee, J. P. (2005). *Before every child is left behind: How epistemic games can solve the coming crisis in education* (WCER Working Paper No. 2005-7). University of Wisconsin-Madison, Wisconsin Center for Education Research.

Shaffer, D. W., Squire, K. D., Halverson, R., & Gee, J. P. (2005). Video games and the future of learning. *Phi Delta Kappan*, *87*(2), 105–111.

Squire, K., & Klopfer, E. (2007). Augmented reality simulations on handheld computers. *Journal of the Learning Sciences*, *16*(3), 371–413.

Squire, K. D. (2002). Rethinking the role of games in education. *Game Studies, 2*(1). Retrieved October 2008 from http://www.gamestudies.org/0102/squire/

Compilation of References

Abd. Rahman Daud. (2000). Kefahaman terhadap konsep penggunaan ICT di dalam bilik darjah (Understanding the use of ICT concepts in the classroom). In *Proceedings of the International Conference on Teaching and Learning* (pp. 609-623).

Abdul Razak, H., & Jamaludin, B. (1998). Penggunaan komputer untuk pengajaran dan pembelajaran di sekolah menengah (The use of computer for teaching-learning in secondary schools). [Journal of Education]. *Jurnal Pendidikan, 23*, 53–64.

ACCENTURE. (2001). *African Virtual University Strategic Review, Final Report*. Boston, USA: Accenture.

Aftab, P. (2006). *Stop, block and tell!* Retrieved February 11, 2009, from http://safety.xanga.com/2006/06/12/stop-block-and-tell/

Aglionby, J. (2002). *English in schools divides Malaysia: reports on an ethnic revolt over education.* Retrieved September, 26, 2002, from http://guardian.co.uk

Ahmad Hanizar, A. H., Muhammad, Z. M. Z., Wong, S. L., & Hanafi, A. (2005). The Taxonomical Analysis of Science Educational Software in Malaysian Smart Schools. [MOJIT]. *Malaysian Online Journal of Instructional Technology, 2*(2), 106–113.

Ahtee, M., & Varjola, I. (1998). Students' understanding of chemical reaction. *International Journal of Science Education, 20*(3), 305–316. doi:10.1080/0950069980200304

Ainley, J., Kos, J., & Nicholas, M. (2008). *Participation in Science, Mathematics and Technology in Australian Education* (ACER Research Monograph No. 63). Melbourne, Australia: ACER.

Ainsworth, S. (1999). The functions of multiple representations. *Computers & Education, 33*, 131–152. doi:10.1016/S0360-1315(99)00029-9

Akilagpa, S. (2004). *Challenges Facing African Universities.* Accra, Ghana: Association of African Universities. AT&T. (2008). *AT&T Connect©. Conferencing for the Entreprise.* Retrieved March 2, 2009, from http://www.interwise.com/emc_solution_eroom.html

Albirini, A. (2006). Teachers' attitudes toward information and communication technologies: the case of Syrian EFL teachers. *Computers & Education, 47*, 373–398. doi:10.1016/j.compedu.2004.10.013

American Association for the Advancement of Science. (1989). *Project 2061 - Science for all Americans.* Retrieved March 18, 2009, from http://www.aaas.org/

American Association for the Advancement of Science. (1993). *Benchmarks for science literacy.* New York: Oxford University Press.

American Library Association. (1989). *Presidential Committee on Information Literacy. Final Report.* Chicago: American Library Association. Retrieved October 16, 2008, from http://www.ala.org/ala/mgrps/divs/acrl/publications/whitepapers/presidential.cfm

Anderson, C. A., & Bushman, B. J. (2001). Effects of violent games on aggressive behavior, aggressive cognition, aggressive affect, physiological arousal, and prosocial behavior: A meta-analytical review of the scientific literature. *Psychological Science, 12*, 353–359. doi:10.1111/1467-9280.00366

Anderson, N., Lankshear, C., Courtney, L., & Timms, C. (2008). 'Because it's boring, irrelevant and I don't like computers': Why female secondary school students avoid professionally-oriented ICT subjects. *Computers & Education, 50*(4), 1304-1318. Retrieved February 11, 2009, from http://www.sciencedirect.com/science?_ob=ArticleURL&_udi=B6VCJ-4MY0T-NR-2&_user=972264&_rdoc=1&_fmt=&_orig=search&_sort=d&view=c&_acct=C000049659&_version=1&_urlVersion=0&_userid=972264&md5=432f65a0b2421d22a5b452835904c080

Anderson, R. E., & Dexter, S. (2005). School technology leadership: An empirical investigation of prevalence and effect. *Educational Administration Quarterly, 41*(1), 49–82. doi:10.1177/0013161X04269517

Angelis, C., & Valanides, N. (2009). Epistemological and Methodological Issues for the Conceptualisation, Development and Assessment of ICT-TPCK: Advances in Technological Pedagogical Content Knowledge (TPCK). *Computers & Education, 52*(1), 154–168. doi:10.1016/j.compedu.2008.07.006

Anstey, M. (2002). *Literate futures: Reading*. Coorparoo, DC., Queensland: State of Queensland, Department of Education.

Anstey, M., & Bull, G. (2004). *Literate futures. Professional development.The teaching of reading for a multiliterate world. Years 7 - 12*. Brisbane, Queensland: Literate Futures, Curriculum Strategy Branch, Department of Education, The State of Queensland.

Anstey, M., & Bull, G. (2006). *Teaching and learning multiliteracies. Changing times, changing literacies*. Kensington Gardens, South Australia: Australia Literacy Educators' Association in conjunction with the International Reading Association, Newark DE.

Antoniette, A., & Cantoia, M. (2000). To see a painting versus to walk in a painting: An experiment on sense making through virtual reality. *Computers & Education, 34*(3-4), 213–223. doi:10.1016/S0360-1315(99)00046-9

APEC, Asian and the Pacific Regional Bureau for Education. (2004). *Integrating ICTS in Education: Lessons Learned: A collective Case Study of Six Asian Countries*. Bangkok, Thailand: UNESCO

Apple Inc. (2008). *GarageBand*. Retrieved February 15, 2009, from http://docs.info.apple.com/article.html?path=GarageBand/4.0/en/6624.html

Apple Inc. (2009). *iLife*. Retrieved February 15, 2009, from http://www.apple.com/ilife/

Aspyr Studios. (2007). *iQuiz Maker*. Retrieved February 15, 2009, from http://www.iquizmaker.com

Asselin, M. M., & Lee, E. A. (2002). 'I wish someone had taught me': Information literacy in a teacher education programmes. *Teacher Librarian, 30*(2), 10–17.

Association for ICT in Education. (2001). *Teaching ICT volume 1 issue 2*. Retrieved February 15, 2009, from http://acitt.digitalbrain.com/acitt/web/resources/pubs/Journal%2002/whiteboards.htm

Audacity. (2009). Retrieved 15th February 15, 2009, from http://audacity.sourceforge.net

Australian Government Department of Education. Employment and Workplace Relations. (2006). *Australian school innovation in science, technology and mathematics (ASISTM): BirdNet Creating and online community for far north Queensland students*. Retrieved February 11, 2009, from http://project.asistm.edu.au/special/successful.asp?r=8&state=QLD

Australian Government Department of the Environment and Heritage. (2005). *Educating for a sustainable future: A national environmental education statement for Australian schools*. Retrieved February 2, 2009, from http://www.environment.gov.au/education/publications/sustainable-future.html

Australian Government Net Alert. (2007). *Cyber bullying. Information sheet*. Retrieved January 2, 2009, from http://www.netalert.gov.au

Australian Government, Department of Education, Science and Training. (2006). *The SiMERR National Survey on Science, ICT and Mathematics Education in Rural and Regional Australia*. Retrieved February 2, 2009, from http://www.une.edu.au/simerr/pages/projects/1nationalsurvey/

Australian Government. (2007). *A teacher's guide to Internet safety: How to teach Internet safety in our schools*. Retrieved February 11, 2009, from http://www.netalert.

gov.au/__data/assets/pdf_file/0019/1819/01427-A-Teachers-Guide-to-Internet-Safety.pdf

Australian Information & Communications Technology in Education Committee (AICTEC). (2008). *Research Report: Interoperability standards across the Australian education & training sector*. Retrieved October 7, 2008, from http://www.aictec.edu.au

Aviram, R., & Eshet-Alkalai, Y. (2006). Towards a theory of digital literacy: Three scenarios for the next steps. *European Journal of Open Distance E-Learning*. Retrieved December 29, 2008, from http://www.eurodl.org/materials/contrib/2006/Aharon_Aviram.htm

AVU. (2001). *AVU Equipment Technology*. Nairobi, Kenya: African Virtual University.

AVU. (2004). *The AVU Business Plan (2004–2009)*. Nairobi, Kenya: African Virtual University.

AVU. (2005). *Background Information*. Nairobi, Kenya: African Virtual University.

AVU. (2006). *Vice-Chancellors' Meeting Report*. Nairobi, Kenya: African Virtual University.

Azevedo, R. (2004). Using hypermedia as a metacognitive tool for enhancing student learning? The role of self-regulated learning. *Educational Psychologist, 40*(4), 199–209. doi:10.1207/s15326985ep4004_2

Azian, T. S. A. (2006). *Deconstructing Secondary Education: The Malaysian Smart School Initiative*. Paper presented at the 10th SEAMEO INNOTECH International Conference on Learning for Life: Creating Endless Possibilities in Secondary Educatio*n*, Pearl Hall, SEAMEO INNOTECH, Philippines.

Baker, L. (1991). Metacognition, Reading and Science Education. In C.M. Santa & D.E. Alvermann (Eds.), *Science Learning: Processes and Applications*. Newark: International Reading Association, IRA.

Bakhtin, M. M. (1986). *Speech genres and other late essays* (V. McGee, Trans.). Austin: University of Texas Press. Retrieved March 30, 2009, from http://pubpages.unh.edu/~jds/BAKHTINSG.htm

Barber, T. X. (1993). *The human nature of birds*. London: St Martins Press.

Barnea, N., & Dori, Y. (1999). High-school chemistry students' performance and gender differences in a computerised molecular modelling learning environment. *Journal of Science Education and Technology, 8*(4), 257–271. doi:10.1023/A:1009436509753

Barnett, M., Squire, K., Higgenbotham, T., & Grant, J. (2004). Electromagnetism Supercharged! In *Proceedings of the 2004 International Conference of the Learning Sciences*. Los Angeles: UCLA Press.

Barraclough, A., & Guymer, I. (1998). Virtual reality-A role in environmental Engineering education? *Water Science and Technology, 38*(11), 303–310. doi:10.1016/S0273-1223(98)00668-4

Barrett, T. (2008). *Interesting ways to use your interactive whiteboard*. Retrieved February 15, 2009, from http://docs.google.com/Presentation?docid=dhn2vcv5_106c9fm8j&hl=en_GB

Bartle, R. (1983). A Voice from the Dungeon. *Practical Computing, December,* 126-130.

Basler, D. (2009). *Baslercast Physics*. Retrieved February 15, 2009, from http://www.aasd.k12.wi.us/staff/baslerdale/Ppodcasts.asp

Baudrillard, J. (1988). Simulacra and simulations. In M. Poster (Ed.). *Jean Baudrillard. Selected writings* (pp. 166-184). Oxford, UK: Stanford University Press/Polity Press.

Beard, C., & Wilson, J. P. (2006). *Experiential learning: A best practice handbook for educators and trainers*. London: Kogan Page.

Bearne, E., Clark, C., Johnson, A., Manford, P., Mottram, M., & Wolstencroft, H. (2007). *Reading on screen. Research report*. Leicester, UK: United Kingdom Literacy Association.

Beavis, C. (1997). Computer games, culture and curriculum. In I. Synder (Ed.), *Page to screen. Taking literacy into the electronic era* (pp. 234-255). St Leonards, NSW: Allen and Unwin.

BECTA. (2006). *Safeguarding children in a digital world: Developing a strategic approach to e-safety*. Retrieved January 5, 2009, from http://publications.becta.org.uk/display.cfm?resID=25933

BECTA. (2007). *Signposts to safety: teaching e-safety at Key Stages 3 and 4*, Retrieved February 15, 2009, from http://publications.becta.org.uk/display.cfm?resID=32424&page=1835

BECTA. (2008). *Harnessing Technology Review 2008: The role of technology and its impact on education*. Retrieved January, 1, 2009, from http://publications.becta.org.uk/display.cfm?resID=37346

Bell, P., Hoadley, C. M., & Linn, M. C. (2004). Design-based research in education. In M. C. Linn, E. A. Davis, & P. Bell (Eds.), *Internet environments for science education* (pp. 73-85). Mahway, NJ: Lawrence Erlbaum Associates.

Bentley, D., & Watts, M. (1989). *Learning and Teaching in School Science: Practical Alternatives*. Milton Keynes, UK: Open University Press.

Bereiter, C., & Scardamalia, M. (1987). *The psychology of written composition*. Hillsdale, NJ: Lawrence Erlbaum Associates.

Berkenkotter, C., & Huckin, T. N. (1995). *Genre Knowledge in Disciplinary Communication: Cognition/Culture/Power*. Mahwah, NJ: Lawrence Erlbaum Associates.

Bernama. (2005). *Teach Science, Maths in mother tongue: Chinese movement. Daily Express, Wednesday, October 12, 2005*. Retrieved December 20, 2008, from http://www.dailyexpress.com.my/news.cfm?NewsID=37720

Berryhill, A. H. (2007). *A correlational study of adoption of instructional technology by higher education faculty and their social communications network*. Unpublished doctoral dissertation, Mississippi State University.

Berson, I. R., & Berson, M. J. (2003). Digital literacy for effective citizenship. *Social Education*, *67*(3), 164–167.

Bijker, W. E. (1992). The social construction of fluorescent lighting or how an artifact was invented in its diffusion stage. In W.E. Bijker & J. Law (Eds.), *Shaping technology/building society. Studies in sociotechnical change* (pp. 75-104). Cambridge, MA: MIT Press.

Bijker, W. E., & Law, J. (Eds.). (1992). *Shaping technology/building society. Studies in sociotechnical change*. Cambridge, MA: MIT Press.

Bilal, D. (2005). Children's information seeking and the design of digital interfaces in the affective paradigm. *Library Trends*, *54*(2), 197–208. doi:10.1353/lib.2006.0013

Bilal, D., & Kirby, J. (2002). Differences and similarities in information seeking: Children and adults as web users. *Information Processing & Management*, *38*, 649–670. doi:10.1016/S0306-4573(01)00057-7

Blizzard Entertainment. (2004). *World of Warcraft* [computer game]. Designers: Pardo, Kaplan & Chilton.

Bodner, G. M. (1986). Constructivism – A theory of knowledge. *Journal of Chemical Education*, *63*(10), 873–878.

Boellstorff, T. (2008). *Coming of Age in Second Life: An Anthropologist Explores the Virtually Human*. Princeton, NJ: Princeton University Press.

Bore, A. (2006). Creativity, continuity and context in teacher education: Lessons from the field. *Australian Journal of Environmental Education*, *22*(1), 31–38.

Bradley, J. (2009). *DNA spooling experiment*. Retrieved February 15, 2009, from http://scienceguyinatie.blogspot.com/2009/02/videocast-dna-spooling-experiment.html

Brainiac: Science Abuse. (2004). *"Alkali Metals". 2004-09-02. No. 1, season 2*. Retrieved from http://video.google.com/videoplay?docid=-2134266654801392897

Brandt, L., Elen, J., Hellemans, J., Heerman, L., Couwenberg, I., Volckaert, L., & Morisse, H. (2001). The impact of concept mapping and visualization on the learning of secondary school chemistry students. *International Journal of Science Education*, *23*(12), 1303–1313. doi:10.1080/09500690110049088

Bransford, J. D., Brown, A. L., & Cocking, R. R. (1999). *How people learn: Brain, mind, experience, and school*. Washington, DC: National Academy Press.

Branson, J. (1991). Gender, Education and Work. In D. Corson (Ed.), *Education for Work: Background to Policy and Curriculum*. Clevedon, UK: Open University Press/Multilingual Matters Ltd.

Braund, M., & Reiss, M. (2006). Towards a More Authentic Science Curriculum: The contribution of out-of-school learning. *International Journal of Science Education*, *28*(12), 1373–1388. doi:10.1080/09500690500498419

Braund, M., & Reiss, M. J. (2004). *Learning science outside the classroom*. London: Routledge.

Braund, M., Reiss, M., Tunnicliffe, S. D., & Moussouri, T. (2008). *Getting the most from 'Out-of-School' learning in science: What should teachers know?* Retrieved July 15, 2008, from http://www.ase.org.uk/htm/conferences/atsi/pdf/F9.pdf

Bray, O. (2008). *Nintendo DS in Education - Harper Collins 100 Classic Book Collection*. Retrieved February 15, 2009, from http://olliebray.typepad.com/olliebraycom/2008/12/nintendo-ds-in-education-harper-collins-100-classic-book-collection.html

Bray, O. (2009). *Computer Games Based Learning: Sharing good practice between Musselburgh, East Lothian and Sydney, Australia*. Retrieved February 15, 2009, from http://olliebray.typepad.com/olliebraycom/2009/01/computer-games-based-learning-sharing-good-practice-between-musselburgh-east-lothian-and-sydney-australia.html

Briano, R., Midoro, V., & Trentin, G. (1997). Computer mediated communication and online teachers training in environmental education. *Journal of Information Technology for Teacher Education*, *6*(2), 127–146.

Brittain, S., Glowacki, P., Van Ittersum, J., & Johnson, L. (2006). Podcasting Lectures: Formative evaluation strategies helped identify a solution to a learning dilemma. *Educase Quarterly, 29*, 3. Retrieved February 24, 2009, http://connect.educause.edu/Library/EDUCAUSE+Quarterly/PodcastingLectures/39987

Brooks, J. G., & Brooks, M. G. (1999). *In search of Understanding: The Case for Constructivist Classrooms*. Alexandria, VA: Association for Supervision and Curriculum Development.

Brosseau, J., Potvin, M. J., & Rouleau, I. (2007). Aging affects motor skill learning when the task requires inhibitory control. *Developmental Neuropsychology*, *32*(1), 597–613.

Brown, A. (1992). Design experiments: Theoretical and methodological challenges in creating complex interventions in classroom settings. *Journal of the Learning Sciences*, *2*(2), 141–178. doi:10.1207/s15327809jls0202_2

Brown, C., & Krumholz, L. R. (2002). Integrating information literacy into the science curriculum. *College & Research Libraries*, *63*(2), 111–123.

Brown, J. S. (1994). Borderline Issues: Social and Material Aspects of Design. *Human-Computer Interaction*, *9*, 3–36. doi:10.1207/s15327051hci0901_2

Bruce, S. P., & Bruce, B. C. (2000). Constructing images of science: people, technologies, and practices. *Computers in Human Behavior*, *16*(3), 241–256. doi:10.1016/S0747-5632(00)00004-2

Brunton, M., & Coll, R. K. (2005). Enhancing Technology Education by Forming Links with Industry: A New Zealand Case Study. *International Journal of Science and Mathematics Education*, *3*, 141–166. doi:10.1007/s10763-004-1516-1

Bull, J., & McKenna, C. (2000). CAA Centre update. In *Proceedings of the 4th International Computer Assisted Assessment Conference*, Loughborough. Retrieved January 26, 2007, from http://www.caaconference.com/pastConferences/2000/proceedings/jbull.pdf

Bull, J., & McKenna, C. (2004). *Blueprint for Computer-Assisted Assessment*. London: RoutledgeFalmer.

Bundy, A. (Ed.). (2004). *Australian and New Zealand information literacy framework. Principles, standards and practice*. Adelaide: Australian and New Zealand Institute for Information Literacy (ANIIL).

Burden, J. (2005). The new science curriculum at key stage 4. *Education in Science Education, 213*(10-12).

Burger, N. (2000). *Vorstellungen von Schülern über Elektrochemie - eine Interviewstudie [Students conceptions about electro-chemistry an interview study]*. Unpublished doctoral dissertation, University of Dortmund.

Burkman, E., & Gagne, R. M. (Eds.). (1987). *Factors Affecting Utilization. Instructional technology: Foundations*. Hillsdale, NJ: Lawrence Erlbaum.

Butler, D. (2003). *The impact of computer-based testing on student attitudes and behaviour. The Technology Source. January/February*. Retrieved November 7, 2006, from http://ts.mivu.org/default.asp?show=article&id=1013

Butler, G. (2005). Who's in our classrooms? *Teacher*, *156*, 6–9.

Butts, B., & Smith, R. (1987). What do student perceive as difficult in H.S.C. chemistry? *Australian Science Teachers' . Journal*, *32*(4), 45–51.

Buxton, B. (2007). *Sketching User Experiences: Getting the Design Right and the Right Design*. Redmond, WA: Morgan Kaufmann Publishers.

Bybee, R. W. (1997). *Achieving scientific literacy: From purposes to practices*. Portsmouth, NH: Heinemann.

Bybee, R. W., Powell, J. C., & Trowbridge, L. W. (2008). *Teaching secondary school science. Strategies for developing scientific literacy* (9th ed.). Columbus, OH: Pearson Education/Merrill Prentice Hall.

Callava, T. (2007). *Mainstream social science faculty uses and attitudes toward information technologies*. Unpublished doctoral dissertation, Capella University.

Capri, A. (2004). Matter: States of matter. *Visionlearning, CHE, 3*(1). Retrieved November 1, 2008, from http://www.visionlearning.com/library/module_viewer.php?mid=120

Cassady, J. C., & Johnson, R. E. (2002). Cognitive test anxiety and academic performance. *Contemporary Educational Psychology, 27*(2), 270–295. doi:10.1006/ceps.2001.1094

Cavas, B., Cavas, P., Karaoglan, B., & Kışla, T. (2009). A Study on Science Teachers' Attitudes toward Information and Communication Technologies in Education, *The Turkish Online Journal of Educational Technology, 8*(2). Retrieved January 1, 2009, from http://www.tojet.net

Cavas, P., Cavas, B., Kışla, T., & Karaoglan, B. (2008). The use of ICT in science education: a case study. In *Proceedings of the XIII. IOSTE Symposium*, Aydin, Turkey.

CEOP. (2007). *ThinkUKnow - Teachers and Trainers Area*. Retrieved February 15, 2009, from http://www.thinkuknow.co.uk/teachers/

Chan, A., & Lee, M. J. W. (2005). An MP3 a day keeps the worries away: Exploring the use of podcasting to address preconceptions and alleviate pre-class anxiety amongst undergraduate information technology students. In D.H.R. Spennemann & L. Burr (Eds.), *Good Practice in Practice: Proceedings of the Student Experience Conference* (pp. 58-70). Wagga Wagga: Charles Sturt University. Retrieved February 24, 2009, from http://www.csu.edu.au/division/studserv/sec/papers/chan.pdf

Chan, F.-M. (2002). *ICT in Malaysian Schools: Policy and Strategies*. Paper presented at a Workshop on the Promotion of ICT in Education to Narrow the Digital Divide, Tokyo Japan. Center for the Research and Support of Educational Practice. Retrieved October 30, 2008, from http://www.comminit.com/redirect.cgi?m=ac050d863535736c87376cac1c064d92

Chan, F.-M. (2006). *Case study one on ICT integration into education in Malaysia: The Malaysian Smart School Project*. Retrieve November 19, 2006, from http://www2.unescobkk.org/education/ict/resources/JFIT/schoolnet/case-studies/Malaysia_ICT.doc

Chapman, J. H. (2000). *The Final Report of the Research into Male Participation: Male Participation and the Lifelong Learning Agenda*. Northern College, Barnsley.

Cheal, C. (2007). Second Life: hype or hyperlearning? *Horizon, 15*(4), 204–210. doi:10.1108/10748120710836228

Chien, I., Cook, W., & Harding, J. (1948). The field of action research. *The American Psychologist, 3*, 43–50. doi:10.1037/h0053515

Chittaro, L., & Ranon, R. (2007). Web3D Technologies in Learning, Education and Training: Motivations, Issues, Opportunities. *Computers & Education, 49*(1), 3–18. doi:10.1016/j.compedu.2005.06.002

Chong, C. K., Sharaf, H., & Daniel, J. (2005). A Study on the Use of ICT in Mathematics Teaching. [MOJIT]. *Malaysian Online Journal of Instructional Technology, 2*(3), 43–51.

Christensen, H. (2001). What cognitive changes can be expected with normal ageing? *The Australian and New Zealand Journal of Psychiatry, 35*(6), 768–775. doi:10.1046/j.1440-1614.2001.00966.x

Chronicle, U. N. (2002). Partnerships: 'E-Learning for Life'. *UN Chronicle Online Edition, 39*(2). Retrieved November 28, 2008, from http://www.un.org/Pubs/chronicle/2002/issue2/0202p22_elearning.html

Chung, J. S., & Neuman, D. (2007). High school students' information seeking and use for class projects. *Journal of the American Society for Information Science and Technology, 58*(10), 1503–1517. doi:10.1002/asi.20637

Clariana, R., & Wallace, P. (2002). Paper-based versus computer-based assessment: Key factors associated with the test mode effect. *British Journal of Educational Technology, 33*(5), 593–602. doi:10.1111/1467-8535.00294

Clark, C. S. (1993). TV Violence. *CQ Researcher, 3*(12), 167–187.

Clarke, S. (2005). *Formative Assessment in the Secondary Classroom*. London: Hodder Murray

Collins, T. (2009). *Maths DS Lite Project Adventures from Campie Primary 6b*. Retrieved February 15, 2009, from http://edubuzz.org/blogs/campiep6b/2009/01/23/maths-ds-lite-project/

Colvin, M. (2006). *Govt plans digital TV shake-up*. Australian Broadcasting Commission. Retrieved March 16, 2006, from http://www.abc.net.au/pm/content/2006/s1591580.htm

Conroy, S. (2007). *Labor's plan for cyber-safety (fact sheet)*. Retrieved January 5, 2009, from http://www.alp.org.au/download/now/labors_plan_for_cyber_safety.pdf

Convery, A. (2009). The pedagogy of the impressed: How teachers become victims of technological vision. *Teachers and Teaching, 15*(1), 25-41. Retrieved March 1, 2009, from http://dx.doi.org/10.1080/13540600802661303

Cope, B., & Kalantzis, M. (Eds.). (2000). *Multiliteracies. Literacy learning and the design of social futures*. London: Routledge.

Costello, P. J. M. (2003). *Action research*. London: Continuum.

Cottrell, J. (1997). The diffusion of innovations: Applying change theory to academic computing. In *Proceedings of the ACM SIGUCCS*, Monterey, CA.

Couros, A. (2008). *What is a PLN? Or, PLE vs. PLN?* Retrieved February 15, 2009, from http://educationaltechnology.ca/couros/1156

CQU. (2009). *Information literacy at CQU Library*. Retrieved January 5, 2009, from http://www.library,cqu.edu.au Curriculum Corporation for MCEETYA. (2008). *Digital education- Making change happen. Learning in an ONLINE world*. Retrieved October 7, 2008, from http://www.icttaskforce.edna.edu.au/icttaskforce/go

Creanor, L., Trinder, K., Gowan, D., & Howells, C. (2006). *The learner experience of e-learning*. Retrieved April 6, 2007, from http://www.jisc.ac.uk/uploaded_documents/LEX%20Final%20Report_August06.pdf

Creative Commons. (2009). Retrieved February 15, 2009, from http://creativecommons.org

Crocodile Clips. (2008). *Yenka Electricity*. Retrieved February 15, 2009, from http://www.yenka.com/en/Yenka_Electricity/

Crosby, B. (2008). *One reason we blog - finding and clarifying misconceptions* Retrieved February 15, 2009, from http://learningismessy.com/blog/?p=550

Cuban, L. (2001). *Oversold and Underused: Reforming Schools Through Technology, 1980-2000*. Cambridge, MA: Harvard University Press.

Czaja, S. J. (1998). Gerontology Case Study: designing a computer-based communication system for older adults. In V. J. Berg Rice (Ed.), *Ergonomics in Health Care and Rehabilitation*. Boston: Butterworth-Heinemann.

Damanpour, F. (1992). Organizational size and innovation. *Organization Studies, 13*(3), 375–402. doi:10.1177/017084069201300304

Data Harvest Ltd. (2008). Retrieved February 15, 2009, from http://www.dataharvest.co.uk/distributor/files/documents/sensors/3143/3143_ds038_2_force.pdf

De Jong, O., & Treagust, D. F. (2002). The teaching and learning of electrochemistry. In J. K. Gilbert, O. De Jong, R. Justi, D. F. Treagust, & J. H. Van Driel (Eds.), *Chemical Education: Towards research-based practice* (pp. 317-338). Dordrecht, The Netherlands: Kluwer.

De Posada, J. M. (1997). Conceptions of high school students concerning the internal structure of metals and their electric conduction: structure and evolution. *Science Education, 81*(4), 445–367. doi:10.1002/(SICI)1098-237X(199707)81:4<445::AID-SCE5>3.0.CO;2-C

Deary, I. J., Lawrence, J., Whalley, H. L., Crawford, J. R., & Starrc, J. M. (2000). The Stability of Individual Differences in Mental Ability from Childhood to Old Age: Follow-up of the 1932 Scottish Mental Survey. *Intelligence, 28*(1), 49–55. doi:10.1016/S0160-2896(99)00031-8

Deboer, G. E. (2000). Scientific literacy: Another look at its historical and contemporary meanings and its relationship to science education reform. *Journal of Research in Science Teaching, 37*(6), 582–601. doi:10.1002/1098-2736(200008)37:6<582::AID-TEA5>3.0.CO;2-L

Deci, E. L., & Ryan, R. M. (2000). The "what" and "why" of goal pursuits: Human needs and self-determination of behavior. *Psychological Inquiry, 11*, 227–268. doi:10.1207/S15327965PLI1104_01

DEEWR. (2009). *Digital Education Revolution. Australian Government Policy.* Retrieved January 6, 2009, from http://www.digitaleducationrevolution.gov.au

Dembo, S. (2008). *Be a better blogger in just 30 days.* Retrieved February 15, 2009, from http://www.teach42.com/2008/10/23/be-a-better-blogger-in-just-30-days/

Department for Education and Science. (2003). *Fulfilling the Potential: Transforming Teaching and Learning Through ICT in Schools.* London: DfES.

Design-Based Research Collective (DBRC). (2003). Design-based research: An emerging paradigm for educational inquiry. *Educational Researcher, 32*(1), 5–8. doi:10.3102/0013189X032001005

Design-Based Research Collective. (2003). Design-Based Research: An Emerging Paradigm for Educational Inquiry. *Educational Researcher, 32*(1), 5–8. doi:10.3102/0013189X032001005

Diallo, B., & Rasagu, P. (2008). *Elsevier Connect Interview.* Nairobi, Kenya: African Virtual University.

Diallo, B., & Wangeci, C. (2008). *Brief of the AfDB/AVU Multinational Support Project.* Nairobi, Kenya: African Virtual University.

Diaper, D., & Stanton, N. (Eds.). (2004). *The Handbook of Task Analysis for Human-Computer Interaction.* London: Lawrence Erlbaum Associates.

DiSessa, A. A. (2000). *Changing Minds: Computer, Learning, and Literacy.* Cambridge, MA: MIT Press.

DiSessa, A. A., & Sherin, B. (2000). Meta-representation: An introduction. *The Journal of Mathematical Behavior, 19*(4), 385–398. doi:10.1016/S0732-3123(01)00051-7

Dodge, B. (1997). *Some thoughts about WebQuests.* Retrieved January 5, 2009, from http://webque.sdsu.edu/about_webquests.html

Dodge, T., Barab, S., Stuckey, B., Warren, S., Heiselt, C., & Stein, R. (2008). Children's Sense of Self: Learning and Meaning in the Digital Age. *Journal of Interactive Learning Research, 19*(2), 225–249.

Dooley, K. E., & Murphrey, T. P. (2000). *How the perspectives of administrators, faculty and support units impact the rate of distance education adoption.* Retrieved January 1, 2009, from http://www.westga.edu/~distance/ojdla/winter34/dooley34.html

Downs, M., Ariss, S. M. B., Grant, E., Keady, J., Turner, S., & Bryans, M. (2006). Family carers' accounts of general practice contacts for their relatives with early signs of dementia. *Dementia, 5*(3), 353–374. doi:10.1177/1471301206067111

Doyle. (2009). *Our Kids are being Jobbed.* Retrieved February 15, 2009, from http://doyle-scienceteach.blogspot.com/2009/01/are-kids-are-being-jobbed.html

Dresang, E. T. (2005). The information-seeking behavior of youth in the digital environment. *Library Trends, 54*(2), 178–196. doi:10.1353/lib.2006.0015

Drexler, W. (2008). *Networked Student.* Retrieved February 15, 2009, from http://www.youtube.com/watch?v=XwM4ieFOotA

Driver, R., Squires, A., Rushworth, P., & Wood-Robinson, V. (1994). *Making Sense of Secondary Science: Research into Children's Ideas.* London: Routledge Falmer.

Durant, J. R. (1993). What is scientific literacy? In J. R. Durant & J. Gregory (Eds.), *Science and culture in Europe* (pp. 129-137). London: Science Museum.

Durkheim, E. (1922). Education et Sociologie. In A. Giddens (Ed.), *Emile Durkheim Selected Writings.* Cambridge, UK: Cambridge University Press.

Dwyer, D. (1994). Apple classrooms of tomorrow: What we've learned. *Educational Leadership, 51*(7), 4–10.

Edelson, D. C. (2002). Design research: What we learn when we engage in design. *Journal of the Learning Sciences, 11*, 105–121. doi:10.1207/S15327809JLS1101_4

Edtechroundup. (2008). *Edtechroundup podcast episode 4.* Retrieved February 15, 2009, from http://www.edtechroundup.com/2008/05/05/edtechroundup-podcast-episode-4/

Education Testing Service (ETS). (2005). *Beyond Technical Competence: Literacy in Information and Communica-*

tion Technology (An Issue Paper from ETS Princeton). NJ: ETS. Retrieved June 15, 2009, from http://www.ets.org/Media/Tests/ICT_Literacy/pdf/ICT_Beyond_Technical_Competence.pdf

Eilks, I. (2003). Students' understanding of the particulate nature of matter and some misleading illustrations from textbooks. *Chemistry in Action*, *69*, 35–40.

Eilks, I., Möllering, J., & Valanides, N. (2007). Seventh-grade students' understanding of chemical reactions – Reflections from an action research interview study. *Eurasia Journal of Mathematics . Science and Technology Education*, *4*(3), 271–286.

Eilks, I., Ralle, B., Markic, S., Pilot, A., & Valanides, N. (2006). Ways towards research-based science teacher education. In I. Eilks & B. Ralle (Eds.), *Towards research-based science teacher education* (pp. 179-184), Aachen, Germany: Shaker.

Eilks, I., Witteck, T., & Pietzner, V. (2009). A critical discussion of the efficacy of using visual learning aids from the Internet to promote understanding, illustrated with examples explaining the Daniell voltaic cell. *Eurasia Journal of Mathematics . Science and Technology Education*, *5*(2), 145–152.

Ellis, J. D., & Kuerbis, P. J. (1985). *Development and validation of essential computer literacy competencies for science teachers*. Paper presented at the Annual Meeting of NARST, French Lick Springs, IN.

Ellsworth, J. B. (2000). Surviving change: A survey of educational change models. *Syracuse, NY: ERIC Clearinghouse on Information and Technology.* (ERIC Document Reprodcution Service No: ED 443 417)

Elluminate Inc. (2008). *Elluminate Live! Participant Quick Reference Guide*. Retrieved March 2, 2009, from http://www.elluminate.com/support/index.jsp

Emslie, P. (n.d.). *Safety resources*. Retrieved February 11, 2009, from http://education.qld.gov.au/learningplace/cx/docs/resourcelist/safety.doc

Enger, S., & Yager, R. (2001). *Assessing student understanding in science: a standards-based K-12 handbook*. Thousand Oaks, CA: Corwin Press.

Erkkila, K. (1996). *Enterprise Education in the Case of Finland*. Paper presented at the World Congress of Comparative Education Societies, Sydney, Australia.

Eshet-Alkalai, Y. (2004). Digital literacy: A conceptual framework for survival skills in the digital era. *Journal of Educational Multimedia and Hypermedia*, *13*(1), 93–106.

EU. (2004). *Europe needs more scientists!* Brussels: European Commission, Directorate-General for Research, High Level Group on Human Resources for Science and Technology in Europe. Retrieved November 7, 2007, from http://www.tekes.fi/EU/fin/6po/tutkijaliikkuvuus/gago_report_final_en.pdfwww.tekes.fi/EU/fin/6po/tutkijaliikkuvuus/gago_report_final_en.pdf

EU. (2005). *Europeans, science and technology. Eurobarometer 224*. Brussels: The Directorate General Press and Communication of the European Commission. Retrieved November 7, 2007, from http://ec.europa.eu/public_opinion/archives/ebs/ebs_224_report_en.pdf

Evans, A. (2008). *Nintendo DS in the class - What's good beginning to look like?* Redbridge Primary ICT Consultant. Retrieved February 15, 2009, from http://redbridgeprimaryit.blogspot.com/2008/05/nintendo-ds-in-class-whats-good.html

Falk, J., & Storksdieck, M. (2005). Using the Contextual Model of Learning to Understand Visitor Learning from a Science Center Exhibition. *Science Education*, *89*, 744–778. doi:10.1002/sce.20078

Falvo, D. (2008). Animations and simulations for teaching and learning molecular chemistry. *International Journal of Technology in Teaching and Learning*, *4*(1), 68–77.

Faseyitan, S. O., & Hirschbuhl, J. (1992). Computers in university instruction: What are the significant variables that influence adoption? *Interactive Learning International*, *8*(3), 185–194.

FCCS. (1994). *Framework curriculum for the comprehensive school (in Finland)*. Helsinki: State Printing Press and National Board of Education.

Fehring, H., & Green, P. (Eds.). (2001). *Critical literacy. A collection of articles from the Australian Literacy Educators' Association*. Newark, DE: International Reading Association.

Fierce Wireless Daily. (2008). *2008 Asia - Telecoms, Mobile and Broadband in Malaysia and Philippines: Executive Summary (Posted July 14, 2008).* Retrieved on December 20, 2008, from http://www.fiercewireless.com/

Fildes, J. (2006). Podcast numbers cut through hype. *BBC New Channel Website.* Retrieved April 2, 2009, from http://news.bbc.co.uk/1/hi/technology/4885010.stm

Finkelhor, D., Mitchell, K., & Wolak, J. (2000). *Online Victimization: A report on the Nation's Youth.* National Center for Missing and Exploited Children Bulletin, Alexandria, VA, US Department of Justice, Washington, DC.

Fisch, K. (2006). *Did You Know?* Retrieved February 15, 2009, from http://thefischbowl.blogspot.com/2006/08/did-you-know.html

Fisher, V. F. (2005). *Rogers' diffusion theory in education: the implementation and sustained use of innovations introduced staff development.* Unpublished PhD Thesis, University of Nebraska, Lincoln, NE.

Fleer, M., & Robbins, J. (2004). "Yeah that's what they teach you at uni, its just rubbish": The participatory appropriation of new cultural tools as early childhood student teachers move from a developmental to a sociocultural framework for observing and planning. *Journal of Australian Research in Early Childhood Education, 11*(1), 47–62.

Fleer, M., Jane, B., & Hardy, T. (2007). *Science for children: Developing a personal approach to teaching* (3rd ed.). Australia, NSW: Pearson Education.

Fokkema, T., & Knipscheer, K. (2007). Escape loneliness by going digital: A quantitative and qualitative evaluation of a Dutch experiment in using ECT to overcome loneliness among older adults. *Aging & Mental Health, 11*(5), 496–504. doi:10.1080/13607860701366129

Ford, C. (2007). Boys are doctors; girls are nurses: Sustaining the stereotypes of career aspirations. *Redress,* (December), 2-7.

Foster, T. C. (1999). Involvement of hippocampal synaptic plasticity in age-related memory decline. *Brain Research. Brain Research Reviews, 30*(3), 236–249. doi:10.1016/S0165-0173(99)00017-X

Friedman, V., & Lennartz, S. (2008). *Screencasting: How to start, tools and guidelines Smashing Magazine.* Retrieved February 15, 2009, from http://www.smashingmagazine.com/2008/08/19/screencasting-how-to-start/

Friere, P. (1970, 1993). *Pedagogy of The Oppressed* (rev. ed.) (M. B. Ramos, Trans.). New York: Penguin Books.

Frost & Sullivan. (2005). *Benchmarking of the Smart School integrated solution: Educating hearts and minds.* Malaysia: Multimedia Development Corporation.

Fullan, M. G. (1992). *Successful school improvement.* Buckingham, UK: Open University Press.

Funk, J. B. (1993). Reevaluating the impact of video games. *Clinical Pediatrics, 32*(2), 86–90. doi:10.1177/000992289303200205

G8 Heads of State. (2000). *G8 Okinawa Communiqué.* Retrieved on December 30, 2006, from http://en.g8russia.ru/g8/history/okinawa2000/4/

Gardner, H. (1983). *Frames of mind. The theory of multiple intelligence (Tenth Anniversary Edition).* New York: Basic books.

Gardner, H. (1993a). *Creating minds.* New York: Basic Books.

Gardner, H. (1993b). *Multiple Intelligences: The theory in practice.* New York: Basic Books.

Gardner, H. (1998). A multiplicity of intelligences. *Scientific America presents: Exploring intelligences, 94*(4), 18-23.

Gardner, H. (1999). *Intelligences reframed: Multiple intelligences for the 21st century.* New York: Basic Books.

Garnett, P. J., & Treagust, D. F. (1992a). Conceptual difficulties by senior high school students of electrochemistry: electric circuits and oxidation-reduction equations. *Journal of Research in Science Teaching, 29*(2), 121–142. doi:10.1002/tea.3660290204

Garnett, P. J., Garnett, P. J., & Hackling, M. W. (1995). Students' alternative conceptions in chemistry: a review of research and implications for teaching and learning. *Studies in Science Education, 25,* 69–95. doi:10.1080/03057269508560050

Gazit, E., Yair, Y., & Chen, D. (2005). Emerging Conceptual Understanding of Complex Astronomical Phenomena by Using a Virtual Solar System. *Journal of Science Education and Technology, 14*(5-6), 459–470. doi:10.1007/s10956-005-0221-3

Gee, J. P. (2003). High score education [Electronic Version]. *Wired, 11.* Retrieved June 3 2004, from http//www.wired.com/wired/archive/11.05/view.html?pg=1

Gee, J. P. (2003). *What Video Games Have to Teach Us About Learning and Literacy.* New York: Palgrave Macmillan.

Gee, J. P. (2004). *Situated Language and Learning: A Critique of Traditional Schooling.* London: Routledge.

Gee, J. P. (2008). Learning and games. In K. Salen (Ed.), *The ecology of games: connecting youth, games, and learning* (pp. 21-40). Cambridge, MA: MIT Press.

Giddens, A. (1984). *The Constitution of Society: Outline of the Theory of Structure.* Berkeley CA: University of California Press.

Gill, M. (1993, August 17). Literacy - always a slippery customer for wordsmiths. *Age*, 13.

Global, O. E. C. D. Science Forum. (2005). *Papers presented in the conference Declining enrolment in science and technology in November 2005.* Retrieved November 7, 2007, from http://www.oecd.org/document/52/0,3343,en_2649_34319_35398132_1_1_1_1,00.html

Goldacre, B. (2006a, July 15). Sky's limit for big bangs. *The Guardian.* Retrieved November 23, 2008, from http://www.guardian.co.uk/science/2006/jul/15/badscience.uknews

Goldacre, B. (2006b, July 22). Smile while you're faking it. *The Guardian.* Retrieved November 23, 2008, from http://www.guardian.co.uk/science/2006/jul/22/badscience.uknews

Gomez, M.-A., Pozo, J.-I., & Sanz, A. (1995). Students' ideas on conservation of matter: effects of expertise and context variables. *Science Education, 79*(1), 77–93. doi:10.1002/sce.3730790106

Gooch, M., Rigano, D., Hickey, R., & Fien, J. (2008). How do primary pre-service teachers in a regional Australian university plan for teaching, learning and acting in environmentally responsible ways? *Environmental Education Research, 14*(2), 175–186. doi:10.1080/13504620801951715

Goodrum, D., Hackling, M., & Rennie, L. (2001). *The Status and Quality of Teaching and Learning of Science in Australian Schools.* Canberra, Australia: Department of Education, Training and Youth Affairs.

Gordon, C. A. (2000). The effects of concept mapping on the searching behavior of tenth-grade students. *School Library Media Research, 3.* Retrieved January 15, 2009, from http://www.ala.org/ala/mgrps/divs/aasl/aaslpubsandjournals/slmrb/slmrcontents/volume32000/mapping.cfm

Gray, T. (2008). *Alkali Metal Bangs and videos.* Retrieved November 23, 2008, from http://www.theodoregray.com/PeriodicTable/AlkaliBangs/

Griffin, J. (2004). Research on Students and Museums: Looking More Closely at the Students in School Groups. *Science education, 88*, 59–70. doi:10.1002/sce.20018

Griffin, M. M. (1995). You can't get there from here: Situated learning, transfer and map skills. *Contemporary Educational Psychology, 20*, 67–87. doi:10.1006/ceps.1995.1004

Grosslight, L., Unger, C., Jay, E., & Smith, C. (1991). Understanding models and their use in science: conceptions of middle and high school students and experts. *Journal of Research in Science Teaching, 28*(9), 799–822. doi:10.1002/tea.3660280907

Groundwater-Smith, S., Ewing, R., & Le Cornu, R. (2006). *Teaching challenges and dilemmas* (3rd ed.). Australia: Thomson Learning.

Gruenewald, D. A., & Smith, G. A. (2008). Creating a movement to ground learning in place. In D. A. Gruenewald & G.A. Smith (Eds.), *Place-based education in the global age* (pp. 345-358). New York: Lawrence Erlbaum Associates.

Haleblian, R. (2007). *DsLibris - an ebook reader for the Nintendo DS.* Retrieved February 15, 2009, from http://rhaleblian.wordpress.com/dslibris-an-ebook-reader-for-the-nintendo-ds/

Halliday, M. A. K. (1993). Towards a language based theory of learning. *Linguistics and Education, 5*, 93–116. doi:10.1016/0898-5898(93)90026-7

Halliday, M. A. K. (1994). *An introduction to functional grammar* (2nd ed.). London: Edward Arnold.

Halliday, M. A. K., & Hasan, R. (1985). *Language, Context and Text: aspects of language in a social semiotic perspective*. Geelong, Australia: Deakin University Press.

Hamilton, J., & Thompson A. (1992). *The adoption and diffusion of an electronic network for education*. (ERIC Document Reproduction Service No. ED 347 991)

Hancock, J. (Ed.). (1999). *Teaching literacy using information technology. A collection of articles from the Australian Literacy Educators' Association*. Carlton South, Victoria, Australia: ALEA in conjunction with the IRA Newark, DE.

Hanewald, R. (2008). Confronting the Pedagogical Challenge of Cyber Safety. *Australian Journal of Teacher Education, 33*(3), 1–16.

Hansen, K. F. (2006). *Fælles Mål i skolens hverdag - hvordan? Inspirationsmateriale til kommuner og skoler om implementering af Fælles Mål i skolens hverdag* [Common goals in schools – how? Inspiration material for municipalities and schools on the implementation of common goals in schools]. Copenhagen: Ministry of Education, Department of Primary, Lower Secondary and General Adult Education (Ministry of Education's Handbook series 13).

Harlen. (1999). *Effective Teaching of Science: A Review of Research. SCRE*. Retrieved February 15, 2009, from http://www.scre.ac.uk/pdf/science.pdf

Harris, S., Kington, A., & Lee, B. (2001). ICT and innovative pedagogy: examples from case studies in two schools collected as part of the Second Information Technology in Education Study (SITES) in England. In *Proceedings of the British Educational Research Association Annual Conference*, University of Leeds. Retrieved from http://www.leeds.ac.uk/educol/documents/00001906.htm

Harrison, A. (2000). *How do teachers and textbook writers model scientific ideas for students?* Paper presented at the NARST Annual Meeting, New Orleans, LA.

Hartman, G. (2008). *Twitter4Teachers*. Retrieved February 15, 2009, from http://twitter4teachers.pbwiki.com

Hasan, R. (2005). Language and society in a system functional perspective. In R. Hasan, C. Matthiessen, & J.J. Webster (Eds.), *Continuing discourse on Language: A functional perspective. Volume 1*. London: Equinox.

Havelock, R. G., & Huberman, A. M. (1977). *Solving educational problems*. Paris: UNESCO.

Hawken, S. (2008). Lockie Leonard multi-literacy DVD-ROM and resource package. Australian Children's Television Foundation. *Literacy Learning: The Middle Years, 16*(3), 67–68.

Haythorne, E. (1991). *On Earth To Make The Numbers Up*. Leeds, UK: Yorkshire Arts Circus.

Hazen, R. (2002). *Why Should You Be Scientifically Literate?* Retrieved December 29, 2008, from http://www.actionbioscience.org/newfrontiers/hazen.html

Hazzard, B., & Badger, J. (n.d.). *SMARTboard Lessons Podcast*. Retrieved February 15, 2009, from http://pdtogo.com/smart/

Healey, A. (1999). *Children reading in a post-typographic age: Two case studies*. Unpublished doctoral dissertation, Queensland University of Technology, Brisbane.

Healy, A. (Ed.). (2008). *Multiliteracies and diversity in education. New pedagogies for expanding landscapes*. South Melbourne, Australia: Oxford.

Heath, A. (1987). Class in the Classroom. *New Society, 81*, 13–15.

Hedegaard, M. (2001). Learning through acting within societal traditions: Learning in classrooms. In M. Hedegaard (Ed.), *Learning in classrooms: A cultural-historical approach* (pp. 15-35). Aarhus, Denmark: Aarhus Unwood Press.

Hektner, J. M., Schmidt, J. A., & Csicksentmihalyi, M. (2007). *Experience sampling method: Measuring the quality of everyday life*. London: Sage.

Henderson, R. (2004). Recognising difference: One of the challenges of using a multiliteracies approach. *Practically Primary, 9*(2), 11–14.

Heppell, S. (2008). *Learning to Change/Changing to Learn*. Pearson Foundation. Retrieved February 15, 2009, from http://www.pearsonfoundation.org/pg5.6.html

Herring, J. E. (2006). A critical investigation of students' and teachers' views of the use of information literacy skills. *School Library Media Research, 9*. Retrieved January 15, 2009, from http://www.ala.org/ala/mgrps/divs/aasl/aaslpubsandjournals/slmrb/slmrcontents/volume9/informationliteracy.cfm

Hickey, R. (2007). Understanding how children learn science. In V. Dawson & G. Venville (Eds.), *The Art of teaching primary science* (pp. 43-61). Crows Nest, NSW, Australia: Allen & Unwin.

Hickey, R. L., & Anderson, N. (2003). The science education website: Making judgements about science understandings. In C.P. Constantinou, & Z.C. Zacharia (Eds.), *Computer-based Learning in Science: Vol. 1. New Technologies and Their Applications in Education* (pp. 746-753). Nicosia: University of Cyprus.

Hickey, R. L., & Whitehouse, H. (Eds.). (2008). *BirdNet2008. Cairns, Australia: James Cook University*. Retrieved February 11, 2009, from http://www.soe.jcu.edu.au/birdnet2008/

Higgins, A. H. (1981). Distance Education and Pupils: From Horseback to Satellite. *Education in Rural Australia, 10*(2), 31–36.

Hill, D. (1988). Misleading illustrations. *Research in Science Education, 18*(1), 290–297. doi:10.1007/BF02356607

Hill, S., Comber, B., Louden, W., Rivalland, J., & Reid, J. (1998). *100 children go to school: Connections and disconnections in literacy development in the year prior to school and the first year of school*. Canberra, ACT: Commonwealth Department of Employment, Education,Training and Youth Affairs.

Hodgkinson, H. L. (1988). Action Research—a critique. In S. Kemmis & R. McTagggert (Eds.), *The Action Research Reader* (3rd ed.). Geelong, Australia: Deakin University Production Unit.

Hofer, B. K., & Pintrich, P. R. (1997). The development of epistemological theories: Beliefs about knowledge and knowing and their relation to learning. *Review of Educational Research, 67*(1), 88–140.

Hoffman, J. L., Wu, H. K., Krajcik, J. S., & Soloway, E. (2003). The nature of middle schools learners' science content understandings with the use of on-line resources. *Journal of Research in Science Teaching, 40*(3), 323–346. doi:10.1002/tea.10079

Hoggart, R. (1957). *The Uses of Literacy*. London: Penguin.

Hoggart, R. (1995). *The Way We Live Now*. London: Pimlico.

Hollingworth, S., Allen, K., Hutchings, M., Abol Kuyok, K., & Williams, K. (IPSE). (2008). *Technology and school improvement: reducing social inequity with technology?* Coventry, UK: Becta. Retrieved January 1, 2009, from http://partners.becta.org.uk/index.php?section=rh&catcode=_re_rp_02&rid=14541

Holloway, R. E. (1997). Diffusion and adoption of educational technology: A critique of research design. In D. H. Jonassen (Ed.), *Handbook of research for educational communications and technology* (pp. 1107-1133). New York: Simon & Schuster Macmillan.

House of Lords Select Committee on Science and Technology. (2000). *Science and society*. London: HMSO.

Huberman, M., & Miles, M. (1984). *Innovation up close*. New York: Plenum.

Huk, T. (2007). Who benefits from learning with 3D models? The case of spatial ability. *Journal of Computer Assisted Learning, 22*(6), 392–404. doi:10.1111/j.1365-2729.2006.00180.x

Hurt, J. S. (1979). *Elementary Schooling and The Working Class*. London: Routledge and Kegan Paul.

Hytch, T. (2008). Transforming pedagogies with new technologies. The power of protest. *Engineers Australia, 43*(2), 47–50.

Intel. (2005). *3 million teachers help students learn to develop 21st century skills*. Retrieved January 2, 2006, from http://www.intel.com/pressroom/archive/releases/20051116edu.htm

International ICT Literacy Panel. (2002). *Digital transformation: A framework for ICT literacy (A report of the International ICT Literacy Panel)*. Princeton, NJ: Educational Testing Service. Retrieved February 24, 2009, http://www.ets.org/Media/Research/pdf/ictreport.pdf

Iske, S., Klein, A., Kutscher, N., & Otto, H.-W. (2008). Young People's Internet Use and Its Significance for Informal Education and Social Participation. *Technology, Pedagogy and Education*, *17*(2), 131–141. doi:10.1080/14759390802116672

ITEA. (2000). Standards for technological literacy. Content for the study of technology. *Technology Teacher*, *58*(5), 8–13.

Jacobsen, M. (1997). *Bridging the gap between early adopters' and mainstream faculty's use of instructional technology*. Calgary, Alberta, Canada: University of Calgary. (ERIC Document Reproduction Service No. ED423785)

Jamieson, J., & Steele, G. (2008). *Radioactivity Experiments*. Dunfermline, UK: Scottish Schools Equipment Research Centre.

Janssens, J. (2002). Kahootzing around in Year 3. *Practically Primary*, *7*(3), 10–11.

Jarman, R., & McClune, B. (2007). *Developing scientific literacy. Using news media in the classroom*. Maidenhead, Berkshire, England: Open University Press, McGraw-Hill Education.

Jasinski, M. (2007). *Innovate and integrate: Embedding innovative practices*. Canberra, Australia: Australian Flexible Learning Framework and the Australian Government Department of Education, Science and Training. Retrieved February 11, 2009, from http://innovateandintegrate.flexiblelearning.net.au/docs/Innovate_and_Integrate_Final_26Jun07.pdf

Jelfs, A., & Whitelock, D. (2000). The Notion of presence in virtual learning environments: What makes the environment "real". *British Journal of Educational Technology*, *31*(2), 135–152. doi:10.1111/1467-8535.00145

Jenkins, E. W. (1994). Scientific literacy. In T. Husen & T. N. Postlethwaite (Eds.), *The International Encyclopedia of Education* (Vol. 9, pp. 5345-5350). Oxford, UK: Pergamon Press

Jenkins, E. W. (1997). Scientific and technological literacy: Meanings and rationales. In E. W. Jenkins (Ed.), *Innovations in science and technology education* (Vol. 6, pp. 11-42). Paris: UNESCO.

Jenkins, M. (2006). Supporting students in e-learning. In A. Martin & D. Madigan (Eds.), *Digital literacies for learning* (pp. 162-171). London: Facet Publishing.

Johnson, M. K., Reeder, J. A., Raye, C. L., & Mitchell, K. J. (2002). Second thoughts versus second looks: An age-related deficit in selectively refreshing just-active information. *Psychological Science*, *13*(1), 64–67. doi:10.1111/1467-9280.00411

Johnson, P. (1998). Progression in children's understanding of a 'basic' particle theory: a longitudinal study. *International Journal of Science Education*, *20*(4), 393–412. doi:10.1080/0950069980200402

Johnson, W. L., Rickel, J. W., & Lester, J. C. (2000). Animated pedagogical agents: Face-to face interaction in interactive learning environments. *International Journal of Artificial Intelligence in Education*, *11*, 47–78.

John-Steiner, V., & Mahn, H. (1996). Sociocultural approaches to learning and development: A Vygotskian framework. *Educational Psychologist*, *31*(3/4), 191–206. doi:10.1207/s15326985ep3103&4_4

Johnstone, A. H. (1991). Why is science difficult to learn? Things are seldom what they seem. *Journal of Computer Assisted Learning*, *7*(2), 75–83. doi:10.1111/j.1365-2729.1991.tb00230.x

Jonassen, D. H. (1994). Thinking technology. *Educational Technology*, *33*(4), 34–37.

Jonassen, D. H. (2002). Learning as activity. *Educational Technology*, *42*(2), 45–51.

Jones, G. M., Howe, A., & Rua, J. (2000). Gender differences in students' experiences, interests, and attitudes toward science and scientists. *Science Education*, *84*(2), 180–192. doi:10.1002/(SICI)1098-237X(200003)84:2<180::AID-SCE3>3.0.CO;2-X

Jones, S. (2003). *Let the games begin: Gaming technology and entertainment among college students*. Retrieved June 3, 2004, from http//www.pewinternet.org/

Jordan, K. (2008). 'But it doesn't count, sir' - A conversation about using electronic discussion in VCE English. *Engineers Australia*, *43*(2), 59–62.

Justi, R. S., & Gilbert, J. K. (2002a). Science teachers' knowledge about and attitudes towards the use of models and modelling in learning science. *International Journal of Science Education, 24*(12), 1273–1292. doi:10.1080/09500690210163198

Justi, R. S., & Gilbert, J. K. (2002b). Models and modelling in chemistry education. In J. K. Gilbert, O. De Jong, R. Justi, D. F. Treagust, & J. H. Van Driel (Eds.), *Chemical Education: Towards research-based practice* (pp. 47-68). Dordrecht, The Netherlands: Kluwer.

Kalantzis, M., & Cope, B. (2004). *Learning by design*. Altona, Victoria: Common Ground Publishing.

Kalantzis, M., Cope, B., & Fehring, H. (2002). Multiliteracies: Teaching and learning in the new communications environment. *PEN, 133*(March).

Karaoglan, B., Cavas, B., Cavas, P., & Kı la, T. (2007). *Fen Bilgisi Öğretmenlerinin Bilgi ve İletişim Teknolojilerini Kullanma Bilgi ve Becerilerinin Araştırılmasına ve Geliştirilmesine Yönelik Bir Araştırma*, Araştırma Raporu, SOBAG- SBB-104K034, Tübitak.

Karpov, Y. V. (2003). Vygotsky's doctrine of scientific concepts. Its role for contempary education. In A. Kozulin, B. Gindis, V.S. Ageyev, & S.M. Miller (Eds.), *Vygotksy's educational theory in cultural context* (pp. 65-82). NewYork: Cambridge University Press.

Kay, K. (2008). *Learning to Change/Changing to Learn*. Pearson Foundation. Retrieved February 15, 2009, from http://www.pearsonfoundation.org/pg5.6.html

Keddie, N. (1971). Classroom Knowledge. In M. F. D. Young (Ed.), *Knowledge and Control*. London: Collier-MacMillan.

Kelly, R. M. (2005). *Exploring how animations of sodium chloride dissolution affect students' explanations*. Unpublished doctoral dissertation, University of Northern Colorado, Greeley, CO.

Khatoon, B. bt. A. K. (2007). Malaysia's Experience in Training Teachers to Use ICT. In *ICT in Teacher Education: Case Studies from the Asia-Pacific Region* (pp. 10-12). Bangkok, Thailand: UNESCO.

Kibby, M. W. (2000). What will be the demands on literacy in the workplace in the next millennium? *Reading Research Quarterly, 35*(3), 380–381.

Kiili, K. (2005). Digital game-based learning: Towards an experiential gaming model. *The Internet and Higher Education, 8*, 13–24. doi:10.1016/j.iheduc.2004.12.001

Kirner, G., Kirner, C., & Kawamoto, A. L. S. Canta˜o, J., Pinto, A., & Wazlawick, R. S. (2001). Development of a collaborative virtual environment for educational applications. In S. Diehl & M. V. Capps (Eds.), *Proceedings of the 6th international conference on 3D web technology* (pp. 61-68). New York: ACM Press.

Kışla, T., Çavaş, P., Çavaş, B., & Karaoglan, B. (2008). Turkish Science Teachers' Attitudes toward ICT in Education. In *Proceedings of the International Computer and Instructional Technologies Symposium,* Aydin, Turkey.

Kitt, J., Davies, N. G., & Gillam, J. A. (1983). *School of the Air by Satellite. A Study of the Improvement of Distance Education in North-west Queensland Using the Australian Communications Satellite System*. (Evaluative Report) Queensland Dept. of Education, Brisbane (Australia). Abstract ED247883 from ERIC database.

Klug, B. (2002). *The Plagiarism Checker*. Retrieved February 15, 2009, from http://www.dustball.com/cs/plagiarism.checker/

Knapp, C. E. (2008). Place-based curricular and pedagogical models. In D.A. Gruenewald & G.A. Smith (Eds.), *Place-based education in the global age* (pp. 5-28). New York: Lawrence Erlbaum Associates.

Koehler, M. J., Mishra, P., & Yahya, K. (2007). Tracing the development of teacher knowledge in a design seminar: integrating content, pedagogy and technology. *Computers & Education, 49*(3), 740–762. doi:10.1016/j.compedu.2005.11.012

Kolsto, S. D. (2001). Scientific literacy for citizenship: Tools for dealing with the science dimension of controversial socioscientific issues. *Science Education, 84*(3), 291–310. doi:10.1002/sce.1011

Kozma, R., & Russell, J. (2005a). Students becoming chemists: developing representational competence. In J.

K. Gilbert (Ed.), *Visualization in science education* (pp. 121-145). Dordrecht, The Netherlands: Springer.

Kozma, R., & Russell, J. (2005b). Multimedia learning of chemistry. In R. Mayer (Ed.), *The Cambridge handbook of multimedia learning* (pp. 409-428). New York, NY: Cambridge University Press.

Krutsch, J. (2007). *Are you building a creepy treehouse?* Retrieved February 15, 2009, from http://technagogy.learningfield.org/2007/11/19/are-you-building-a-creepy-treehouse/

Kuhlthua, C. C. (2004). *Seeking meaning: A process approach to library and information services* (2nd ed.). Westport, CT: Libraries Unlimited.

Kuhn, D. (2002). What is scientific thinking and how does it develop? In U. Goswami (Ed.), *Blackwell Handbook of Childhood Cognitive Development* (pp. 371-393). Malden, MA: Blackwell Publishing.

Kuhn, D. (2005). *Education for Thinking*. Cambridge, MA: Harvard University Press.

Kuitunen, H., & Meisalo, V. (1988). Science and technology education and industry. In C. Layton (Ed.), *Innovations in science and technology education 2*. Paris: UNESCO, 141-154.

Laaser, W. (1986). Some didactic aspects of audiocassettes in distance education. *Distance Education*, *7*(1), 143–152. doi:10.1080/0158791860070110

Lacina, J. (2005). Media Literacy and Learning. *Childhood Education*, *82*(2), 118–120.

Lacina, L. J. (1984). *The determination of computer competencies needed by classroom teachers.* Dubuque, IA: Loras College. (ERIC Document Reproduction Service No. ED264 831)

Langsford, S. (2002). Museums as resources for science teaching. *Australian Primary & Junior Science Journal*, *18*(1), 17–19.

Lankshear, C., & Knobel, M. (2003). *New Literacies: Changing Knowledge and Classroom Learning.* Buckingham, UK: Open University Press.

Lankshear, C., Snyder, I., & Green, B. (2000). *Teachers and techno-literacy: Managing literacy, technology and learning in schools.* St. Leonards, NSW: Allen & Unwin.

Lantolf, J. P. (2003). Intrapersonal communication and internalisation in the second language classroom. In A. Kozulin, B. Gindis, V. S. Ageyev, & S. M. Miller (Eds.), *Vygotsky's educational theory in cultural context* (pp. 349-370). Cambridge, UK: Cambridge University Press.

Larrabee, G. J., & Crook, T. H. (1994). Estimated prevalence of age-associated memory impairment derived from standardized tests of memory function. *International Psychogeriatrics*, *6*, 95–104. doi:10.1017/S1041610294001663

Lau, B. T., & Chia, H. S. (2008). Exploring The Extent of ICT Adoption Among Secondary School Teachers in Malaysia. *International Journal of Computing and ICT Research*, *2*(2), 19-36. Retrieved from http://www.ijcir.org/volume2-number2/article3%2019-36.pdf

Lau, K.-C. (2009). A critical examination of PISA's assessment on scientific literacy [Published Preprint Online]. *International Journal of Science and Mathematics Education*, 1-28. Retrieved March 18, 2009, from http://springerlink.com.ezproxy.lib.rmit.edu/content/x6620h1156810417/fulltext.pdf

Laugksch, R. C. (2000). Scientific literacy: A conceptual overview. *Science Education*, *84*(1), 71–94. doi:10.1002/(SICI)1098-237X(200001)84:1<71::AID-SCE6>3.0.CO;2-C

Lave, J., & Wenger, E. (1991). *Situated Learning. Legitimate peripheral participation.* Cambridge, UK: University of Cambridge Press.

Lavonen, J., Juuti, K., Byman, R., Uitto, A., & Meisalo, V. (2006). Job characteristics found important for their future career choice by ninth grade students. In S. Yoong, M. Ismail, A. Nurulazam, F. Salleh, F. S. Fook, L. C. Sam, & M. N. L. Yan (Eds.), *Proceedings of the XII IOSTE SYMPOSIUM: Science and Technology Education in the Service of Humankind,* School of Educational Studies, Universiti Sains Malaysia, Penang, Malaysia.

Law, J. (1991). Power, Discretion and Strategy. J. Law (Ed.), *A Sociology of Monsters: Essays on Power, Technology and Domination* (Sociological Review Monograph 38). London: Routledge.

Law, J., & Bijiker, W. E. (1992). Postscript: Technology, stability, and social theory. In W.E. Bijker & J. Law (Eds.), *Shaping technology/building society. Studies in sociotechnical change* (pp. 290-308). Cambridge, MA: MIT Press.

Lee, H. (2007). Instructional design of web-based simulations for learners with different levels of spatial ability. *Instructional Science, 35*, 467–479. doi:10.1007/s11251-006-9010-5

Lee, J. (2009). Properly Using ICT in the Classroom. *Education in Malaysia Blog*. Retrieved February 3, 2009, from http://educationmalaysia.blogspot.com/2009/02/properly-using-ict-in-classroom.html

Lee, O., Eichinger, D. C., Anderson, C. W., Berkheimer, G. D., & Blakeslee, T. S. (1993). Changing middle school students' conceptions of matter and molecules. *Journal of Research in Science Teaching, 30*(3), 249–270. doi:10.1002/tea.3660300304

LeFever, L. (2008). *Twitter in Plain English*. Retrieved February 15, 2009, from http://www.commoncraft.com/twitter

Leonard, J. (1993). *Cutting to the root of ageism: The effect of mental stimulation on the elderly*. OH: The Union Institute.

Lepper, M. R., & Chabay, R. W. (1985). Intrinsic motivation and instruction: Conflicting views on the role of motivational processes in computer-based education. *Educational Psychologist, 20*(4), 217–230. doi:10.1207/s15326985ep2004_6

Lester, J. C., Converse, S. A., Kahler, S. E., Barlow, S. T., Stone, B. A., & Bhoga, R. S. (1997). The persona effect: Affective impact of animated pedagogical agents. In S. Pemberton (Ed.), SIGCHI conference on human factors in computing systems (pp. 21-27). New York: ACM Press.

Lethbridge, D., & Davies, G. (1995). An initiative for the development of creativity in science and technology (CREST): An interim report on a partnership between schools and industry. *Technovation, 15*(7), 453–465. doi:10.1016/0166-4972(95)96594-J

Li, Q. (2004). *Cyber-bullying in school: Nature and extent of adolescents' experience*. Retrieved January 5, 2009, from http://www.ucalgary.ca/~qinli/publication/cyberbully_aera05%20.html

Limniou, M., Roberts, D., & Papadopoulos, N. (2008). Full immersive virtual environment CAVETM in chemistry education. *Computers & Education, 51*, 584–593. doi:10.1016/j.compedu.2007.06.014

Lin, S. (2002). Piaget's developmental stages. In B. Hoffman (Ed.), *Encyclopedia of Educational Technology*. Retrieved January 1, 2009, from http://coe.sdsu.edu/eet/Articles/piaget/start.htm

Linden Labs. (2003). *Second Life*. Retrieved from http://lindenlab.com/

Loughran, J., Milroy, P., Berry, A., & Gunstone, R. (2001). Documenting science teachers' pedagogical content knowledge through PaP-eRs. *Research in Science Education, 31*(2), 289–307. doi:10.1023/A:1013124409567

Loukomies, A., Lavonen, J., & Juuti, K. (2009). *The Motivational Features of an Industry Site Visit*. Paper submitted to ESERA 2009 Conference. Istanbul.

Lowe, I. (2006). *Sustainable natural resource management. Inaugural Rick Farley Lecture 2006, Sydney Conservatorium of Music, Sydney, NSW, Australia*. Retrieved February 11, 2009, from http://www.acfonline.org.au/uploads/res/RFL1-Ian_Lowe.pdf

LTScotland. (2008). *Curriculum for Excellence*. Retrieved January 1, 2009, from http://www.ltscotland.org.uk/curriculumforexcellence/outcomes/

Ludington, J. (2005). *Windows media enanced podcast*. Retrieved February 15, 2009, from http://www.jakeludington.com/project_studio/20051004_windows_media_enhanced_podcast.html

Luft, O. (2008). *Podcasts popularity surges in the US*. Retrieved February 20, 2009, from http://www.guardian.co.uk/media/2008/aug/29/podcasting.digitalmedia?gusrc=rss

Lundblad, J. (2003). A review and critique of Rogers' diffusion of innovation theory as it applies to organizations . *Organization Development Journal, 21*(4), 50–64.

Lyons, T., Cooksey, R., Panizzon, D., & Pegg, J. (2005). *Science, ICT and Mathematics education in rural and regional Australia: The SiMERR national survey*. Canberra, Australia: University of New England and the Australian Government Department of Education, Science and Training.

Macdonald, J. (2004). Developing competent e-learners: The role of assessment. *Assessment & Evaluation in Higher Education, 29*(2), 215–226. doi:10.1080/0260293042000188483

MacKenzie, D., & Wajcman, J. (Eds.). (1985). *The social shaping of technology*. Milton Keynes, UK: Open University Press.

Mackenzie, S. (2008a). *Smart scale diagrams*. Retrieved February 15, 2009, from http://blog.mrmackenzie.co.uk/2008/06/16/smart-scale-diagrams/

Mackenzie, S. (2008b). *Is there really "dead time" in the school year?* Retrieved February 15, 2009, from http://blog.mrmackenzie.co.uk/2008/04/07/is-there-really-dead-time-in-the-school-year/

Mackenzie, S. (2008c). *Youtube solutions to latest higher hw*. Retrieved February 15, 2009, from http://mrmackenzie.co.uk/2008/12/16/youtube-solutions-to-latest-higher-hw/

Mackenzie, S. (2008d). *iPod my Physics*. Retrieved February 15, 2009, from http://blog.mrmackenzie.co.uk/2008/11/05/ipod-my-physics/

Madden, A. D., Ford, N. J., & Miller, D. (2007). Information resources used by children at an English secondary school. Perceived and actual level of usefulness. *The Journal of Documentation, 63*(3), 340–358. doi:10.1108/00220410710743289

Madden, A. D., Ford, N. J., Miller, D., & Levy, P. (2006). Children's use of the internet for information-seeking. What strategies do they use, and what factors affect their performance? *The Journal of Documentation, 62*(6), 744–761. doi:10.1108/00220410610714958

Magnussen, R. (2007). Games as a Platform for Situated Science Practice. In S.de Castell & J. Jenson (Eds.), *Worlds in Play: International Perspectives on Digital Games Research*. New York: Peter Lang.

Magnussen, R. (2009). *Representational Inquiry in Science Learning Games*. Unpublished doctoral dissertation, Copenhagen: Danish School of Education, University of Aarhus. Retrieved from http://www.dpu.dk/site.aspx?p=6604&init=rma&msnr=3&lang=eng

Magnussen, R., & Jessen, C. (2006). Naturfaglig Praksis og Spil-lignende Læring [Science praxis and game-like learning]. *MONA: Matematik og naturfagsdidaktik*, (2), 7-26.

Mahathir, M. (1991). *Malaysia: The way forward*. Kuala Lumpur: National Printing Department.

Mahathir, M. (1998). *The Way Forward - Vision 2020*. London: Weidenfield and Nicholson Publications.

Malaysia, Government of. (1996). *Seventh Malaysia Plan: 1996-2000*. Putrajaya, Malaysia: The Economic Planning Unit, Prime Minister's Department.

Malaysia, Government of. (2006a). *Ninth Malaysia Plan: 2006-2010*. Putrajaya, Malaysia: The Economic Planning Unit, Prime Minister's Department.

Malaysia, M. S. C. (2008). *What is MSC Malaysia / Facts & Figures. National Roll Out - MSC Malaysia Phase 2: MSC Malaysia NEXT LEAP (2004-2010)*. Retrieved December 20, 2008, from http://www.mscmalaysia.my/topic/12066956321774

Malaysia, Ministry of Education. (1997a). *The Malaysian Smart School: A Quantum Leap*. Kuala Lumpur, Malaysia: Ministry of Education.

Malaysia, Ministry of Education. (1997b). *The Malaysian Smart School: A Conceptual Blueprint*. Kuala Lumpur, Malaysia: Ministry of Education, 11 July 1997. Retrieved November 28, 2008, from http://www.msc.com.my/smartschool/downloads/blueprint.pdf

Malaysia, Ministry of Education. (1997c). *The Malaysian Smart Schools Implementation Plan*. Kuala Lumpur, Malaysia: Smart School Project Team, Ministry of Education. Retrieved on November 28, 2008 from http://www.ppk.kpm.my/special/sslmPlan.pdf

Malaysia, Ministry of Education. (1997d). *The Smart School Concept Request for Proposals for Teaching and Learning Materials*. Kuala Lumpur, Malaysia: Ministry of Education, Malaysia.

Malaysia, Ministry of Education. (1997e). *The Smart School Concept Request for Proposals for a Smart School Management System*. Kuala Lumpur, Malaysia: Ministry of Education, Malaysia.

Malaysia, Ministry of Education. (1997f). *The Smart School Concept Request for Proposals for a Smart School Assessment System*. Kuala Lumpur, Malaysia: Ministry of Education, Malaysia.

Malaysia, Ministry of Education. (1997g). *The Smart School Concept Request for Proposals for a Smart School Technology Infrastructure*. Kuala Lumpur, Malaysia: Ministry of Education, Malaysia.

Malaysia, Ministry of Education. (1997h). *The Smart School Concept Request for Proposals for Systems Integration*. Kuala Lumpur, Malaysia: Ministry of Education.

Malaysia, Ministry of Education. (2001a). *A report on the collaborative monitoring of the Smart School Pilot Project for 2000*. Kuala Lumpur, Malaysia: Schools Division, Ministry of Education. (Translated title: Original in the Malay language).

Malaysia, Ministry of Education. (2001b). *A report on the collaborative monitoring of the Smart School Pilot Project for 2001*. Kuala Lumpur, Malaysia: Schools Division, Ministry of Education. (Translated title: Original in the Malay language).

Malaysia, Ministry of Education. (2003a). *Educational development plan 2001-2010*. Kuala Lumpur, Malaysia: Ministry of Education.

Malaysia, Ministry of Education. (2003b). *Benchmarking of the Smart School Integrated Solution*. Kuala Lumpur, Malaysia: Educational Technology Division, Ministry of Education.

Malaysia, Ministry of Education. (2003c). *User Acceptance and Effectiveness of the Smart School Integrated Solution*. Kuala Lumpur, Malaysia: Educational Technology Division, Ministry of Education. (Translated title: Original in the Malay language).

Malaysia, Ministry of Education. (2003d). *Education development plan: 2001-2010*. Kuala Lumpur: Ministry of Education.

Malaysia, Ministry of Education. (2006b). *Pelan Induk Pembangunan Pendidikan 2006-2010* (*Blueprint for Education Development 2006-2010*). Putrajaya, Malaysia: Ministry of Education.

Malone, T. W. (1981). Toward a theory of intrinsically motivating instruction. *Cognitive Science*, *4*, 333–369.

Malone, T. W., & Lepper, M. R. (1987). Making learning fun: A taxonomy of Intrinsic motivations for learning. In R. E. Snow & M. J. Farr (Eds.), *Apptitude, Learning, and Instruction. Volume 3: Conative and Affective Process Analyses* (pp. 223-250). Hillsdale, NJ: Lawrence Erlbaum Associates Inc.

Mareeba Wetland Foundation. (2009). Mareeba savanna and wetlands reserve. Retrieved February 11, 2009, from http://www.mareebawetlands.org/index.html

Markauskaite, L. (2006). Towards an integrated analytical framework of information and communications technology literacy: from intended to implemented and achieved dimensions. *Information Research, 11*(3). Retrieved December 29, 2008, from http://InformationR.net/ir/113/paper252.html

Markic, S., & Eilks, I. (2006). Cooperative and context-based learning on electrochemical cells in lower secondary science lessons – A project of Participatory Action Research. *Science Education International*, *17*(4), 253–273.

Marks, A. (2000a). Lifelong Learning and The Breadwinner Ideology. *Educational Studies*, *26*(3), 304–320. doi:10.1080/03055690050137123

Marks, A. (2000b). Men's Health in the Era of "Crisis" Masculinities - A Health Policy Role for UK Lifelong Learning? *Adults Learning*, 11–13.

Marks, A. (2003). Welcome to the new Ambivalence: Reflections on the current and historical antagonism between the working class male and higher education. *British Journal of Sociology of Education, (Carfax), 24*(1), 84-94.

Marks, A. (2005). A Post- Gesellschaft-Gemeinshaft: A Sociological account of the 'PIPS' Project. In S. Rodrigues (Ed.), *International Perspectives of Teacher Professional Development* (pp. 149-163). New York: Nova Science Publications.

Martin, A. (2005). DigEuLit – a European Framework for Digital Literacy: a Progress Report. *Journal of eLiteracy, 2*, 130-136.

Martin, A., & Madigan, D. (Eds.). (2006). *Digital literacies for learning*. London: Facet Publishing.

Martin, L. (2004). An Emerging Research Framework for Studying Informal Learning and Schools. *Science Education*, *88*, 71–82. doi:10.1002/sce.20020

Matthews. (2009). The Latest Doomed Pedagogical Fad: 21st-Century Skills. *Washington Post*. Retrieved February 15, 2009, from http://www.washingtonpost.com/wp-dyn/content/article/2009/01/04/AR2009010401532.html

Mayer, R. E. (2001). *Multimedia learning*. New York: Cambridge University Press.

Mayer, R. E. (2003). The promise of multimedia learning using the same instructional design methods across different media. *Learning and Instruction*, *13*(2), 125–140. doi:10.1016/S0959-4752(02)00016-6

Mayer, R. E. (2004). Should there be a three-strikes rule against pure Discovery learning? The case for guided methods of instruction. *The American Psychologist*, *59*(1), 14–19. doi:10.1037/0003-066X.59.1.14

Mayer, R. E. (2005). *The Cambridge Handbook of Multimedia Learning*. New York: Cambridge University Press.

Mayes, J. T., & Fowler, C. J. H. (2006). Learners, learning literacy and the pedagogy of e-learning. In A. Martin & D. Madigan (Eds.), *Digital Literacies for Learning* (pp. 26-33). London: Facet Publishing.

McAtee, W., & Zani, T. L. (1975). *The Education of Isolated Children in Western Australia* (Research Report). Education Department of Western Australia, West Perth.

MCEETYA and MCVTE. (2008a, June). *Charter of principles for cross-sectoral collaboration and interoperability across the Australian education and training sector*. Retrieved October 7, 2008, from http://www.aictec.edu.au

MCEETYA and MCVTE. (2008b, June). *Joint Ministerial statement on information and communications technologies in Australian education and training: 2008-2011*. Retrieved October 8, 2008, from http://www.aictec.edu.au

McEneaney, E. H. (2003). The Worldwide Cachet of Scientific Literacy. *Comparative Education Review*, *47*(2), 217–219. doi:10.1086/376539

McFarlane, A., Triggs, P., & Wan, Y. (2008). *Researching mobile learning: interim report*, Coventry, UK: Becta. Retrieved January 1, 2009, from http://partners.becta.org.uk/index.php?section=rh&catcode=_re_rp_02&rid=14204

McGinn, M. K., & Roth, W. M. (1999). Preparing students for competent scientific practice: Implications of recent research in science and technology studies. *Educational Researcher*, *28*(3), 14–24.

McKay, E. (2008). *The human-dimensions of human-computer interaction. Balancing the HCI equation* (Vol. 3). Nieuwe Hemweg, Amsterdam: IOS Press.

McKenzie, J. (1998). Grazing the Net. [from http://www.fno.org/text/grazing.html]. *Phi Delta Kappan*, *79*(September), 26–31. Retrieved January 5, 2009.

MDeC Smart School Department. (2005). *Malaysian Smart School Roadmap 2005 – 2020: An Educational Odyssey – A consultative paper on the expansion of the smart school initiative to all schools in Malaysia*. Cyberjaya: Multimedia Development Corporation. Retrieved October 30, 2008, from http://www.msc.com.my/smartschool/downloads/roadmap.pdf

MDeC, Multimedia Development Corporation. (2004). *Multimedia Super Corridor Impact Survey 2004: Performance of MSC-Status Companies in phase 1 (2003)*. Cyberjaya. Multimedia Development Corporation. Retrieved November 28, 2008, from http://cbdd.wsu.edu/edev/Nigeria_ToT/tr510/documents/UNDPAPDIP_ICT_in_Education.pdf

Medlin, B. D. (2001). *The factors that may influence a faculty member's decision to adopt electronic technologies in instruction*. Unpublished doctoral dissertation, Faculty of the Virginia Polytechnic Institute and State University.

MHS Resources Sdn Bhd. (2005). *Preventive Maintenance Report (PM01) dated 17 Feb. 2005*.

Microsoft. (2007a). *Partners in Learning Progress Report 2007. October, 2007*. Retrieved November 28, 2008, from http://download.microsoft.com/download/2/c/5/2c5f1568-368a-45bf-9a1f-7b38935ae7cc/PiL_Report_complete.pdf

Microsoft. (2007b). *Press Release: Ministry of Education and Microsoft launch Generasi-M Kuala Lumpur, December 10, 2007*. Retrieved November 28, 2008, from (http://www.microsoft.com/malaysia/press/archive2007/linkpage4363.mspx

Millar, R., & Osborne, J. (Eds.). (1998). *Beyond 2000: Science education for the future*. London: Kings College.

Miller, D. J., & Robertson, D. P. (in press). Using a games console in theprimary classroom: Effects of 'Brain Training' programmesme oncomputation and self-esteem. *British Journal of Educational Technology*.

Ministry of Education. (2008). *Fælles Mål*. [Common goals]. Retrieved June 26, 2008, from http://www.faellesmaal.uvm.dk/

Mishra, P., & Koehler, M. J. (2006). Technological Pedagogical Content Knowledge: a framework for teacher knowledge. *Teachers College Record*, *108*(6), 1017–1054. doi:10.1111/j.1467-9620.2006.00684.x

Mistler-Jackson, M., & Songer, N. B. (2000). Student motivation and Internet technology: Are students empowered to learn science? *Journal of Research in Science Teaching*, *37*(5), 459–479. doi:10.1002/(SICI)1098-2736(200005)37:5<459::AID-TEA5>3.0.CO;2-C

Moby. (2007). *Film music*. Retrieved February 15, 2009, from http://mobygratis.com/film-music.html

MOE Intel. (2007). *MOE Intel School Adoption Project Phase I: Project Report*. Malaysia, Ministry of Education and Intel Malaysia.

Moersch, C. (2002). *Beyond hardware: Using existing technology to promote higher level thinking*. Eugene, OR: ISTE.

Mohd Jasmy, A. R., Ismail, M. A., & Norsiati, R. (2003). Tahap kesediaan penggunaan perisian kursus di kalangan guru Sains dan Matematik. (The level of readiness in using courseware among science and mathematics teachers). In *Prosiding Konvensyen Teknologi Pendidikan ke 16 (Proceedings of the 16th Education Technology Convention)* (pp. 372-380).

Mort, P. R., & Cornell, F. G. (1941). *American schools in transition: How our schools adapt their practices to changing needs, a study of Pennsylvania*. New York: Teachers College.

MSC. Multimedia Super Corridor, Malaysia. (1996). *Malaysia's Multimedia Super Corridor (MSC) Background. What Is The Multimedia Super Corridor?* Retrieved April, 1 1996, from http://www.jaring.my/~webmster/msia-new/rnd/msc.html

Mullins, B. (2006). *The Moon Books Project*. Retrieved February 15, 2009, from http://www.moonbooks.info

Muncie, J. (1999). *Youth and Crime: A Critical Introduction*. London: Sage.

Murcia, K. (2005a). *Science for the 21st Century: Teaching for scientific literacy in the primary classroom*. Paper presented at the CONASTA54 University of Melbourne.

Murcia, K. (2005b). Science in the newspaper: A strategy for developing scientific literacy. *Teaching Science*, *51*(1), 40–42.

Murnane, J. S. (2007). Retirement, Mobility and the Internet. In A. Tatnall, J. B. Thompson, & H. Edwards (Eds.), *Education, Training and Lifelong Learning, IFIP WG 3.6 and 3.4 Joint Working Conferences*, Prague. Heidelberg, Australia: Heidelberg Press.

Murnane, J. S. (2008). Age, mobility and eMail. *Journal of Assistive Technologies*, *2*(4), 16–25.

Murphy, J. (2008). *Micro-blogging for science & technology libraries*. Retrieved February 15, 2009, from http://eprints.rclis.org/12752/2/Murphy_Microblogging.pdf

Murray, C. (1990). *The Emerging British Underclass*. London: Institute of Economic Affairs.

Murray, C. (1991). *Underclass: The Crisis Deepens*. London: Institute of Economic Affairs.

Nakhleh, M. B. (1992). Why some students don't learn chemistry. *Journal of Chemical Education*, *69*(3), 191–196.

National Curriculum Board. (2009). *National science curriculum framing paper*. Retrieved January 13, 2009, from http://www.ncb.org.au/verve/_resources/National_Science_Curriculum_-_Framing_Paper.pdf

Nations, U. (2005). *World Summit on the Information Society: Tunis commitment*. New York: United Nations.

NCCBE. (2004). *National Core Curriculum for Basic Education 2004* [Electronic Version]. Helsinki: National Board of Education. Retrieved from http://www.oph.fi/english/page.asp?path=447,27598,37840,72101,72106

Nelson, B. C., & Ketelhut, D. J. (2007). Scientific Inquiry in Educational Multi-user Virtual Environments. *Educational Psychology Review*, *19*, 265–283. doi:10.1007/s10648-007-9048-1

Neville, M. (2004). Brisbane multiliteracies project. *Practically Primary, 9*(2), 20–22.

New London Group. (1996). A pedagogy of multiliteracies: Designing social futures. *Harvard Educational Review, 66*(1), 60–92.

New Straits Times. (2008). *Auditor-General's Report: Hauled up over RM9.56 m 'missing' equipment. 30 August 2008.*

Ng, P., & Yeung, Y. (2000). Implications of Data-logging on A.L. Physics Experiments: A Preliminary Study. *Asia-Pacific Forum on Science Learning and Teaching, 1*(2), Article 5. Retrieved February 15, 2009, from http://www.ied.edu.hk/apfslt/issue_2/phng/phng3.htm

Ng, W. (2006). Web-based Technologies, Technology Literacy and Learning. In L. W. H. Tan & R. Subramaniam (Eds.), *Handbook of Research on Literacy in Technology at the K1-2 Level* (pp. 94-117). Hershey, PA: Idea Group Publishing.

Ng, W. (2008a). Virtual teamwork: students learning about ethics in an online environment. *Journal of Research in Science & Technological Education, 26*(1), 13–29. doi:10.1080/02635140701847421

Ng, W. (2008b). Self-directed learning with web-based sites: How well do students' perceptions and thinking match with their teachers? *Teaching Science, 54*(2), 26–30.

Ng, W., & Gunstone, R. (2002). Students' perceptions of the effectiveness of the World Wide Web as a research and teaching tool in science learning. *Research in Science Education, 32*(4), 489–510. doi:10.1023/A:1022429900836

Ng, W., & Gunstone, R. (2003). Science and computer-based technologies in Victorian government schools: Attitudes of secondary science teachers. *Journal of Research in Science and Technology Education, 21*(2), 243–264. doi:10.1080/0263514032000127266

Nintendo of America, Inc. (1994). *Donkey Kong Country* [Computer Game]. Developer: Rare Ltd.

Nixon, H. (2003). New research literacies for contemporary research into literacy and new media? *Reading Research Quarterly, 38*(3), 407–413.

Nixon, J., Cope, P., McNally, J., Rodrigues, S., & Stephen, C. (2000). University-based initial teacher education: Institutional re-positioning and professional renewal. *International Studies in Sociology of Education, 10*(3), 243-261. Retrieved January 4, 2009, from http://dx.doi.org/10.1080/09620210000200062

Noraini, I., Loh, S. C., Norjoharuddeen, M. N., Ahmad, Z. A. R., & Rahimi, M. S. (2007). The Professional Preparation of Malaysian Teachers in the Implementation of Teaching and Learning of Mathematics and Science in English. *Eurasia Journal of Mathematics, Science & Technology Education, 3*(2), 101-110. Retrieved from http://www.jmste.com

Norris, S. P., & Phillips, L. M. (2003). How literacy in its fundamental sense is central to scientific literacy. *Science Education, 87*, 224–240. doi:10.1002/sce.10066

Northcote, M. (2003). Online assessment in higher education: The influence of pedagogy on the construction of students' epistemologies. *Issues in Educational Research, 13*(1), 66–84.

Novick, S., & Nussbaum, J. (1978). Junior high school pupils' understanding of the particulate nature of matter: an interview study. *Science Education, 62*(3), 273–281. doi:10.1002/sce.3730620303

Nuffield Foundation. (2004). *Ticker-timers for investigating speed.* Retrieved February 15, 2009, from http://www.practicalphysics.org/go/Experiment_266.html

Nugent, C. D. (2007). ICT in the elderly and dementia. *Aging & Mental Health, 11*(5), 473–476. doi:10.1080/13607860701643071

O'Neill, H. M., Pouder, R. W., & Buchholtz, A. K. (1998). Patterns in the diffusion of strategies across organizations: Insights from the innovation diffusion literature. *Academy of Management Review, 23*(1), 98–114. doi:10.2307/259101

Oblinger, D. (2004). The next generation of educational engagement. *Journal of Interactive Media in Education, 8*, 1–18.

OECD, Organization of Economic Co-operation and Development. (2001). *The well-being of nations: The role of human and social capital.* Paris: OECD.

OECD, Organization of Economic Co-operation and Development. (2006). *Are Students Ready for a Technology-Rich World? What PISA Tells us.* Paris: OECD.

OECD. (2001). *Knowledge and skills for life: First results from the OECD Programmesme for International Student Assessment (PISA) 2000.* Retrieved December,17, 2008, from http://cat/lib.rmit.edu.au/vwebv/searchBasic

OECD. (2003). *The PISA 2003 Assessment Framework – Mathematics, Reading, Science and Problem Solving Knowledge and Skills.* Paris: OECD.

OECD. (2006). *Assessing scientific, reading and mathematical literacy.* Paris: OECD.

OECD. (2007). *Assessing Scientific, Reading and Mathematical Literacy: A Framework for PISA 2006.* Paris: OECD Publications.

Ogude, A. N., & Bradley, J. D. (1994). Ionic conduction and electrical neutrality in operating electrochemical cells. *Journal of Chemical Education, 71*(1), 29–34.

Ogude, A. N., & Bradley, J. D. (1996). Electrode processes and aspects relating to cell EMF, current and cell components in EC cells. *Journal of Chemical Education, 73*(12), 1145–1149.

Olitsky, S. (2007). Promoting student engagement in science: Interaction rituals and the pursuit of a community of practice. *Journal of Research in Science Teaching, 44*(1), 33–56. doi:10.1002/tea.20128

Oliver, R., & Tomei, L. (2000). Information and communications technology literacy: getting serious about IT. In *Proceedings of ED-MEDIA 2000. World Conference on Educational Multimedia, Hypermedia and Telecommunications*, VA. Retrieved January 5, 2004, from http://elrond.scam.ecu.edu.au/oliver/2000/emict.pdf

Olsen, P. (2007). The state of Australian birds. *Wingspan Supplement, 14*(4), 1-17. Retrieved February 11, 2009, from http://www.birdsaustralia.com.au/images/stories/downloads/soab/soab2007s.pdf

Ondrejka, C. (2008). Education unleashed: participatory culture, education, and innovation in second life. In K. Salen (Ed.), *The ecology of games: connecting youth, games, and learning* (pp. 229- 251). Cambridge, MA: MIT Press.

Organisation for Economic Co-operation and Development (OECD). (2007). *Programme for International Student Assessment (PISA) 2006 Science competencies for tomorrow's world.* Retrieved February 11, 2009, from http://www.pisa.oecd.org/dataoecd/30/17/39703267.pdf

Ormrod, J. E. (2008). *Human Learning* (5th ed.). Upper Saddle River, NJ: Pearson.

Osborne, J., & Dillon, J. (2008). *Science Education in Europe: Critical Reflections. A Report to the Nuffield Foundation.* London: The Nuffield Foundation.

Osborne, J., Simon, S., & Collins, S. (2003). Attitude towards science: a review of the literature and its implications. *International Journal of Science Education, 25*(9), 1049–1079. doi:10.1080/0950069032000032199

Osborne, R. J., Bell, B. F., & Gilbert, J. K. (1983). Science teaching and children's views of the world. *European Journal of Science Education, 5*(1), 1–14.

Palmer, J. A., & Birch, J. C. (2005). Changing academic perspectives in environment education research and practice: Progress and promise. In E. A. Johnson & M. J. Mappin (Eds.), *Environmental Education and Advocacy* (pp. 114-136). Cambridge, UK: Cambridge University Press.

Panofsky, C. P., John-Steiner, V., & Blackwell, P. J. (1990). The development of scientific concepts and discourse. In L. C. Moll (Ed.), *Vygotsky and Education: Instructional implications of sociohistorical psychology* (pp. 251-267). New York: Cambridge University Press.

Papert, S. (1980). *Mindstorms.* New York: Basic Books.

Papert, S. (1993). *The children's machine: Rethinking school in the age of the computer.* New York: Basic Books.

Parisot, A. H. (1995). *Technology and teaching: The adaptation and diffusion of technological innovations by a community college faculty.* Unpublished doctoral dissertation, Montana State University.

Parvin, J., & Stephenson, M. (2004). Learning Science at Industrial site. In M. Braund & M. J. Reiss (Eds.), *Learning science outside the classroom* (pp. 129-137). London: Routledge.

Pavio, A. (1986), *Mental respresentations: A dual coding approach.* Oxford, UK: Oxford University Press.

Payne, I. (1983). A Working Class Girl in a Grammar School. In D. Spender & E. Sarah (Eds.), *Learning to Lose: Sexism and Education.* London: the Women's Press.

Pedretti, E., Mayer-Smith, J., & Woodrow, J. (1998). Technology, text and talk: Students' learning in a technology enhanced secondary science classroom. *Science Education, 82*, 569–589. doi:10.1002/(SICI)1098-237X(199809)82:5<569::AID-SCE3>3.0.CO;2-7

Pfundt, H. (1975). Ursprüngliche Erklärungen der Schüler für chemische Vorgänge [Initial students' explanations of chemical phenomena]. *Der Mathematische und Naturwissenschaftliche Unterricht, 28*(3), 157–162.

Pfundt, H. (1982). Vorunterrichtliche Vorstellungen von stofflicher Veränderung [Pre-instructional imaginations of material change]. *Chimica Didactica, 8*, 161–180.

Piaget, J. (1955). *The construction of reality in the child.* Routledge & Keegan Paul: London.

Piaget, J. (1972). *Psychology and epistemology: Towards a theory of knowledge.* London: Penguin University Books.

Picardo, J. (2008a). *Interactive Whiteboard.* Retrieved February 15, 2009, from http://www.boxoftricks.net/?cat=8

Picardo, J. (2008b). *Edmodo: What students think.* Retrieved February 15, 2009, from http://www.boxoftricks.net/?p=432

Picardo, J. (2009). *I teach, therefore you learn...or do you?* Retrieved February 15, 2009, from http://www.boxoftricks.net/?p=863

Pittard, V., Bannister, P., & Dunn, J. (2003). *The big pICTure: The impact of ICT on attainment, motivation and learning.* Nottinghamshire,UK: DfES publishing Department for Education and Skills. Retrieved January 5, 2009, from http://www.dcsf.gov.uk/research/data/uploadfiles/ThebigpICTure.pdf

Ploetzner, R., Bodemer, D., & Neudert, S. (2008). Successful and less successful use of dynamic visualizations. In R. Lowe & W. Schnotz (Eds.), *Learning with Animation – Research Implications for Design* (pp. 71-91). New York, NY: Cambridge University Press.

Podshow. (2005). *Podsafe Music Network.* Retrieved February 15, 2009, from http://music.podshow.com

Poland, R., La Velle, L. B., & Nichol, J. (2003). The virtual field station (VFS): Using a virtual reality environment for ecological fieldwork in A'-level biological studies--Case Study 3. *British Journal of Educational Technology, 34*(2), 215–231. doi:10.1111/1467-8535.00321

Posner, G. J., Strike, K. A., Hewson, P. W., & Gertzog, W. A. (1982). Accommodation of a scientific conception: toward a theory of conceptual change. *Science Education, 66*(2), 211–227. doi:10.1002/sce.3730660207

Prain, V., & Waldrip, B. (2006). An exploratory study of teachers' and students' use of multi-modal representation of concepts in primary science. *International Journal of Science Education, 28*(15), 1843–1866. doi:10.1080/09500690600718294

Preece, J. (1994). *Human-computer interaction.* Harlow, UK: Addison-Wesley.

Premkumar, G., & Ramamurthy, K. (1994). Implementation of electronic data interchange: an innovation diffusion perspective. *Journal of Management Information Systems, 11*(2), 157–186.

Prensky, M. (2001). *Digital Game-Based Learning.* New York: McGraw-Hill.

Prensky, M. (2001). *Digital Natives, Digital Immigrants - A New Way to Look at Ourselves and Our Kids.* Retrieved March 15, 2006, from http://www.marcprensky.com/writing/Prensky%20-%20Digital%20Natives,%20Digital%20Immigrants%20-%20Part1.pdf

Prensky, M. (2002). Not only the lonely: Implications of 'social' online activities for higher education. *Horizon, 10*(4), 1–10.

Prensky, M. (2005). Digital natives, digital immigrants. *Gifted, 135*, 29–31.

Promethean Inc. (2006). *Promethean Planet - Teacher Feature.* Retrieved February 15, 2009, from http://feeds.feedburner.com/PrometheanPlanetTeacherFeature

Queensland College of Teachers. (2006). *Professional Standards.* Retrieved February 11, 2009, from http://www.qct.edu.au/Publications/ProffesionalStandards/ProfessionalStandardsForQldTeachers2006.pdf

Raizen, S., Sellwood, P., Todd, R., & Vickers, M. (1995). *Technology education in the classroom*. San Francisco, CA: Jossey-Bass Publishers.

Ramasundaram, V., Grunwald, S., Mangeot, A., Comerford, N. B., & Bliss, C. M. (2005). Development of an environmental virtual field laboratory. *Computers & Education*, *45*, 21–34. doi:10.1016/j.compedu.2004.03.002

Rennie, L. J., Goodrum, D., & Hackling, M. (2001). Science teaching and learning in Australian schools: Results of a national study. *Research in Science Education*, *31*(4), 455-498. Retrieved January 15, 2009, from http://www.ingentaconnect.com/content/klu/rise/2001/00000031/00000004/00383551

Richardson, W. (2006). *Blogs, Wikis, Podcasts and other powerful web tools for classrooms*. Thousand Oaks, CA: Corwin Press

Ricketts, C., & Wilks, S. J. (2002). Improving student performance through computer-based assessment: insights from recent research. *Assessment & Evaluation in Higher Education*, *27*(5), 475–479. doi:10.1080/0260293022000009348

Riddell, S., Baron, S., & Wilson, A. (2001). Gender and Post-School Experience. In B. Francis & C. Skelton (Eds.), *Investigating Gender: Contemporary Perspectives in Education*. Buckingham, UK: Open University Press.

Rinn, W. E. (1998). Mental decline in normal aging: a review. *Journal of Geriatric Psychiatry and Neurology*, *1*(3), 144–158. doi:10.1177/089198878800100304

Rippon, N. (2005). *AVU Capacity Enhancement Programmes (ACEP) Phase 1*. Nairobi, Kenya: African Virtual University.

RIT Center for Multidisciplinary Studies. (2008). *Survey of internet and at-risk behaviors*. Retrieved February 15, 2009, from http://www.rrcsei.org/RIT%20Cyber%20Survey%20Final%20Report.pdf

Rivard, L. P. (1994). A Review of Writing to Learn in Science: Implications for Practice and Research. *Journal of Research in Science Teaching*, *31*, 969–983. doi:10.1002/tea.3660310910

RMIT. (2002). *Facilitators Training Manuel*. Melbourne, Australia: Royal Melbourne of Technology.

RMIT. (2009a). *Information literacy and RMIT University library*. Retrieved January 5, 2009, from http://www.rmit.edu.au/library

RMIT. (2009b). *Library skills for foundation studies*. Retrieved January 8, 2009, from http://rmit.libguides.com/content.php?pid=6688&sid=42212#106369

RMIT. (2009c). *Library research skills for education students*. Retrieved January 9, 2009, from http://rmit.libguides.com/content.php?pid=4587

Roberts, D. A. (2007). Scientific literacy/Science literacy. In S. K. Abell & N. G. Lederman (Eds.), *Handbook of research on science education* (pp. 729-780). Mahwah, NJ: Lawrence Erlbaum Associates.

Robertson, D. (2007). *Dr. Kawashima's Brain Training - Outcomes and Evaluation*. Dundee, Scotland: Learning and Teaching. Retrieved February 15, 2009, from http://www.ltscotland.org.uk/ictineducation/gamesbasedlearning/sharingpractice/braintraining/outcomes.asp

Rodrigues, S. (2006). *Pupil-appropriate contexts in science lessons: The relationship between themes, purpose and dialogue*. Retrieved January 2, 2009, from http://dx.doi.org/10.1080/02635140600811544

Rodrigues, S. (Ed.). (2005). *International models of teacher professional development in science education: Changes influenced by politics, pedagogy and innovation*. New York: Nova Academic Press.

Rodrigues, S., Marks, A., & Steel, P. (2003). Developing science and ICT pedagogical content knowldge: A model of continuing professional development. *Innovations in Education and Teaching International*, *40*(4), 386-394. Retrieved January 21, 2009, from http://dx.doi.org/10.1080/1470329032000128413

Rogers, E. M. (1983). *Diffusion of innovations* (3rd ed.). New York: Free Press, Collier Macmillan.

Rogers, E. M. (1995). *Diffusion of innovations* (4th ed.). New York: The Free Press.

Rogers, E. M. (2003). *Diffusion of Innovations* (5th ed.). New York: Free Press.

Rogers, G. (1999). Reflections on Teaching Remote and Isolated Children. *Education in Rural Australia*, *9*(2), 65–68.

Rogoff, B. (1995). Observing sociocultural activity on three planes: Participatory appropriation, guided participation, and apprenticeship. In J. V. Wertsch, P. Del Rio, & A. Alvarez (Eds.), *Sociocultural studies of mind* (pp. 139-165). Cambridge, UK: Cambridge University Press.

Rogoff, B. (1998). Cognition as a collaborative process. In D. Kuhn & R. S. Siegler (Eds.), *Handbook of Child Psychology* (5th ed.) (Vol. 2, pp. 679-744). New York: John Wiley.

Rogoff, B. (2003). *The cultural nature of human development*. Oxford, UK: Oxford University Press.

Rönnberg, L. (1998). Quality of life in nursing-home residents: an intervention study of the effect of mental stimulation through an audiovisual programme. *Age and Ageing*, *27*, 393–397. doi:10.1093/ageing/27.3.393

Rosnaini, M. (2006). *Kesediaan teknologi maklumat dan komunikasi asas dalam pendidikan guru-guru sekolah menengah (Basic ICT readiness in the education of secondary school teachers)*. Unpublished doctoral dissertation, Universiti Kebangsaan Malaysia.

Rosnaini, M., & Mohd, A. Hj. I. (2008). Factors Influencing Ict Integration In The Classroom: Implications To Teacher Education. In *Proceedings of the 2008 EABR & TLC Conferences*, Salzburg, Austria.

Ross, M. K. (2006). *Bridging the gap: a multi-case study of the adoption and implementation of instructional technology in higher education*. Unpublished doctoral dissertation, Vanderbilt University.

Roth, W.-M. (2000). Autobiography and science education: An introduction. *Research in Science Education*, *30*(1), 1–12. doi:10.1007/BF02461649

Rothman, A. (2003). *America online 'youth wired' survey finds kids are online an average of four days a week; nearly 20% go online every day. Survey published 28th September 2003*. Retrieved January 10, 2009, from http://corp.aol.com/press

Russell, T., & McGuigan, L. (2001). Promoting understanding through representational redescription: an illustration referring to young pupils' ideas about gravity. In D. Psillos, P. Kariotoglou, V. Tselfes, G. Bisdikian, G. Fassoulopoulos, E. Hatzikraniotis, & E. Kallery (Eds.), *Science education research in the knowledge-based society. Proceedings of the Third International Conference of the ESERA* (pp. 600-602). Thessaloniki, Greece: Aristotle University of Thessaloniki.

Ryan, B., & Gross, N. (1943). The Diffusion of Hybrid Seed Corn in Two Iowa Communities. *Rural Sociology*, *8*(1), 15–24.

Rychen, D. S., & Salganik, L. H. (2003). A holistic model of competence. In D. S. Rychen & L. H. Salganik (Eds.), *Key Competencies for a Successful Life and a Well-Functioning Society* (pp. 41-62). Cambridge, MA: Hogrefe & Huber Publishers.

Ryder, J. (2001). Identifying science understanding for functional scientific literacy. *Studies in Science Education*, *36*, 1–44. doi:10.1080/03057260108560166

Sa'ari, J.R., Wong, S.L., & Roslan, S. (2005). In-service Teachers' Views toward Technology and Teaching and their Perceived Competency toward Information Technology. *Journal of Technology*, *43*(E), 1-14.

Sadik, A. (2005). Factors influencing teachers' attitudes towards personal use and schools use of computers: New evidence from a developing nation. *Evaluation Review*, *2*(1), 1–29.

Sahin, I. (2006). *Instructional computer use by COE faculty in Turkey: Application of Diffusion of Innovation*. Unpublished doctoral dissertation, Iowa State University, Ames, Iowa.

Sahin, I., & Thompson, A. (2006). Using Rogers' Theory to Interpret Instructional Computer Use by COE Faculty. *Journal of Research on Technology in Education*, *39*(1), 81–104.

Salbiah, I. (2008). *ICT in the Classroom: A Malaysian Perspective*. Educational Technology Division, Ministry of Education, Malaysia. Retrieved November 28, 2008, from http://www.moe.gov.my/43seameocc/download/MALAYSIA-%20ICT%20and%20School%20Linkages.pdf

Saljo, R. (1998). Thinking with and through: The role of psychological tools and physical artifacts in human learning and cognition. In D. Faulkner, K. Littleton, & M. Woodhead (Eds.), *Learning relationships in the classroom* (pp. 54-66). London: Routledge.

Samak, Z. A. (2006). *An exploration of jordanian english language teachers'attitudes, skills, and access as indicator of information and communication technology integration in Jordan.* Unpublished doctoral dissertation, Florida State University.

Sanderson, F. (1998). Teaching Experience in a School of the Air. *Education in Rural Australia, 8*(2), 31–33.

Sandholtz, J. H., Ringstaff, C., & Dwyer, D. (1997). *Teaching with technology: Creating student-centered classrooms.* New York: Teachers College Press. (ERIC Document Reproduction Service No. ED 402 923)

Sanford, N. (1970). Whatever happened to action research? *The Journal of Social Issues, 26*(4), 3–23.

Sanger, M. J., & Greenbowe, T. J. (2000). Addressing student misconceptions concerning electron flow in electrolyte solutions with instruction including computer animations and conceptual change strategies. *International Journal of Science Education, 22*(5), 521–537. doi:10.1080/095006900289769

Savery, J. R. (2002). Faculty and student perceptions of technology integration in teaching. *The Journal of Interactive Online Learning, 1*(2), 1-16. Retrieved February 11, 2009, from http://www.ncolr.org/jiol/issues/PDF/1.2.5.pdf

Sax, L. (2007). *Boys Adrift: The Five Factors Driving the Growing Epidemic of Unmotivated Boys and Underachieving Young Men.* New York: Basic Books.

Schiefele, U. (1999). Interest and learning from text. *Scientific Studies of Reading, 3*, 257–279. doi:10.1207/s1532799xssr0303_4

Schnotz, W., & Bannert, M. (2003). Construction and interference in learning from multiple respresentations. *Learning and Instruction, 13*(2), 117–123. doi:10.1016/S0959-4752(02)00015-4

Schreiner, C., & Sjøberg, S. (2005). Et meningsfullt naturfag for dagens ungdom? [A meaningful school science for today's youth?]. *Nordina: Nordic Studies in Science Education, 2*, 18–35.

Schultze, U., & Rennecker, J. (2007). Reframing online games. Synthetic worlds as media for organizational communication. In K. Crowston, S Sier, & E. Wynn (Eds.), *IFIP International Federation of Information Processing, Virtuality and Virtualization* (Vol. 236, pp. 335-351). Boston: Springer.

Schunk, D. H. (2008). *Learning theories: An educational perspective* (5th ed.). Upper Saddle River, NJ: Pearson.

Schwartz, N., Andersen, C., Hong, N., Howard, B., & McGee, S. (2004). The influence of metacognitive skills on learners' memory of information in a hypermedia environment. *Journal of Educational Computing Research, 31*(1), 77–93. doi:10.2190/JE7W-VL6W-RNYF-RD4M

Scottish Association for Mental Health. (2008). *Cyberbullying.* Retrieved February 15, 2009, from http://www.respectme.org.uk//cyberbullying/cyberbullying/cyberbullying.html

Seals, C. D., Clanton, K., Agarwal, R., Doswell, F., & Thomas, C. M. (2008). Lifelong learning: Becoming computer savvy at a later age. *Educational Gerontology, 34*(12), 1055–1069. doi:10.1080/03601270802290185

Sega of America, Inc. (1994). *Sonic the Hedghog* [Computer Game]. Developer: Sonic Team.

Segalowitz, N. (1977). Psychological Perspective on Bilingual Education. In B. Spolsky & R. L. Coopewr (Ed.), *Frontiers of Bilingual Education.* Rowley, MA: Newbury House Publishers.

Selwyn, N., & Facer, K. (2007). *Beyond the digital divide: Rethinking digital inclusion for the 21st century.* Futurelab. Retrieved February 15, 2009, from http://www.futurelab.org.uk/resources/documents/opening_education/Digital_Divide.pdf

Shaffer, D. W. (2006). Epistemic frames for epistemic games. *Computers & Education, 46*(3), 223–234. doi:10.1016/j.compedu.2005.11.003

Shaffer, D. W. (2007). *How Computer Games Help Children Learn.* New York: Palgrave Macmillan.

Shaffer, D. W., & Gee, J. P. (2005). *Before every child is left behind: How epistemic games can solve the coming crisis in education* (WCER Working Paper No. 2005-7). University of Wisconsin-Madison, Wisconsin Center for Education Research.

Shaffer, D. W., Squire, K. D., Halverson, R., & Gee, J. P. (2005). Video games and the future of learning. *Phi Delta Kappan, 87*(2), 105–111.

Shamos, M. H. (1995). *The myth of scientific literacy.* New Brunswick, NJ: Rutgers University Press.

Shapira, N., Barak, A., & Gal, I. (2007). Promoting older adults' well-being through Internet training and use. *Aging & Mental Health, 11*(5), 477–484. doi:10.1080/13607860601086546

Shapiro, A. (1999). The relationship between prior knowledge and interactive overviews during hypermedia-aided learning. *Journal of Educational Computing Research, 20*(2), 143–167. doi:10.2190/BCKU-F3AC-CNPW-M44E

Shapiro, W. L., Roskos, K., Cartwright, G. P., Hirschbuhl, J. J., & Bishop, D. (Eds.). (1995). Technology-enhanced learning environments. *Computers in Education.* Guilford, CT: Dushkin Publishing Group/Brown and Benchmark Publishers.

Sharifah, M., & Lewin, K. M. (Eds.). (1993). *Insights into science education planning and policy priorities in Malaysia.* UNESCO, Paris: International Institute for Education Planning (IIEP).

Shen, S. P. (1975). Science Literacy and the Public Understanding of Science. In S. B. Day (Ed.), *Communication of Scientific Information* (pp. 44-52). Basel, Switzerland: Karger.

Sheremetov, L., & Arenas, A. G. (2002). EVA: an interactive Web-based collaborative learning environment . *Computers & Education, 39*(2), 161–182. doi:10.1016/S0360-1315(02)00030-1

Sherman, R. R. (1991). Vocational Education and Democracy. In D. Corson (Ed.), *Education for Work: Background to Policy and Curriculum.* Clevedon, UK: Open University Press.

Shim, K. C., Park, J. S., Kim, H. S., Kim, J. H., Park, Y. C., & Ryu, H. I. (2003). Application of virtual reality technology in Biology education. *Journal of Biological Education, 37*(2), 71–74.

Shulman, L. S. (1987). Knowledge and teaching – foundation of the new reform. *Harvard Educational Review, 57*(1), 1–22.

Simons, S. (2002). Participatory online environmental education at the Open University, UK. In W. D. Filho (Ed.), *Teaching sustainability at universities: Towards curriculum greening.* Bern Switzerland: Peter Lang.

Siraj-Blatchford, J. (1997). *Learning technology, science and social justice: An integrated approach for 3-3 year olds.* Nottingham, UK: Education Now Publishing Co-operative.

Sjøberg, S. (1997). Scientific literacy and school science – Arguments and second thoughts. In S. Sjøberg & E. Kallerud (Eds.), *Science, technology and citizenship* (pp. 9-28). Oslo, Norway: NIFU Rapport 10/97.

Sjøberg, S. (2000). *Science And Scientists: The SAS-study. Cross-cultural evidence and perspectives on pupils interests, experiences and perceptions. - Background, Development and Selected Results.* Department of Teacher Education and School Development, Acta Didactica 1/2000, University of Oslo. Retrieved November 7, 2007, from http://folk.uio.no/sveinsj/SASweb.htm

Small, S. A. (2001). Age-related memory decline. Current concepts and future directions. *Archives of Neurology, 56,* 360–364. doi:10.1001/archneur.58.3.360

Small, S. A., Stern, Y., Tang, M., & Mayeux, R. (1999). Selective decline in memory function among healthy elderly. *Neurology, 52,* 1392–1399.

Smith, J. A., & Osborn, M. (2003). Interpretative Phenomenological Analysis. In: J. A. Smith (Ed.). *Qualitative Psychology: A Practical Guide to Research Methods.* London: Sage Publications.

Smith, M. K. (1999). Informal Learning. In *Encyclopedia of Informal Education pub.* London: InFed.org. Retrieved February 15, 2009, from http://www.infed.org/biblio/inf-lrn.htm

Smith, P., Mahdavi, J., Carvalho, M., & Tippet, N. (2006). *An investigation into cyber bullying, its forms, awareness and impact, and the relationship between age and gender in cyber bullying* (A Report to the anti-Bullying Alliance). London. UK. Retrieved January 5, 2009, from http://www.anti-bullyingalliance.org/

Smith, P., Rudd, P., & Coghlan, M. (NFER). (2008). *Harnessing Technology: Schools Survey 2008 Report 1:*

Analysis, Coventry, UK: Becta. Retrieved from http://partners.becta.org.uk/index.php?section=rh&catcode=_re_rp_02&rid=15952

Snyder, I. (1999). Using information technology in language and literacy education: An introduction. In J. Hancock (Ed.), *Teaching literacy using information technology. A collection of articles from the Australian Literacy Educators' Association* (pp. 11-30). Carlton South, Victoria, Australia: ALEA.

Snyder, I. (Ed.). (1997). *Page to screen. Taking literacy into the electronic era*. St Leonards, NSW: Allen & Unwin.

Soon, L. (2008). *Creating quizzes for the phone*. Retrieved February 15, 2009, from http://video.google.com/videoplay?docid=-2479360146328027324&ei=_qFrSd-mTM5-QiQLPhvyqBQ&q=2479360146328027324

Sprotte, J. A., & Eilks, I. (2007). *Introducing the particulate nature of matter – Results from a case study on experienced German Science Teachers' PCK of models and modelling*. Paper presented at the 6th ESERA Conference, Malmoe, Sweden.

Squire, K. D. (2002). Rethinking the role of games in education. *Game Studies, 2*(1). Retrieved October 2008 from http://www.gamestudies.org/0102/squire/

Squire, K., & Klopfer, E. (2007). Augmented reality simulations on handheld computers. *Journal of the Learning Sciences, 16*(3), 371–413.

Stacey, E. (1999). Collaborative Learning in an Online Environment. *Journal of Distance Education, 14*(2), 14–33.

Stavy, R. (1990). Children's conception of changes in the state of matter: from liquid (or solid) to gas. *Journal of Research in Science Teaching, 30*(3), 247–266. doi:10.1002/tea.3660270308

Steen, K., Brooks, D., & Lyon, T. (2006). The Impact of Virtual Manipulatives on First Grade Geometry Instruction and Learning. *Journal of Computers in Mathematics and Science Teaching, 25*(4), 373–391.

Suara, K. (2009). *March 7 demo against teaching Science, Maths in English*. Retrieved January 21, 2009, from http://www.suarakeadilan.com/sk/english/2009/01/1960

Suh, J., & Moyer, P. S. (2007). Developing Students' Representational Fluency Using Virtual and Physical Algebra Balances. *Journal of Computers in Mathematics and Science Teaching, 26*(2), 155–173.

Sum, S., Mathews, M. R., Pourghasem, M., & Hughes, I. (2008). Internet technology and social capital: How the Internet affects seniors' social capital and wellbeing. *Journal of Computer-Mediated Communication, 14*(1), 202–220. doi:10.1111/j.1083-6101.2008.01437.x

Surry, D. W., & Farquahr, J. D. (1997). Diffusion theory and instructional technology. *Journal of Instructional Science and Technology, 2*(1), 24–36.

Taber, K. (2008). Towards a curricular model of the nature of science. *Science & Education, 17*(2-3), 179–218. doi:10.1007/s11191-006-9056-4

Takkunen, C. L. (2008). *Learning to teach with technology: New teachers' perspectives on using educational technology: After participating in a PT3 grant initiative*. Unpublished doctoral dissertation, Capella University.

Tarde, G. (1903). *The Laws of Imitation* (E. C. Parson, Trans.). New York: Holt.

Tella, A., Tella, A., Toyobo, O. M., Adika, L. O., & Adeyinka, A. A. (2007). An Assessment of Secondary School Teachers Uses of ICT's: Implications for further Development of ICT's Use in Nigerian Secondary Schools. *The Turkish Online Journal of Educational Technology, 6*(3). Retrieved January 1, 2009, from http://www.tojet.net

Teoh, B. (2005). Revisiting Malaysia's Smart School Pilot Project. Monday, October 24, 2005 http://www.it-sideways.com/2005_10_01_archive.html

Thayer, W. R., & Wolf, W. C. Jr. (1984). The generalizability of selected knowledge diffusion/utilization know-how: A case of educational practice. *Knowledge, 5*(4), 447–467.

The Malaysian Smart School Portal. (n.d.) Retrieved from http://www.msc.com.my/smartschool/whatis/index.asp

The Schome Community. (2007). *The schome-NAGTY Teen Second Life Pilot Final Report: a summary of key findings and lessons learnt*. Milton Keynes, UK: The Open University. Retrieved July 20, 2007, from http://kn.open.ac.uk/public/document.cfm?docid=9851

The Smart School Roadmap 2005-2020: An Educational Odyssey. (n.d.). *A consultative paper on the expansion of the Smart School initiative to all schools in Malaysi October2005*. Retrieved from http://www.csdms.in/gesci/pdf/MALAYSIA-SMARTSCHOOLS-roadmap.pdf

The Star Online. (2008). *Vernacular schools prefer Maths, Science in mother tongue*. Retrieved November 1, 2008 from http://thestar.com.my/news/story.asp?file=/2008/11/1/nation/2431447&sec=nation

The State of Queensland (Department of Education, Training and the Arts). (2003). *Management Internet Service*. Retrieved February 11, 2009, from http://education.qld.gov.au/schools/mis/

The State of Queensland (Department of Education, Training and the Arts)/ (2009a). *Conditions for including personal information on publicly accessible school internet websites (part of ICT-PR-004)*. Retrieved February 11, 2009, from http://education.qld.gov.au/strategic/eppr/ict/ictpr004/conditions.pdf

The State of Queensland (Department of Education, Training and the Arts). (2009b). *Smart Classrooms Professional Development Framework*. Retrieved February 11, 2009, from http://education.qld.gov.au/smartclassrooms/strategy/tsdev_pd.html

Thomson, S., & De Bortoli, L. (2008). *Exploring scientific literacy: How Australia measures up. The PISA 2006 survey of students' scientific reading and mathematical literacy skills*. Camberwell, Victoria, Australia: Australian Council for Educational Research Ltd.

Thomson, S., & De Bortoli, L. (2009). PISA in brief from Australia's perspective. Highlights from the full Australian report: Exploring scientific literacy: How Australia measures up. *The PISA 2006 assessment of students' scientific, reading and mathematical literacy skills*. Retrieved January 10, 2009, from http://www.acer.edu.au/ozpisa/reports.html

Tomlinson, D., Coulter, F., & Peacock, J. (1985). *Teaching and Learning at Home: Distance Education and the Isolated Child* (Research Series No. 4). Queensland Dept of Education, Brisbane (Australia). Abstract ED262927 from ERIC database.

Treder, M. (2004). *Jolt to the system: the transformative impact of nanotechnology*. Retrieved December 29, 2008, from http://www.crnano.org/Speech%20%20Troy%20%20Mike%20Treder,%20CRN.ppt

Tse, M. M. Y., Choi, K. C. Y., & Leung, R. S. W. (2008). E-health for older people: The use of technology in health promotion. *Cyberpsychology & Behavior*, *11*(4), 475–479. doi:10.1089/cpb.2007.0151

Tsui, C.-Y., & Treagust, D. (2004). Motivational aspects of learning genetics with interactive multimedia. *The American Biology Teacher*, *66*(4), 277–285. doi:10.1662/0002-7685(2004)066[0277:MAOLGW]2.0.CO;2

Turner, S. (2005). *Answering the Temperature regulation problem*. Retrieved February 15, 2009, from http://www.esf.edu/EFB/turner/termite/The%20temperature%20problem%202.html

Tyack, D., & Cuban, L. (1995). *Tinkering toward utopia*. Cambridge, MA: Harvard University Press.

Tynjälä, P. (1999). Towards Expert Knowledge? A Comparison Between a Constructivist and a Traditional Learning Environment in University. *International Journal of Educational Research*, *31*(5), 357–442. doi:10.1016/S0883-0355(99)00012-9

Tytler, R. (2008). *Re-imagining Science Education: Engaging students in science for Australia's future* (Australian Education Review Number 51). Camberwell, Victoria, Australia: Australian Council for Educational Research.

Tytler, R., & Symington, D. (2006). Redesigning science teacher education to reflect the nature of contemporary science. In *Proceedings of the annual meeting of the National Association for Research in Science Teaching*, San Francisco. Retrieved February 11, 2009, from http://www.deakin.edu.au/alt/edsmf/steme/?q=user/36

UNDP. (2006). *E-Learning for Life (Apple Malaysia)*. United Nations Development Programme Regional Centre Bangkok: Asia-Pacific Development Information Programme. Retrieved November 28, 2008, from http://www.apdip.net/projects/undp/my03/view

UNESCO. (1990). *Compendium of statistics on illiteracy - 1990 Edition* (Statistical Reports and Studies No. 31). Paris: UNESCO, Division of Statistics on Education, Office of statistics.

UNESCO. (2003a). *ICT Policies of Selected Countries is the Asia-Pacific*. Retrieved November 28, 2008 from http://www.unesco.org

UNESCO. (2003b). *Meta-survey on the use of technologies in education in Asia and the Pacific*. Paris: UNESCO

UNESCO. (2003c). *Trends in the use of ICT in Asia and the Pacific. 29 June 2003*. Retrieved November 30, 2008, from http://www2.unescobkk.org/education/ict/v2/info.asp?id=11012

UNESCO. (2004a). *Integrating ICT into Education: A Collective Case Study of Six Asian Countries - Indonesia, Malaysia, Philippines, Singapore, South Korea, Thailand.* UNESCO Asia and Pacific Regional Bureau for Education. Retrieved November 28, 2008, from http://unesdoc.unesco.org/images/0013/001355/135562e.pdf

UNESCO. (2004b). *The Malaysian Smart School Project. Case Study One on ICT Integration into Education in Malaysia*. UNESCOBKK. Retrieved on November 28, 2008, from http://www2.unescobkk.org

UNIVERSALIA. (2005). *Evaluation of the African Virtual University, Volume I, Final Report*. Montréal, Canada: UNIVERSALIA

University of Colorado. (2008). *PhET Circuit Construction Ki.t* Retrieved February 15, 2009, from http://phet.colorado.edu/simulations/sims.php?sim=circuit_construction_kit_dc_only

University of Nottingham. *The Periodic Table of Videos*. (2008). Caesium. Retrieved November 23, 2008 from http://www.periodicvideos.com/videos/055.htm

Unsworth, L. (2001). *Teaching multiliteracies across the curriculum: Changing contexts of text and image in classroom practice*. Buckingham, UK: Open University Press.

Usun, S. (2004). Factors Affecting the Application of Information and Communication Technologies (ICT) in Distance Education. *Turkish Online Journal of Distance Education, 5*(1), Retrieved January 1, 2009, from http://tojde.anadolu.edu.tr/tojde13/articles/usun.html

Van de Ven, A., Poole, M. S., & Angle, H. (Eds.). (1989). Methods for Studying Innovation Processes. In *Research on the Management of Innovation: The Minnesota Studies*. New York: Oxford University Press.

Van der Veer, R., & Valsiner, J. (1993). *Understanding Vygotsky: A quest for synthesis*. Oxford, UK: Blackwell.

Van Driel, J. H., & Verloop, N. (1999). Teachers' knowledge of models and modelling in science. *International Journal of Science Education, 21*(11), 1141–1153. doi:10.1080/095006999290110

Van Driel, J. H., Verloop, N., & De Vos, W. (1998). Developing science teachers' Pedagogical Content Knowledge. *Journal of Research in Science Teaching, 35*(6), 673–695. doi:10.1002/(SICI)1098-2736(199808)35:6<673::AID-TEA5>3.0.CO;2-J

VCAA. (2005). *General Achievement Test (GAT) 2005*. Retrieved January 3, 2009, from http://www.vcaa.vic.edu/vce/exams/gat/gat.html

VCAA. (2006). *General Achievement Test (GAT) 2006*. Retrieved January 3, 2009, from http://www.vcaa.vic.edu.au/vce/exams/gat/gat.html

VCAA. (2009). *General Achievement Test (GAT)*. Retrieved January 5, 2009, from http://www.vcaa.vic.edu.au/vce/exams/gat/aboutgat.html

Victorian Curriculum Assessment Authority (VCAA). (2005a). *Victorian Essential Learning Standards (VELS). Revised 10th December 2005.*

Victorian Curriculum Assessment Authority (VCAA). (2005b). *Victorian Essential Learning Standards. Revised Edition 10th December 2005*. Retrieved January 2, 2009, from http://vels.vcaa.vic.edu.au/links/standards.html#5

Vincent, T. (2008). *iPods in Education* (Learning in Hand Podcast). Retrieved February 15, 2009, from http://learninginhand.com/ipod/clickwheel.html

Vygotsky, L. (1978). *Mind in society: The development of higher psychological processes*. Cambridge, MA: Harvard University Press.

Vygotsky, L. (1997). The history of the development of higher mental functions. In R. W. Rieber (Ed.), *The collected works of L. S. Vygotsky. Vol. 4* (M. J. Hall & R. W. Rieber, Trans.).New York: Plenum Press.

Vygotsky, L. S. (1962). *Thought and language*. Cambridge, MA: MIT Press.

Vygotsky, L. S. (1987). Thinking and speech. In R. W. Rieber & A. S. Carton (Eds.), *The collected works of L. S. Vygotsky, Vol. 1, Problems of general psychology* (N. Minick, Trans.) (pp. 39-285). New York: Plenum Press.

Wainer, J., Dwyer, T., Dutra, R. S., Covic, A., Magalhaes, V. B., & Ferreria, L. R. R. (2008). Too Much Computer and Internet Use is Bad for Your Grades, Especially if You are Young and Poor: Results from the 2001 Brazillian SAEB. *Computers & Education, 51*(4), 1417–1429. doi:10.1016/j.compedu.2007.12.007

Wajcman, J. (1991). *Feminism confronts technology.* London: Allen and Unwin.

Waldrip, B., Prain, V., & Carolan, J. (2006). Learning junior secondary science through multi-modal representations. *Electronic Journal of Science Education, 11*(1), 87–107.

Walker, D. J., Topping, K., & Rodrigues, S. (2008). Student reflections on formative e-assessment: expectations and perceptions. *Learning, Media and Technology, 33*(3), 221–234. doi:10.1080/17439880802324178

Wallace, R. M. (2002). The Internet as a site for changing practice: the case of Ms Owens. *Research in Science Education, 32*(4), 465–487. doi:10.1023/A:1022477832695

Wallace, R., Kupperman, J., Krajcik, J., & Soloway, E. (2000). Science on the Web: Students on-line in a sixth grade classroom. *Journal of the Learning Sciences, 9*(1), 75–104. doi:10.1207/s15327809jls0901_5

Waller, T. (2007). ICT and Social Justice: Educational technology, global capital and digital divides. *Journal for Critical Educational Policy Studies, 5*(1). Retrieved March 1, 2009, from http://www.jceps.com

Walters, M. (2007). *An Investigation of Adopting, Adapting and Integrating of Information and Communication Technology (ICT) and Incorporating the Explicit Teaching of Thinking Skills Across the Curriculum.* Unpublished doctoral dissertation, RMIT University, Melbourne.

Walters, M., & Fehring, H. (2008). *Integrating ICT and thinking skills - have schools moved on?* Unpublished journal article. RMIT University.

Wandersee, J. H., Mintzes, J. J., & Novak, J. D. (1994). Research on alternative conceptions in science. In D. L. Gabel (Ed.), *Handbook of research in science teaching and learning* (pp. 177-210). New York, NY: Macmillan.

Wang, L. (2006). *Information literacy courses- a shift from a teacher-centred to a collaborative learning environment.* Paper presented at the 4th International Lifelong Learning Conference: Partners, Pathways, and Pedagogies, Yeppoon, Queensland.

Warford, M. K. (2005). Testing a Diffusion of Innovations in Education Model (DIEM). The *Innovation Journal . The Public Sector Innovation Journal, 10*(3), 32.

Warlick, D. F. (2005). *Classroom Blogging: A teacher's guide to the blogosphere.* Raleigh, NC: The Landmark Project.

Waters, S. (2007). *31 Day Project.* Retrieved February 15, 2009, from http://aquaculturepda.wikispaces.com/blogs1

Webster. (2008). *The Podcast Consumer Revealed 2008.* Retrieved February 20, 2009, http://www.edisonresearch.com/home/archives/2008/04/the_podcast_con_1.php

Wei, K. (2007). Sharing knowledge in global virtual teams. How do Chinese team members perceive the impact of national cultural differences on knowledge sharing? In K. Crowston, S. Sier, & E. Wynn (Eds.), *IFIP International Federation of Information Processing, Virtuality and Virtualization* (Vol. 236, pp. 251-265). Boston: Springer.

Weller, M. (2004). Learning objects and the e-learning cost dilemma. *Open Learning, 19*(3), 293–302. doi:10.1080/0268051042000280147

Wenger, E. (1998). *Communities of practice: Learning, meaning and identity.* Cambridge, UK: Cambridge University Press.

Wertsch, J. V. (1990). The voice of rationality in a sociocultural approach to mind. In L. C. Moll (Ed.), *Vygotsky and education: Instructional implications of sociocultural psychology* (pp. 111-126). New York: Cambridge University Press.

Wertsch, J. V. (1990). The voice of rationality in a sociocultural approach to mind. In L. C. Moll (Ed.), *Vygotsky and education: Instructional implications of sociocultural psychology* (pp. 111-126). New York: Cambridge University Press.

White, H., McConnell, E., Clipp, E., Branch, L. G., Sloane, R., Pieper, C., & Box, T. L. (2002). A randomized controlled trial of the psychosocial impact of providing internet training and access to older adults. *Aging & Mental Health, 6*(3), 213–221. doi:10.1080/13607860220142422

Whitehouse, H. (2007). *Final project report for BirdNet: Creating an online community for far north Queensland students.* Unpublished report for the Australian School Innovation in Science, Technology and Mathematics (ASISTM). Cairns, Australia: James Cook University.

Whitehouse, H. (2008). "EE in cyberspace, why not?" Teaching, learning and researching tertiary pre-service and in-service teachers' environmental education online. *Australian Journal of Environmental Education, 24*, 11–21.

Whitehouse, H., & Hickey, R. L. (Eds.). (2007). *BirdNet2007. Cairns, Australia: James Cook University.* Retrieved February 11, 2009, from http://www.soe.jcu.edu.au/birdnet2007/

Whittle, P., & Maharjan, D. (2000). Scientific and technological literacy for sustainable development in the 21st Century. In *Proceedings to the International symposium BioEd2000: The challenge of the next century.* Paris. Retrieved January 5, 2009, from http://www.iubs.org/cbe/pdf/whittle.pdf

Wikipedia. (2008). *Enhanced podcast.* Retrieved February 15, 2009, from http://en.wikipedia.org/wiki/Enhanced_podcast

Wikipedia. (2008a). *Brainiac: Science Abuse Alkali metal experiment with forged results.* Retrieved November 20, 2008, from http://en.wikipedia.org/wiki/Brainiac:_Science_Abuse

Wikipedia. (2008b). *Viewer Special Threequel episode of MythBusters.* Retrieved November 23, 2008, from http://en.wikipedia.org/wiki/MythBusters_(2008_season)#Episode_114_.E2.80.93_.22Viewer_Special_Threequel.22

Wikipedia. (2009). *Nintendo DS Homebrew.* Retrieved February 15, 2009, from http://en.wikipedia.org/wiki/Nintendo_DS_homebrew

Willard, R. (2008). *The Tech Teachers show #83.* Retrieved February 15, 2009, from http://thetechteachers.blogspot.com/2008/08/tech-teachers-show-83.html

Williamson, V. M., & Abraham, M. R. (1995). The effects of computer animation on the particulate mental models of college chemistry students. *Journal of Research in Science Teaching, 32*(5), 521–534. doi:10.1002/tea.3660320508

Willis, P. (1977). *Learning to Labour: How Working Class Kids Get Working Class Jobs.* London: Gower Press.

Wilson, T. D. (2006). 60 years of the best in information research. On user studies and information needs. *The Journal of Documentation, 62*(6), 658–670. doi:10.1108/00220410610714895

Wolak, J., Mitchell, K., & Finkelhor, D. (2006). *Online victimization of youth: five years later* (National Center for Missing & Exploited Children bulletin - #07-06-025). Alexandria, VA.

Woolnough, B. E. (1996). Changing Pupils' Attitudes to Careers in Science. *Physics Education, 31*(5), 301–308. doi:10.1088/0031-9120/31/5/020

World Bank. (2003). *ICT and MDGs: A World Bank perspective.* Washington, DC: World Bank.

Wray, D., & Lewis, M. (1995). *Writing frames: scaffolding children's non-fiction writing in a range of genres.* University of Reading: Reading & Language Information Centre.

Wray, D., & Lewis, M. (1997). *Extending Literacy.* London: Routledge.

Wright, R. E., Palmar, J. C., & Kavanaugh, D. C. (1995). The importance of promoting stakeholder acceptance of educational innovations. *Education, 115*, 628–633.

Wynne, M. E., & Cooper, W. L. (2007). *Power up: The campaign for digital inclusion.* Retrieved December 28, 2008, from http://www.digitalaccess.org/pdf/White_Paper.pdf

Ya'acob, A., Nor, N., & Azman, H. (2005). Implementation of the Malaysian Smart School: An Investigation of Teaching-Learning Practices and Teacher-Student Readiness. *Internet Journal of e-Language Learning & Teaching, 2*(2), 16-25.

Yang, E.-M., Greenbowe, T. J., & Andre, T. (2004). The effective use of an interactive software programmes to reduce students' misconceptions about batteries. *Journal of Chemical Education, 81*(4), 587–595.

Yates, B. L. (2001). *Applying diffusion theory: Adoption of media literacy programmes in schools*. Paper presented at Instructional and Developmental Communication Division, International Communication Association Conference, Washington, DC, Retrieved February 8, 2008, from http://www.westga.edu/~byates/applying.htm

Yates, J., & Orlikowski, W. J. (1992). Genres of Organizational Communication: A Structurational Approach to Studying Communication and Media. *Academy of Management Review*, *17*, 299–326. doi:10.2307/258774

Yelland, N. (2001). *Teaching and learning with information and communication technologies (ICT) for numeracy in the early childhood and primary years of schooling*. Canberra: Department of Education, Training and Youth Affairs.

Yin, D. Y. (2008). *Hong Kong students' performance in scientific literacy. HKPISA Centre*. Retrieved March 10, 2009, from http://www.fed.cuhk.edu.hk/~hkpisa/events/2006/events2006_20071210.htm

Yoder, M. B. (1999). A productive and thought-provoking use of the Internet. *Learning and Leading with Technology*, *26*(7), 6–11.

Yoong, S. (2005). *Crossing Linguistic & Cultural Borders: Problems and Challenges in the Teaching Science and Mathematics in English*. Keynote paper presented at TEACH 2005 symposium on the Teaching of Science and Mathematics in English, Universiti Sarawak Malaysia, UNIMAS.

Yoong, S. (2007a). *Smart Schools for Science & Technology: The Malaysian Experience*. Keynote paper presented at the Turkev School of Science & Technology Symposium, Izmir, Turkey.

Yoong, S. (2007b). *Using ICT in Chemistry Teaching: Reactivity of Alkali Metals. Chemistry Teaching Method, PGT 211E: PPIP ONLINE*. Retrieved from http://ppip.usm.my

Yoshioka, T., & Herman, G. (1999). *Genre Taxonomy: A Knowledge Repository of Communicative Actions*. Retrieved January 1, 2009, from http://ccs.mit.edu/papers/pdf/wp209.pdf

Zacharia, Z. C. (2007). Comparing and combining real and virtual experimentation: an effort to enhance students' conceptual understanding of electric circuits. *Journal of Computer Assisted Learning*, *23*, 120–132. doi:10.1111/j.1365-2729.2006.00215.x

Zakaria, Z. (2001). *Factors related to information technology implementation in the Malaysian Ministry of Education Polytechnics*. Unpublished doctoral dissertation, Virginia, Polytechnic Institute and State University, Blacksburg.

Zeidler, D. L., Sadler, T. D., & Simmons, M. L. (2005). Beyond STS: A Research-based Framework for Socioscientific Issues Education. *Science Education*, *89*(3), 357–377. doi:10.1002/sce.20048

Zhao, Y., & Tella, S. (2002). From the special issue editors. *Language Learning & Technology*, *6*(3), 2–5.

Zulkifli, M. (2008). Teachers Lack Good English Command. *Skor Career Blog*. Retrieved September 8, 2008, from http://skorcareer.com.my/blog/teachers-lack-good-english-command/2008/09/08/

Zumbach, J., Schmitt, S., Reimann, P., & Starkloff, P. (2006). Learning Life Sciences: Design and Development of a Virtual Molecular Biology Learning Lab. *Journal of Computers in Mathematics and Science Teaching*, *25*(3), 281–300.

About the Contributors

Susan Rodrigues, is a Professor of Science Education at the University of Dundee. She is the Director of the Professional Doctorate programme at the School of Education, Social Work and Community Education, University of Dundee. She started as a teacher in chemistry and physical education. She received her Ph.D. in Chemical Education from the University of Waikato in New Zealand.

* * *

Bulent Cavas is an instructor at Dokuz Eylul University, Faculty of Education, Buca, Izmir, Turkey. He currently teaches science and technology courses and is involved with several national and international studies & projects on education. In addition to these academic activities, he plays an advisory role to master students. These duties combine with his special interest in science and technology education. Since October 2009, he works on the Interests and Recruitments in Science and Technology as a post doc study at Middle East Technical University in Ankara, Turkey. Further information can be found about Bulent Cavas from his web page: http://people.deu.edu.tr/bulent.cavas

Pinar Huyuguzel Cavas is a research assistant at department of elementary education of Ege University in Izmir, Turkey. She completed her master degree on the elementary science education. She is still doing her Ph.D on the elementary teachers' scientific literacy level and their competence on the science teaching. She has over 20 national and international publications on the science and technology education.

John Cripps Clark is a researcher and educator at present based in Melbourne. His research interests include: primary science education, teacher professional development and activity theory. At present I teach pre-service primary teachers to teach science and I have taught science and science education at primary, secondary and tertiary levels.

Bakary Diallo is the CEO/Rector of the African Virtual University Dr. Bakary Diallo has been working in the education sector for the past 20 years as a secondary school teacher, an academic, a consultant, a project administrator and a researcher. He joined the African Virtual University (AVU), an Intergovernmental Organization based in Nairobi Kenya, which specializes in Open Distance and Electronic Learning in August 2005. Dr Bakary has a PhD and Masters degree from the University of Ottawa, Ontario, Canada. He also holds a Bachelor's Degree from the Universite Cheikh Anta Diop, Dakar Senegal and has a Certificat d'Aptitude au Professorat from INSEPS/Universite Cheikh Anta

Diop, Dakar Senegal. Prior to joining the AVU, he worked at the University of Ottawa as a part-time Lecturer at the Faculty of Education from July 2001 to July 2005, and as a Consultant of Integration of ICT in Education, at the Center for University Teaching. He taught at the Secondary Level in Senegal from 1988 to 1997 before joining the University of Ottawa in 1997. Dr Diallo is fully bilingual (French and English).

Paul Edwards has a body of research in applying contemporary theories of learning to adult education. His passion is using collaborative techniques (both carbon based and silicon) to bring people together and build a shared learning environment to enhance and maximize learning outcomes. He has worked as a University lecturer, IT consultant, super computer programmer, and is currently consulting in business process engineering.

Ingo Eilks, University of Bremen, Bremen, Germany, was appointed a professor and has held the chair for Chemistry education at the University of Bremen since 2004. He is a fully-trained grammar school teacher for Chemistry and Mathematics. His main research interests are the development of Participatory Action Research for research and continuous professional development in science education, cooperative learning in chemistry teaching, the socio-critical and problem-oriented approach to chemistry teaching, empirical research on science teachers' pedagogical content knowledge, attitudes and beliefs, and the development of modern chemistry curricula, teaching materials and textbooks.

Heather Fehring has been writing and researching for many years in the area of children's literacy development specifically literacy assessment and reporting. Her PhD research involved investigating the influences on teachers' judgement of students' literacy development. Heather is currently the CI in an Australian Research Council Linkage Grant team – November 2005 entitled 'Comparative Analysis of Accrued Benefits of the Trade and Bachelor degree Graduates over the First Ten Years of their Working Life'. Among her publications is a jointly authored book called Keying into Assessment: Case Studies, Strategies, Classroom Management and a co-edited book entitled Critical Literacy: A Collection of Articles from the Australian Literacy Educators' Association. A comprehensive curriculum vitae can be found on her professional web site: http://raws.adc.rmit.edu.au/~e51502/

Therrezinha Fernandes is the project coordinator of the Programme Informatique de Laval à l'Afrique Francophone (PILAF) program within the African Virtual University (AVU). Working thightly with the partner institutions, the students, l'Université Laval and the Association of Universities and Colleges of Canada, Dr. Fernandes is at the core center of the PILAF project. She joined the AVU in July 2007. Prior to the AVU, Dr. Fernandes was a doctorate student in the computer science field in l'Université Laval. She thus knows very well the PILAF program delivered to the students. She also already was very engaged for the cause of "education for all", volountering for example for UNICEF. Dr Fernandes holds a Ph.D. in Computer Science from l'Université Laval and has completed graduate level coursework in international management at l'Université Laval.

Marilyn Fleer is a Professor of Early Childhood Education at Monash University and Research leader for the Centre for Childhood Studies. Currently she is the World President for the International Society for Cultural Activity Research. Marilyn has taught in universities since 1988. She has over 200 publications, and has work on a range of research projects covering early childhood education, technology education, science education, family studies and Indigenous education. Much of her conceptual work has concentrated on pedagogy, culture and learning in the early years. Recent publications include: Fleer, M., Hedegaard, M., and Tudge, J., (2009) (eds) Constructing childhood: Global-local policies and practices, World Year Book Series, Routledge, New York; Fleer, M., and Hedegaard, M., (2008) Studying children: A Cultural-Historical Approach, Open University Press, UK (with contributions from Pernille and Jytte).

Colette Fortuna started teaching Mathematics in 1979 and held posts of Teacher, Guidance Teacher and Assistant Principal Teacher of Mathematics. In 2003 she was appointed to the University of Dundee to teach on the B.Ed, PGDEP and PGDES programmes. She is the Assessment Convenor for PGDEP and she is currently an Associate Inspector for HMIe.

Simon Gallagher worked for the National Foundation for Educational Research (NFER) between 1993 and 1997 as a researcher and has recently returned as the Acting Head of the Research, Evaluation and Information Department. Since 1997, he has worked for six years at the Training and Development Agency for Schools (TDA), managing the teacher training market in England and as the Head of Education at the Wellcome Trust. He has also worked as an education-policy adviser with a number of professional and statutory bodies.

Elizabeth Gibson has a background of senior management roles in science based government, university and industry organisations. As former Executive Director of the Royal Australian Chemical Institute, the professional association for chemists, she had a strong interest in science in education and secured grants to deliver novel science education programmes to schools.

John Gipps is a professional officer in the Faculty of Education, Monash University. His interests include the history of science, human evolution and how it might be approached in school biology, the uses of data logging and control technology in education, and the effectiveness of school-based science programs for pre-service primary teachers.

Richard Green completed an MSc in Maths and Computer Science, an ME (Electrical) and a PhD focusing on computer vision based real-time simulations for education in sports. While pursuing a PhD at the University of Sydney, Richard won the IEEE CSVT Transactions Journal Best Paper Award. He now lectures in computer science at the University of Canterbury with over 70 refereed publications. He currently heads the Computer Vision Research Lab with teaching and research interests in computer vision and computer game engine based simulations for education.

Michael Grimley is a Senior Lecturer in Education in the School of Educational Studies and Human Development at the University of Canterbury. His research interests are in the enhancement of learning, and in particular as it relates to cognition, motivation, interest, interactivity, new technologies and e-learning. These interests have led him into the study of how technology can be leveraged to improve learning. He currently specializes in the utility of new technologies for education with a keen interest in computer games and virtual worlds for education.

Sheila Henderson is a Senior Lecturer in Mathematics Education at the University of Dundee. She spent 25 years as a secondary mathematics teacher before moving into teacher education.

Ivor Hickey's academic roots are in biology. After graduating in plant sciences his doctoral research was in cancer cell biology. Since then he continued to research in a number of areas of cancer biology, particularly ovarian cancer and the alteration of gene expression patterns in tumours by DNA methylation. His academic roles have included running a B.Sc. in Genetics, extensive support for teachers through the provision of in-service courses and responsibilities for science education provision for students taking B.Ed. degrees. About ten years ago he moved to become fully involved in teacher education. His current research centres on development of novel approaches toward science education in both primary and secondary settings. This is largely focused on the development of The Leonardo Effect, a programme of learning that synchronises art and science education and is being widely used to support teachers address the requirements of new skills-based curricula where creativity is a major factor in children's learning.

Ruth Hickey is a Senior Lecturer at the School of Education, James Cook University, Cairns. Her current projects involve preservice teachers using information and communication technologies to develop a website of interviews with experts in water quality and catchment conservation science; and supporting colleagues to embed sustainability issues, values and solutions into their practice on campus. She has been a classroom teacher, school principal, and educational consultant in both science and English language. Her research interests are in science education, in particular pedagogical content knowledge, and how teachers make judgements about science achievement in students. Her recent work in environmental education is integrated with digital pedagogies and the potential of online communities.

Beverley Jane's career as an educator began with teaching Science and Biology in secondary schools in Victoria, Australia and Penang, Malaysia. As a senior lecturer in primary science and technology education at Monash University she gained an international reputation for her research related to linking science and technology education. Her books (with Fleer) 'Science for Children' and 'Technology for Children' continue to be popular set texts for pre-service teachers across Australia. With a PhD in Technology Education and an interdisciplinary Masters in Theology she blurs the boundaries as she argues for incorporating spirituality in science education. Current research interests involve intergenerational science learning with a focus on grandparents and grandchildren, and children's technological learning through block play in early childhood.

Kalle Juuti (Kalle.Juuti@Helsinki.Fi), PhD is Senior Lecturer in Physics education at the Department of Applied Sciences of Education, University of Helsinki. He is also a secretary of the executive board of the Finnish Mathematics and Science Education Research Association.

Bahar Karaoglan is Professor and Vice Director of International Computer Institute of Ege University, Izmir, Turkey. Turkish Scientific Committee Head in EU MedNet'U (Mediterenean Network of Universities) project. Project leader in several national projects funded by Scientific and Technological Research Council of Turkey (TÜBİTAK), and Ege University Science and Technology Application and Research Center. Partner in Socrates Thematic Network Projects: EIE-Surveyor (Electrical and

Information Engineering in Europe) 225997-CP-1-2005-1-FR-ERASMUS-TNPP and Enhancing Lifelong Learning for the Electrical and Information Engineering Community 142814-LLP-1-2008-1-FR-ERASMUS-ENW.

Roslyn Kerr is a lecturer in Sociology in the School of Social and Political sciences at the University of Canterbury where she teaches the sociology of new everyday technologies. Roslyn holds a BA and an M.Phil in Education, and is currently completing her PhD in Sociology which focuses on the use of technologies in the sport of gymnastics in New Zealand.

Minkee Kim (minkee.kim@helsinki.fi), PhD is a postdoctoral researcher in Physics and Chemistry Education, University of Helsinki, Finland. Ahead of his current position, he has obtained his PhD of science education in Seoul National University, Korea about student attitude toward science examined by structural equation modeling.

Tarık Kısla is working as an instructor at Department of Computer and Instructional Technologies. He gives lectures on information technologies, alghorithms, programming language, databases, web design and networks. He has two master degrees on the educational sciences and computer sciences. He is still working on the Ph.D thesis on Natural Language Processing fields in the International Computer Institute at Ege University. Further information about his publications and projects can be found from http://egitim.ege.edu.tr/~tkisla

Jarkko Lampiselkä (jarkko.lampiselka@helsinki.fi), PhD is Senior Lecturer in Chemistry Education at the Department of Applied Sciences of Education, University of Helsinki. He is currently the Head of the Centre for Mathematics and Science Teaching and Learning at the Department of Applied Sciences of Education.

Antti Laherto (antti.laherto@helsinki.fi), MSc, is PhD student at the Department of Physics, University of Helsinki. He works in Physics Teacher Education and is co-ordinator of "Kondensaattori", the Physics Teaching Resource Center. In his thesis he is studying and developing an informal learning environment on nanoscience.

Jari Lavonen (Jari.Lavonen@Helsinki.Fi), PhD, is Professor of Physics and Chemistry Education and Director of the Subject Teacher Education Section at the Department of Applied Sciences of Education, University of Helsinki, Finland. He is also a Director of the Finnish Graduate School of Mathematics, Physics and Chemistry Education.

Lee Yuen Lew is an assistant professor at Long Island University C.W. Post Campus, New York, USA. She played leadership roles in various curriculum development, teacher professional development, and innovative school projects spearheaded by the Malaysian Ministry of Education (MOE) while teaching science, biology, and mathematics for over 18 years in several Malaysian public schools. A major part of her education, up to a Masters in Science and Math Education was completed in Malaysia. However, as a visiting scholar under a full World Bank/MOE Scholarship, Lee Yuen received her Ph.D. in Science Education from the University of Iowa, USA, which also awarded her with the Franklin D. Stone

International Student Award. She serves as a board member of the International Organization of Science and Technology Education (IOSTE) as its North American representative and is a consulting editor for the magazine, Science Activities.

Anni Loukomies (anni.loukomies@helsinki.fi), M.Ed is PhD student at the Department of Applied Sciences of Education, University of Helsinki. She is also working as a primary school teacher. The title of her thesis is: Design-Based Research on Motivation in an Industry Site Visit in Materials Science Context.

Sinclair Mackenzie is a late entrant to teaching, having spent 15 years in a high technology career including contract research & development at the Irish National Microelectronics Research Centre, University College Cork; process & applications engineering for several multinational semiconductor manufacturing corporations and the role of principal engineer in a start-up Scottish Biophotonics instrumentation company. Since taking up his current post as Teacher of Physics at Thurso High School, Sinclair has continued to draw upon the ICT skills he developed in his previous career to enhance learning and teaching both inside and outside his classroom. He maintains a classroom blog at http://mrmackenzie.co.uk and the Fizzics podcast on the Apple iTunes store. In addition to using web2.0 in his own lessons, Sinclair has provided several INSET sessions for other teachers looking to update their ICT skills in the classroom environment.

Rikke Magnussen is an Assistant Professor at The Danish School of Education, Aarhus University. The subject of her research concerns the social learning dynamics in game-based science learning spaces. She has conducted and published on numerous learning game research and development projects and has been an active part in establishing game research groups at the Danish School of Education where researchers and game developers design and study new types of game-based learning spaces. She has published a number of papers on social learning in game-based learning environments and player transformation of educational games and has been co-developer and main researcher on the IT-supported role-playing game Homicide, a forensic homicide investigation game for cross-subject teaching of math, science and Danish. Rikke Magnussen has a PhD in science learning game technology and holds a M.Sc. in molecular biology and science communication. She has produced science TV documentary for the Danish Broadcast Cooperation and Discovery Canada before doing game studies.

Andrew Marks is a Sociologist who has in the past taught at Liverpool Hope, whilst pursuing his doctorate (on mature students) at Liverpool University, before subsequently taking a Research Fellowship at Stirling University (in the Institute of Education) earlier this decade. He has several publications related to the issues surrounding mature study and Lifelong Learning, but is also interested in the relative position of technology as a social phenomenon

Joe McLuckie is Programme Director for the Masters in Education at Dundee University. He is also IT Teaching and Learning Coordinator for the ESWCE and, as such, was instrumental in Dundee University's choice of standardising on one VLE as an online course delivery system. Additionally, over the last 10 years he has been involved in a multitude of international projects incorporating the use of the Internet to support blended learning courses. These partnerships have involved university collaborators across the following countries: Austria, Norway, Finland, Sweden, Latvia, Lithuania, Estonia, Iceland, Belgium, France, Spain, Italy, Germany, Holland, The Czech State, Switzerland and England. As such he sees himself as a Scot, a Brit, a European and a World Citizen. He strongly believes that the more we communicate the less we confront and sees online systems, on a global scale, as supporting this communication.

Veijo Meisalo (veijo.meisalo@helsinki.fi), PhD, is Professor Emeritus at the Department of Applied Sciences of Education, University of Helsinki, Finland. He has formerly been the Dean of the Faculty and the Director of the Department and is presently a part-time project researcher and research coordinator.

John Murnane teaches about using computers and the Internet in any educational setting. John is a former primary teacher with long standing record in the use of computers in Education. He has taught Primary and Secondary trainees Maths, Maths Method and Computer Science, and from around 1980 he's taught about education and computers. In between all of this he had a stint programming for Ford Australia as he is also a professional computer scientist. For some time he has been preoccupied with Lego Logo, how to introduce beginners to programming Robots, programming for beginners generally, and beginner's programming language design. He also has a great interest in introducing retirees to the world of the Internet.

Wan Ng is Associate Dean (International) and Senior Lecturer in science and technology education in the Faculty of Education at La Trobe University, Australia. Her major research interest is in the use of ICT, including mobile technologies, in education. She has published widely in the areas of science education, gifted education, online learning at both school and tertiary levels and the use of mobile technologies for learning. Wan has co-ordinated several nationally-funded science initiatives and has worked extensively with primary and secondary science teachers in professional learning courses.

Trond Nilsen graduated with an MSc (Hons) in Computer Science at the University of Canterbury and HITLab NZ in 2006 with research into the impact of augmented reality on game design and player motivation. Following this, he worked on research into virtual worlds, games, and education before beginning a PhD at the University of Washington where he hopes to explore how augmented reality and related technologies can support human cognition, memory, and performance. He maintains a blog on his research at: http://www.meme-hazard.org/

Verena Pietzner, University of Koblenz-Landau, Landau, Germany, was appointed a professor of Chemistry education at the University of Koblenz-Landau in Spring 2009. After her teacher education for secondary schools in chemistry and mathematics, she completed her Ph.D. in chemistry education, which researched a concept for computer-based exercises. She then worked as a lecturer at the Technical University of Braunschweig. Her research interests include computer-based learning in chemistry, interdisciplinary learning, and the pedagogical content knowledge of chemistry teachers.

Lorraine Syme Smith worked in the financial sector as a systems programmer. She began a career in education in 1993 as course tutor for a community project providing IT education for women. Moving to Edinburgh's Telford College as a senior lecturer, she then became programme leader for the computing curriculum managing a varied provision including Higher National and National qualifications, special educational needs and community projects. She has also completed a two year secondment as Development Manager at COLEG (Colleges Open Learning Exchange Group). Lorraine is currently a lecturer at the University of Dundee working on the Teaching Qualification for Further Education.

Neil Taylor trained as a teacher of Physics and has taught in several schools in Scotland. He is formerly Principal Teacher of Physics at Auchmuty High,Glenrothes, Scotland, Neil has also worked with Learning & Teaching Scotland as a national development officer in science education 5-14 and national qualifications posy-16. He is currently Programme Director for Secondary Education in the school of education, social work and community education at the University of Dundee.

David Thompson is a PhD researcher at the University of Canterbury, focusing in virtual environments research. He completed an MSc through the HIT Lab NZ, developing multi-touch screen technology, studying collaboration and contributing to interactive exhibition projects. David has a background in professional software engineering, and interests in narratology and indie game development (especially serious/educative gaming).

Sidiki Traore, is the Director of the West African Regional Office, Dakar, Senegal. Mr. Sidiki Traoré holds a Master of Science degree in Applied Linguistics and a Master of Arts degree in Instructional Systems Technology (IST), both from Indiana University in the USA. He also has a Diplôme D'Etudes Approfondies (D.E.A) from the University of Paris 7 in France. Mr. Sidiki Traore, joined the AVU during its Pilot Phase. Mr. Traore contributed his expertise as a seasoned educator, instructional designer, cross-cultural liaison and translator to the development of quality long-distance education for Africa. As a Senior Education specialist, he designed many short courses such as the Massachusetts Institute of Technology (MIT) Java Revolution. He designed and managed the current Certificate Programs in Information Technology in both English and French, Certificate in Journalism, Business English Communication and Certificate in Renewable Energy. As a Site Liaison Officer, he was instrumental in opening the AVU Learning centers in numerous African countries (Anglophone, Francophone and Lusophone countries) such as the Gambia, Guinee –Conakry, Guinee-Bissau, Cameroon, DRC, Malawi, Madagascar; Sudan, Nigeria, Senegal, Cote d'Ivoire. He has been in the forefront of expanding the AVU Network on the African continent.

David Walker is Senior Learning Technologist and head of the Assessment Development & Enhancement Team in the Library and Learning Centre, University of Dundee. He has considerable experience in the fields of e-learning and online assessment and has written on the importance of staff education and understanding the student experience to the achievement of success in these respective areas.

Hilary Whitehouse teaches science and environmental education at undergraduate and postgraduate level and coordinates the Honours program in the School of Education, James Cook University, Cairns. She won a national Australian award for university teaching in 2006 and has coordinated two curriculum innovation projects, BirdNet and Science on the Oval (SOTO). Her current research projects include investigating socio-ecological resilience in primary schools; curriculum change in teacher education in relation to education for sustainability, and conducting oral histories of local environments.

Graham Williamson qualified as a secondary teacher of English in 1972 and in his school career broadened his interests into such areas as Scots language and media studies. Since 1984 he has worked in teacher education, both primary and secondary, and has become increasingly interested in media education as a means to enhance learning across the curriculum. His leisure pursuits are swimming and cinema but not simultaneously.

Torsten Witteck, Engelbert-Kaempfer-Gymnasium, Lemgo, Germany is a teacher of Chemistry and Mathematics in a grammar school. In 2006, he earned his PhD in chemistry education as a member of Ingo Eilks' Participatory Action Research group. The group was performing a study based on cooperative learning in chemistry education with special emphasis on the inclusion of multimedia tools and open experimentation tasks.

Suan Yoong is a Research Fellow at the Sultan Idris University of Education, Malaysia and Deputy Chair of the International Organisation for Science and Technology Education (IOSTE). His bachelor's degree was in Chemistry and Mathematics, and he specialized in science, technology and mathematics education, psychometric, computer-in-education and statistics. He was the Chairman of Higher Degree Program at the School of Educational Studies, University of Science Malaysia. He was also the Project Coordinator of the Malaysian General Science Project at the Malaysian Curriculum Development Center (1974-79). He had won the Outstanding Graduate Student award at Indiana University and was a recipient of the UNESCO Fellowship Award for the 1992-93 Participation Program for research in Psychometric & Testing - as Research Fellow at the California Testing Bureau, Monterey, California, and visiting scholar to the Educational Testing Service at Princeton, New Jersey, the Lindquist Center at University of Iowa and the Psychological Corporation at San Antonio, Texas. He was also a Research Fellow at the University of California at Berkeley and University of Wisconsin at Madison, USA.

Index

K

L

M

N

O

P

R

S

T

U

V

W

Z